Drug Delivery to Tumors

Lesheng Teng • Zhaogang Yang • Chong Li
Editors

Drug Delivery to Tumors

Recent Strategies and Techniques

Springer

Editors
Lesheng Teng
School of Life Sciences
Jilin University
Changchun, China

Zhaogang Yang
School of Life Sciences
Jilin University
Changchun, China

Chong Li
College of Pharmaceutical Sciences
Southwest University
Chongqing, China

ISBN 978-981-19-8929-2 ISBN 978-981-19-8930-8 (eBook)
https://doi.org/10.1007/978-981-19-8930-8

This Springer imprint is published by the registered company Springer Nature Singapore Pte Ltd.
The registered company address is: 152 Beach Road, #21-01/04 Gateway East, Singapore 189721, Singapore

Preface

As early as 1890, the German scientist Paul Ehrlich envisioned a miraculous compound that had a good affinity for pathogenic organisms and was harmless to the human body, which Ehrlich imaginatively compared to "magic bullets." More than a hundred years later, amidst remarkable advances in materials science and drug delivery technology, and iterations of tumor treatment options, scientists are looming further toward this goal. The burden of cancer continues to grow globally, placing tremendous physical, emotional, and economic stress on individuals, families, communities, and health systems. Unfortunately, conventional therapies remain impotent in the face of malignant tumor migration and drug resistance. Rapid, unrestricted tumor growth dramatically alters the tumor microenvironment, and the inter-regulation of its components is a major source of drug resistance in tumors. With such a dilemma as a backdrop, it is logical to develop drug delivery systems that are more tumor-targeted and have low side effects. Today's research on DDS has shifted from passive targeting to active targeting and other emerging tumor-targeted drug delivery strategies. However, the complexity and heterogeneity of tumor biology has led to the realization that a one-size-fits-all drug to cure all tumors may not be coming soon and that tailor-made drugs for specific individuals may be the focus of future research.

In the book *Drug Delivery to Tumors: Recent Strategies and Techniques*, the first chapter provides an overview of the causes and consequences of tumor microenvironment components and the interactions between these components. The aim of this chapter is to illustrate the complexity of this interconnected network and its contribution to chemoresistance, and to substantiate the need for the development of drug delivery strategies that target tumors. The second chapter summarizes the history of drug delivery systems (DDS) and highlights the classification characteristics and applications of third-generation drug delivery technologies. The third chapter discusses different delivery strategies for targeting tumors. Chapter 4 describes commonly used clinical treatments for cancer like chemotherapy, nucleic acid drug, phototherapy and immunotherapy and the impact of different combined therapy between them in cancer treatment, discussing the prospects for application and the direction of improvement of combination therapies. Based on the above four chapters, Chaps. 5, 6, 7, 8, 9, and 10 discuss the research progress and future challenges of six common DDS, including prodrug nanocarrier, smart drug delivery vehicles, cell derived/bionic-drug delivery vehicles, cellular drug delivery vehicles, nanocrystal technology, and microneedle. In this section, the different types of DDS and their subordinate subclasses are discussed in more detail, presenting the advantages, disadvantages, and application scenarios of the different delivery systems.

This book summarizes the different drug delivery strategies and techniques of recent years and discusses their possible clinical applications. We hope that this book will serve as a tool for reviewing the progress of DDS in recent years and will inspire experts from different specialties to work together to advance the development of DDS.

Changchun, China — Lesheng Teng
Changchun, China — Zhaogang Yang
Chongqing, China — Chong Li

Contents

Contents

About the Editors

Lesheng Teng is professor at Jilin University. His research focuses on a highly interdisciplinary combination of nanotechnology, biomaterials, drug delivery, and immunotherapy. He has developed various nanoparticle platforms for delivery of small molecules, proteins, and nucleic acids (e.g., siRNA and mRNA). Dr. Teng has published over 160 peer-reviewed papers and holds over 20 patents/patent applications. Representative papers were published in *Nature Biomedical Engineering*, *Nano Letters*, *Biotechnology Advances*, *ACS Applied Materials & Interfaces*, *Journal of Controlled Release*, *Theranostics*, and *Nanoscale*. He has also received many awards, such as excellent supervisor, and others. Dr. Teng obtained his BS degree and PhD degree from Jilin University.

Zhaogang Yang is professor at the School of Life Sciences, Jilin University. He received his BS and MS degrees from Peking University, China, and completed his PhD degree at the Ohio State University, USA. Dr. Yang has published over 130 peer-reviewed articles in journals such as *Nature Biomedical Engineering*, *Nature Nanotechnology*, and *Nature Communications*, with more than 5,000 citations and an H-index of 47. He is serving on the Young Editorial Board of several journals, including *Asian Journal of Pharmaceutical Sciences*, and *Interdisciplinary Medicine*. Dr. Yang's research focuses on the application of nanotechnology in drug delivery, including novel nanoparticle-based drug delivery systems, gene delivery, and cancer immunotherapy.

Chong Li is professor in the College of Pharmaceutical Sciences at Southwest University (Chongqing, China). He received BS degree from Sichuan University (China) in 2005 and earned his PhD degree in pharmacy from Fudan University in 2010. He is on the Young Editorial Board of the *Acta Pharmaceutica Sinica B*. Prof. Chong Li is a recipient of several awards such as the National Excellent Doctoral Dissertation (2013), Young Pharmaceutical Scientist Award from the Chinese Pharmaceutical Association (2015), and the Youth Scholar of Yangtze River (2018). He has published more than 30 papers in *J Am Chem Soc*, *Angew Chem Int Ed*, *ACS*

Nano, *Nano Lett*, and *Proc Natl Acad Sci USA*. His research interests focus on the design of smart drug delivery systems based on natural products and derivatives (functional peptides and small molecules, etc.), and their application in targeted drug delivery, involving antitumor and anti-infection therapy, and the treatment of central nervous system diseases.

Chapter 1
Cancer Biology and Microenvironment

Yijie Wang, Tongzheng Liu, Shuang Peng, Lai Jin, and Yan Chen

Abstract Malignant transformation and sustained growth of tumors profoundly alter the tumor microenvironment, including angiogenesis, metabolism, immune response, and oxygen levels. In turn, remodeling of the tumor microenvironment can effectively promote tumor progression and hinder drug delivery, which is a prerequisite to the therapeutic efficacy of anticancer drugs. Moreover, the components of the tumor microenvironment mutually regulate each other, which further contribute to the complexity of cancer biology. This chapter aims to overview the causes and outcomes of abnormal tumor vascularization, metabolism reprogramming, immune evasion, endoplasmic reticulum stress and unfolded protein response, and hypoxic conditions in tumors. We will also briefly discuss the complicated interplay between these components of the tumor microenvironment.

Y. Wang
Shandong Provincial Key Laboratory of Animal Resistance Biology, Collaborative Innovation Center of Cell Biology in Universities of Shandong, Center for Cell Structure and Function, Institute of Biomedical Science, College of Life Sciences, Shandong Normal University, Jinan, Shandong, China

T. Liu
School of Pharmacy, Jinan University, Guangzhou, Guangdong, China

S. Peng
School of Medicine, Jinan University, Guangzhou, Guangdong, China

L. Jin
School of Basic Medical Sciences, Nanjing Medical University, Nanjing, Jiangsu, China

Y. Chen (✉)
Shandong Provincial Key Laboratory of Animal Resistance Biology, Collaborative Innovation Center of Cell Biology in Universities of Shandong, Center for Cell Structure and Function, Institute of Biomedical Science, College of Life Sciences, Shandong Normal University, Jinan, Shandong, China

School of Medicine, Jinan University, Guangzhou, Guangdong, China
e-mail: yanchen@sdnu.edu.cn

L. Teng et al. (eds.), *Drug Delivery to Tumors*,
https://doi.org/10.1007/978-981-19-8930-8_1

Keywords Cancer · Angiogenesis · Metabolism reprogramming · Immune evasion · Hypoxia · Microenvironment · Drug delivery

1.1 Formation of Tumor Blood Vessel

The inner layer of blood vessels is comprised of an endothelial cell layer surrounded by a basement membrane, which in turn is surrounded by pericytes. Angiogenesis is a complicated multi-step process in which novel capillary blood vessels can form through vasculogenesis and sprouting angiogenesis. Vasculogenesis is the de novo formation of new blood vessels through induction of vascular progenitor cell proliferation (Lugano et al. 2020), whereas sprouting angiogenesis is the formation of the blood vessels via co-option of the preexisting vasculature (Folkman 2007). The growth of new blood vessels promotes the upregulation of various proangiogenic factors, such as vascular endothelial growth factor (VEGF), fibroblast growth factors (FGFs), angiopoietins (Ang), placental growth factor (PIGF), and some integrins, as well as contributes to the downregulation of some antiangiogenic factors, including angiostatin, interferons, platelet factor 4, endostatin, and inhibitors of metalloproteinase-1, -2, and -3 (Folkman 2007). Malignant cells acquire excessive oxygen and nutrients through the blood circulation system to sustain unlimited proliferation and survival (Wang et al. 2020b; Chen et al. 2020). The progression of a tumor that grows beyond 2 mm in diameter is critically dependent on neoangiogenesis. Rapidly growing tumors, as compared with normal tissues, are heavily vascularized. The structure and function of the nascent primitive vasculatures are usually abnormal due to several features, such as the abrupt change of the blood vessel tortuosity, leakiness, and diameter, leading to limited blood perfusion and diffusion (Chen et al. 2020). Consequently, dysfunction of blood vessels causes tumor hypoxia, blockage of nutrients, and drug delivery.

1.1.1 Vasculogenesis

In response to some cytokines, chemokines, and growth factors produced by tumor and stromal cells, endothelial progenitor cells (EPCs) are recruited to generate new vessels (Fig. 1.1b) (Greenfield et al. 2010). Several stem cell biomarkers, such as CD34, CD133, Nanog, and Oct-4, are expressed in EPCs. Thus, EPCs have the capacity to self-renew, proliferate, and participate in neovascularization (Shin et al. 2005; Chopra et al. 2018). There are two types of EPCs: one is endothelial colony-forming cell (ECFC), which is not derived from the hematopoietic lineage, and another one is hematopoietic-derived myeloid angiogenic cells (MACs). ECFCs are

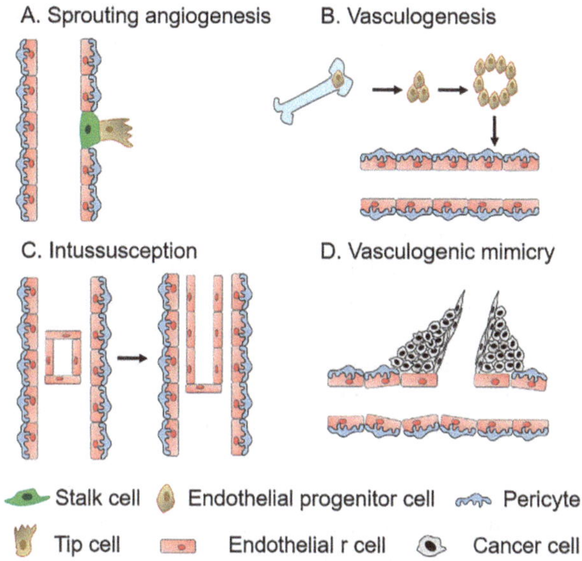

Fig. 1.1 Mechanism of blood vessel formation in the tumor ecosystem. (**a**) Sprouting angiogenesis: a process in which stalk cells and tip cells are fused with an existing blood vessel. (**b**) Vasculogenesis: The endothelial progenitor cells proliferate and are recruited to assemble new blood vessels by forming lumen. (**c**) Intussusception: a preexisting blood vessel is split into two to generate new vasculature. (**d**) Vasculogenic mimicry: malignant tumor cells form vessel-like structures. (Figure adapted from Roberta Lugano et al. (2020) (CC BY 4.0 licensed OA article))

detected in peripheral blood, but not bone marrow, and play a dominant role in vascularization in vivo (Ingram et al. 2004).

Proangiogenic factors stimulate the activation of the matrix metalloprotease 9 (MMP9), which in turn triggers the release of soluble kit-ligand, thereby enhancing the mobility of endothelial and hematopoietic stem cells from the osteoblastic zone to the vascular zone (Heissig et al. 2002). Tumors also secret proangiogenic factors, including chemokines C-C motif ligand 2 (CCL2) and CCL5, the hypoxia-inducible chemokine stromal-derived factor 1 (SDF-1), and adiponectin to promote the tumor bed recruitment of EPCs and enhance neovascularization (Chang et al. 2007; Spring et al. 2005; Xie et al. 2020). Once homed, EPCs proliferate, form lumens, and ultimately assemble into new blood vessels. Hypoxia is a hallmark of tumors. In hypoxic conditions, hypoxia-inducible factor (HIF) augments the transcription of VEGF, PDGF, SDF-1, and C-X-C chemokine receptor type 4 (CXCR4) (Rey et al. 2017). Depletion of HIF impedes EPCs proliferation and differentiation.

It is well known that vasculogenesis is critical for tumor growth and progression because blockage of tumor vasculogenesis using inhibitors of proangiogenic factors will dramatically impair tumor growth. Intriguingly, vasculogenesis also participates in the dissemination and metastasis of tumor cells. EPCs in distant organs secret SDF-1, which interacts with tumor cell-expressed CXCR4; thus, EPC-secreted SDF-1 functions as a chemoattractant of tumor cells (Jin et al. 2012).

Moreover, EPCs enhance tumor cell migration and invasion by activating MMP9 (Kopp et al. 2006). In summary, EPCs are pivotal for tumor growth and metastasis.

1.1.2 Sprouting Angiogenesis

Sprouting angiogenesis is a multi-step process by which new capillaries can bud from preexisting host capillaries (Fig. 1.1a). The formation of sprouts involves three key steps: (1) tip and stalk cell selection, (2) sprout extension, and (3) Lumen formation. Upon induction by proangiogenic factors, endothelial cells that are normally quiescent can sprout and initiate angiogenesis. Sprouting endothelial cells induced by proangiogenic factors are characterized by higher invasion and motility. The leading endothelial cells during the sprouting process are "tip cells," while the highly proliferative endothelial cells located at the stalk of the angiogenic sprout are known as "stalk cells." Tip and stalk cell selection is coordinately regulated by the VEGF and Dll4/Notch pathways (Jakobsson et al. 2009). In response to VEGF, tip cells generate Delta-like-4 (DLL4), which activates Notch signaling in the neighboring endothelial cells, inhibiting tip cell formation (Tammela et al. 2008; Shawber et al. 2007). VEGF stimulates sprouting, branching, invasive phenotype, and numerous filopodia formation in tip cells, thus guiding new blood formation by sensing molecular signals and recruiting stromal cells (Gerhardt et al. 2003). Stalk cells have fewer filopodia as compared to the tip cells and are responsible for tube and branch formation (Gerhardt et al. 2003).

1.1.3 Vascular Mimicry

Vascular mimicry refers to the ability of some malignant cells to form vessel-like structures (Fig. 1.1d) (Maniotis et al. 1999). These aggressive tumor cells dedifferentiate and gain endothelial-like properties. The endothelial-like tumor cells secret collagens IV and VI, proteoglycans, laminin, and heparan sulfate, thus enhancing tubular structure formation and stabilization (Angara et al. 2017). Remodeling of extracellular matrix (ECM), tumor hypoxia, and activation of transmembrane metalloproteinases have been reported to stimulate vascular mimicry (Seftor et al. 2005; Chen et al. 2019). Vascular mimicry has been detected in many human tumors and contributes substantially to tumor growth and progression in several ways. Thus, it is a biomarker for many cancer types (Ren et al. 2014). However, the absence of reliable techniques to distinguish vascular mimicry from normal endothelial cell lining hinders the investigation of vascular mimicry.

1.1.4 Intussusception

Intussusception is briefly described as the formation of new vascular structures where a preexisting vessel splits into two (Fig. 1.1c) (Lugano et al. 2020). It is reported that intussusception occurs after sprouting angiogenesis or vasculogenesis in an endothelial cell proliferation-dependent manner, but this process can also be induced by VEGF, PDGF, and erythropoietin (Zuazo-Gaztelu and Casanovas 2018). However, the underlying molecular mechanisms are still poorly understood. Intussusception has been observed in several types of tumors, such as melanoma, glioma, and colorectal cancer.

1.2 Metabolism Reprogramming

Cancer cells in solid tumors undergo rapid metabolic reprogramming due to limited oxygen and nutrient supply, accumulation of metabolic waste, and undesirable pH in the local tumor microenvironment. Thus, unraveling the molecular mechanism of metabolic pathways will provide new insights into cancer biology and potential therapeutic targets for cancer.

1.2.1 Glucose Metabolism

Cancer cells, even in an aerobic environment, barely use mitochondrial oxidative phosphorylation (OXPHOS). Instead, they mainly rely on aerobic glycolysis, a process where glucose is metabolized to pyruvate and ultimately converted to lactate to supply ATP and intermediate metabolites for proliferation and survival (Fig. 1.2) (Cantor and Sabatini 2012). This phenomenon is known as the Warburg effect. Several studies have demonstrated that cancer cells use glucose as their primary carbon source since the Warburg effect was first described. Once glucose enters the cytoplasm, it is metabolized to generate ATP and NADH for cellular reactions via glycolysis, hexosamine biosynthesis pathway (HBP), or the pentose phosphate pathway (PPP).

Glucose transporter (GLUT), including GLUT1-4, is pivotal for glucose uptake (Fig. 1.2). The expression of GLUT1 is stimulated by several oncoproteins, such as c-Myc, KRAS, YAP, and HIF in many cancers (Barthel et al. 1999; Yun et al. 2009; Osthus et al. 2000; Cox et al. 2018; Luo et al. 2011). The overexpression of YAP and depletion of p53 increase GLUTs expression (Kawauchi et al. 2008; Wang et al. 2015). GLUT1 is overexpressed in most cancers, and its high expression causes poor clinical outcomes (Wilson and Hay 2011). Hexokinase (HK) catalyzes the first reaction in glycolysis, the conversion of glucose to glucose-6-phosphate (G6P), which in turn inhibits HK. Furthermore, HK2 expression is upregulated in cancer

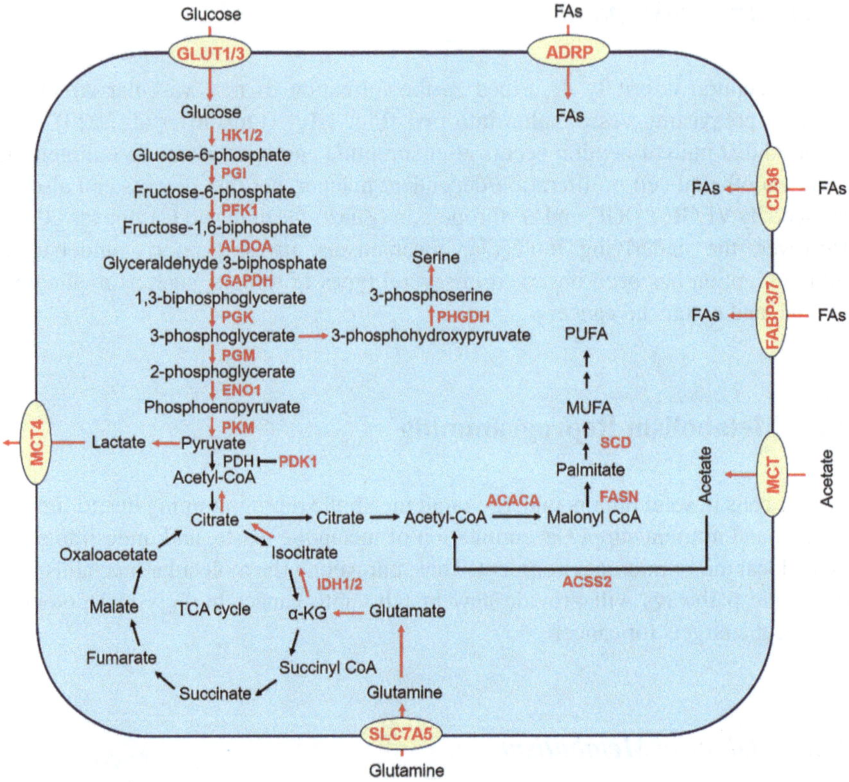

Fig. 1.2 Metabolic reprogramming in tumor cells. Aerobic glycolysis is dramatically increased in cancer cells, whereas glycolytic flux into the TCA cycle is blocked. The glycolytic intermediate 3-phosphoglycerate can be utilized for serine generation. Cancer cells can sustain the TCA cycle and lipogenesis via glutaminolysis. Moreover, fatty acids are accumulated in cancer cells to maintain their proliferation and survival through fatty acid uptake and de novo synthesis. Genes highlighted in bold and red text are upregulated by hypoxia and HIF

cells, leading to increased glycolysis through interaction with the mitochondrial membrane (Hirschey et al. 2015; Hay 2016). HK2 activity is further augmented through the PI3K/AKT pathway by increasing its association with the mitochondrial membrane (Roberts et al. 2013; Gottlob et al. 2001). Glucose analog 2-Deoxyglucose (2-DG) is also phosphorylated by HK. However, phosphorylated 2-DG competes with glucose to bind to HK (Scatena et al. 2008). Thus, 2-DG is an ideal clinical candidate. Unfortunately, its efficacy in killing tumor cells in vivo is very low (Fu et al. 2016). Collectively, GLUT1 and HK2 are attractive potential targets for cancer treatment.

PK muscle isozyme M2 (PKM2), a dominant isoform of pyruvate kinase (PK) in cancer cells, is involved in some cancers, including breast and colon cancer (Cortes-Cros et al. 2013; Israelsen et al. 2013). PKM2 activity is relatively low but can be stimulated by fructose-1, 6-bisphosphate (F-1, 6-BP) and serine. PKM2 is hydroxylated by prolyl hydroxylase 3 (PHD3) to increase its interaction with HIF-1α and subsequently promotes the expression of HIF-1 target genes by recruiting HIF-1 coactivator p300 (Luo and Semenza 2012). PKM2 expression correlates with motility and invasive capacity in different cancer cells due to an increased flux through glycolysis that leads to the generation of methylglyoxal, which activates yes-associated protein (YAP) signaling in breast cancers (Nokin et al. 2016).

The HBP is a branch of glycolysis responsible for the production of a key metabolite, UDP-GlcNAc. Glycosyltransferases use this metabolite in glycosylation reactions of proteins and lipids, as well as glycosylation of UDP-GalNAc and CMP-Neu5Ac (Vasconcelos-Dos-Santos et al. 2015). It has been demonstrated that HBP profoundly affects diverse aspects of tumor biology, including tumor growth and metastasis. The HBP regulates multiple proteins in cancer cells via *O*-linked-β-*N*-acetylglucosamine transferase (OGT), which catalyzes O-GlcNAcylation reactions. In cancer cells, elevated O-GlcNAcylation disrupts the interaction between HIF-1 and von Hippel-Lindau protein (pVHL), thereby promoting the expression of many HIF-1-induced glycolytic enzymes, including GLUT1 and HK2 (Ferrer et al. 2014). O-GlcNAcylation also occurs on transcription factors YAP, c-Myc, and NF-κB and promotes tumorigenesis by increasing their activity (Ma et al. 2013; Chou et al. 1995; Peng et al. 2017). Hence, GlcNAcylation is considered a novel hallmark of cancers.

The PPP pathway is a metabolic pathway parallel to glycolysis. It branches from G6P and generates NADPH as well as ribose 5-phosphorate (R5P). NADPH and R5P are both critical for the proliferation and survival of cells (He et al. 2016). R5P is a precursor for the synthesis of nucleotides. NADPH has a reducing power, which is required for the synthesis of fatty acids, nucleotides, non-essential amino acids (NEAAs), and sterols. NADPH is also essential for the conversion from the oxidized glutathione (GSSG) to the reduced glutathione (GSH). Thus, the PPP signaling is pivotal for maintaining redox homeostasis and protecting cancer cells from oxidative stress. In cancer cells, PI3K/AKT, c-Myc, and mTORC1 pathways and p53 loss-of-function mutation upregulate PPP flux (Patra and Hay 2014; Wamelink et al. 2008). Various enzymes in the PPP pathway, such as G6PD, 6PGD, RPI, and TKT, have been shown to be potential therapeutic targets in cancer treatment (Wang et al. 2016; Shan et al. 2014; Ying et al. 2012; Dasgupta et al. 2018).

1.2.2 Amino Acid Metabolism

Amino acids provide carbons to the TCA cycle, nitrogen to nucleotide synthesis, and redox homeostasis. Therefore, they are essential for cancer cell growth and survival. Glutamine is the most abundant amino acid in plasma (Cantor and Sabatini 2012). It is well-known that cell proliferation relies highly on high concentrations of glutamine supplemented in the cell culture medium. Serine and methionine are also important for tumor growth and progression (Park et al. 2020).

Glutamine is imported into cells via transporters, such as SLC7A5 and SLC1A5, where the latter is the major glutamine transporter in many cancers (Fig. 1.2). Due to glutamine's key roles as nitrogen and carbon donors, SLC1A5 and SLC7A5 are usually activated by many oncogenic drivers, including c-Myc, n-Myc, and mTORC1 (Qing et al. 2012; Ren et al. 2015; Peng et al. 2002). SLC1A5 inhibition by benzylserine, 1-γ-glutamyl-p-nitroanolide (GPNA), and γ-BP blocks tumor growth in lung cancer and melanoma (Hassanein et al. 2015; Wang et al. 2014). SLC7A5 inhibitor 2-aminobicyclo (2, 2, 1)-heptane-2-carboxylic acid (BCH) also restricts tumor growth (Nicklin et al. 2009). After being imported into cells, glutamine is catabolized into many intermediate metabolites, such as glutamate, citrate, pyruvate, lactate, and CO_2. This is known as glutaminolysis (Fig. 1.2). Glutaminase (GLS1/2)-mediated deamination of glutamine into glutamate and ammonia is the first reaction in glutaminolysis. Glutamate will be further converted by glutamate dehydrogenase (GLDH) into NAD(P)H, ammonia, and α-KG. Carbamoyl phosphate synthetase and phosphoribosyl pyrophosphate synthase, two rate-limiting enzymes of the pyrimidine and purine synthesis pathways, incorporate an ammonia group into carbamoyl phosphate (Yang et al. 2017b). Nitrogen from glutamine can be catalyzed by transaminases to generate NEAAs. In cancer cells, c-Myc inhibits microRNA miR-23a and miR-23b, which target GLS mRNA, thereby increasing GLS protein levels (Gao et al. 2009). α-KG produced from glutamine can be used to generate citrate and ATP in the mitochondria of hypoxic cells. Under reductive conditions, isocitrate dehydrogenase 1 (IDH1) and IDH2 function as carboxylases and carboxylate α-KG to generate isocitrate that is used to produce citrate (Fig. 1.2) (Metallo et al. 2011). Citrate translocates from the mitochondria to the cytoplasm and is catalyzed by the ATP citrate lyase (ACLY) enzyme to produce acetyl-CoA. Then, fatty acid synthase (FASN) and other enzymes mediate fatty acid synthesis using acetyl-CoA as substrate (Jiang et al. 2016; Mullen et al. 2011).

Serine is an NEAA that can be taken up from extracellular nutrients or synthesized de novo through the serine synthesis pathway (SSP) (Yang and Vousden 2016). Serine is critical for the growth of certain cancers, such as breast cancer, colorectal cancer, and melanoma, due to its critical roles in one-carbon metabolism and the production of NADPH and glutathione. SSP enzymes are highly expressed in these cancers, and depletion of SSP enzymes blocks tumor growth (Mattaini et al. 2016; Pacold et al. 2016). The first and rate-limiting enzyme in the serine biosynthesis is 3-phosphoglycerate dehydrogenase (PHGDH). It converts 3-phosphoglycerate to 3-phosphohydroxypyruvate (Fig. 1.2) (Amelio et al. 2014). PHGDH expression levels are significantly elevated in multiple cancers, including triple-negative breast cancer (TNBC) and non-small cell lung cancer (NSCLC) (Nowotarski et al. 2013; Locasale et al. 2011; Possemato et al. 2011). PHGDH and other serine biosynthesis enzymes are transcriptionally upregulated by c-Myc (Sun et al. 2015). Several PHGDH inhibitors are undergoing preclinical research, such as CBR-5884, NCT-502, and NCT-503 (Mullarky et al. 2016; Pacold et al. 2016).

Methionine is an essential amino acid that is used for protein synthesis and generation of S-adenosyl-L-methionine (SAM), which is essential for protein and DNA methylation (Ramani and Lu 2017). Methionine adenyltransferase 2 (MAT2) modulates methionine catabolism, and depletion of MAT2 results in cell death (Ikeda et al. 2017; Liu et al. 2007), suggesting that MAT2 is a potential therapeutic target in cancer (Ramani and Lu 2017). However, the MAT2A allosteric inhibitor PF-9366 drives an adaptive and compensatory transcriptional upregulation of MAT2A (Quinlan et al. 2017). The underlying mechanism needs to be further investigated.

1.2.3 Fatty Acid Metabolism

Fatty acids (FAs) play important roles in cell structure maintenance, energy generation and storage, and cell signaling. Elevated endogenous FA synthesis and exogenous FA uptake are common phenotypes in numerous cancer cells, while cancer cells in certain survival situations and some special cancer cells can perform FA catabolism through FA beta-oxidation (FAO). The amount of FAs in cancer cells is coordinately regulated by FA uptake, de novo synthesis, FAO, and lipolysis.

Cancer cells take up exogenous FAs from the surrounding microenvironment through special transporters, including CD36 (also known as FAT), fatty acid transport proteins (FATPs, also known as SLC27), and plasma membrane fatty acid-binding proteins (FABPpm) (Fig. 1.2). The gene and protein levels of these transporters are all elevated in malignant tumors (Su and Abumrad 2009). A major application of elevated uptake of exogenous FAs is their subsequent storage in lipid droplets (LDs), which are dynamic cytoplasmic organelles that regulate the storage and hydrolysis of neutral lipids (Olzmann and Carvalho 2019). LDs play critical roles in regulating lipid and energy homeostasis and preventing lipotoxicity (Olzmann and Carvalho 2019; Petan et al. 2018). High levels of CD36 are correlated with poor clinical outcomes in several cancer types, and depletion of CD36 leads to significant inhibition of tumor growth (Ladanyi et al. 2018; Watt et al. 2019). Hypoxia promotes FA uptake and LD accumulation by stimulating the expression of FABP3, FABP7, and CD36 via HIF-1 (Bensaad et al. 2014; Du et al. 2017; Lisec et al. 2019). These findings suggest that lipid uptake is regulated by the hypoxic tumor microenvironment and HIF, thus providing a rationale for targeting HIF to reduce lipid accumulation in cancer cells.

De novo lipogenesis, the process by which carbon atoms derived from glucose and amino acids are converted into FAs, is primarily controlled by transcriptional factor sterol regulatory element-binding proteins (SREBPs). When cellular lipid concentration is saturated, insulin-induced genes (INSIGs) bind to SREBP-cleavage-activating proteins (SCAPs) and recruit SREBP precursor to the endoplasmic reticulum (ER) (Adams et al. 2004; Sun et al. 2005; Yabe et al. 2002). In contrast, under conditions of low lipid levels, SCAPs facilitate the translocation of SREBPs from the ER to the Golgi, where membrane-bound transcription factor site 1 proteases (MBTPS1 and MBTPS2) cleave SREBPs at the N-terminus (Brown and Goldstein 1997; Yang et al. 2001). The N-terminal fragment is translocated into the nucleus to bind the sterol regulatory elements (SREs) of multiple genes that are involved in lipid de novo synthesis, such as *FASN*, *ACLY*, and *ACC* (Fig. 1.2) (Yokoyama et al. 1993; Bennett et al. 1995; Nakakuki et al. 2014). Intriguingly, the hypoxic region within tumors in vivo displays increased expression and activity of SREBP1 through Akt and HIF-1 (Furuta et al. 2008).

Fatty acid synthesis is an oxygen-consuming process, whereas hypoxia is a hallmark of solid tumors. Hence, cancer cells tend to compensate by increasing exogenous lipid uptake instead of performing de novo synthesis. Not surprisingly, the real condition appears to be far more complex. For instance, hypoxia alone decreases the expression of FA synthesis enzymes, while hypoxia in combination with low-serum conditions causes increased expression of these enzymes (Jung et al. 2012). Additionally, the expression of some FA synthesis enzymes is also affected by cell density (Jung et al. 2012).

FAO mediates FA degradation by cleaving two carbons at a time to generate acetyl-CoA, NADH, and $FADH_2$. Theoretically, cancer cells often decrease FAO to elevate cellular lipid concentration. However, excessive ATP demand and cancer cells undergoing anoikis usually activate FAO (Carracedo et al. 2013; Schafer et al. 2009). HIF-1 suppresses FAO by inhibiting the expression of medium- and long-chain acyl-CoA dehydrogenases (MCAD and LCAD) (Huang et al. 2014). FAs are delivered to FAO via carnitine palmitoyl transferase 1 (CPT1), which converts FAs to FA carnitine (Currie et al. 2013). Current studies of CPT1 in cancer have so far been inconsistent. Consequently, further investigations are necessary to elucidate its role.

1.3 Tumor Immunity

The tumor microenvironment (TME) consists of cancer cells, immune cells, stromal cells, endothelial cells, and cancer-associated fibroblasts. Cancer cells can escape from immune surveillance and immune killing through a variety of mechanisms. In general, the tumor-associated immune cells can be divided into two types: tumor-antagonizing and tumor-promoting immune cells. Effector T cells, natural killer (NK) cells, and dendritic cells (DCs) are tumor-antagonizing immune cells. Tumor-promoting immune cells comprise regulatory T cells (Tregs) and myeloid-derived suppressor cells (MDSCs). Recently, a growing body of study has indicated that the cGAS-STING pathway is critical for tumor growth and progression.

1.3.1 The Tumor-Antagonizing Immune Cells in TME

Human NK cells are derived from multipotent CD34+ hematopoietic progenitors in the bone marrow (Yu et al. 2013). NK cell maturation occurs in the bone marrow as well as in the lymphoid organs. Unlike T cells, NK cell maturation does not require the thymus (Yu et al. 2013; Renoux et al. 2015). Human NK cells are defined as

CD3-CD56+ cells and present 5–15% of circulating lymphocytes (Guillerey et al. 2016). NK cells release perforin and granzymes to induce the apoptosis of tumor cells, suggesting that NK cells are important in tumor immunosurveillance (Smyth et al. 2000; Shimasaki et al. 2020). Expectedly, functionally impaired NK cells lead to worse clinical outcomes in mouse models and humans. Moreover, NK cells secrete pro-inflammatory cytokines and chemokines, such as CCL5, IL-6, and TNF, to enhance antitumor activity (Guillerey et al. 2016). The mechanisms by which NK cells exert tumor immunosurveillance and influence tumor growth remain elusive. NK cell infiltration of tumors is determined by the expression of heparinase. NK cells attract T cell infiltration and elicit inflammatory responses through cytokine and chemokine secretion, although they are not the predominant lymphoid population in tumors. Intriguingly, even though NK cells are critical in tumor immunosurveillance, their killing function is usually inhibited in tumors (Guillerey et al. 2016). NK cells within the TME of breast cancer are immature, which is caused by the reduced expression of DX5, NKp46, perforin, and granzyme B.

DCs are a part of the hematopoietic cell lineage. They mainly function as professional antigen-presenting cells (APCs) to present antigens and activate naive T cells, thus linking both the innate and adaptive immune systems. Most tumors contain only a few mature DCs, and the absence of DCs within tumors is correlated with poor clinical outcomes (Michea et al. 2018; Truxova et al. 2018). TME hamper DC maturation or the activity of DCs to compromise the antitumor immune response. For instance, high levels of CCL2, CXCL1, CXCL5, and VEGF within the TME of human colorectal tumors inhibit DC maturation (Michielsen et al. 2011). DCs interact with Treg cells to suppress CD8+ T cell-mediated antitumor immunity. DCs also interact with NK cells which produce cytokines and chemokines (Bottcher et al. 2018). DC-derived factors, such as IL-12 and IL-15, enhance NK cell functions to eliminate tumor cells (Mahmood et al. 2015). Aside from immune cells, DCs also communicate with cancer cells and stromal cells via multiple pathways, such as Wnt signaling (Cheng et al. 2016; Nusse and Clevers 2017).

1.3.2 The Tumor-Promoting Immune Cells in TME

Treg cells are a dynamic subset of CD4+ T lymphocytes and can be divided into two groups according to their development site: natural Treg (nTreg) and induced Treg (iTreg) (Paluskievicz et al. 2019; Togashi et al. 2019). In addition to CD4+ markers, Foxp3 is a reliable marker for Treg cells (Samstein et al. 2012). Foxp3 is critical for Treg lineage specification and the regulation of the suppressive signal (Samstein et al. 2012). CTLA-4 is an immune checkpoint receptor that is constitutively expressed on the surface of Treg. The interaction between CTLA-4 on Treg and CD80 (B7–1) or CD86 (B7–2) ligands on APCs directly suppresses APC function, attenuates the ability of APCs to activate conventional effector T cells, and induces the activity of indoleamine 2,3-dioxygenase (IDO) in DCs to promote the metabolism of tryptophan amino acid, which is required for the function of effector T cells (Qureshi et al. 2011; Fong and Small 2008; Takahashi et al. 2000).

MDSCs are pathologically activated neutrophils and monocytes with potent immunosuppressive activity. According to their origin, MDSCs are classified into two categories, namely, granulocytic/polymorphonuclear MDSCs (PMN-MDSCs) and monocytic MDSCs (M-MDSCs) (Veglia et al. 2018). The PMN-MDSCs are morphologically similar to N2-polarized neutrophils, whereas M-MDSCs are similar to M2-polarized macrophages (Veglia et al. 2018). Immune suppression by MDSCs involves multiple mechanisms. MDSCs suppress T cell activation by depleting key nutrition factors, including cysteine, cystine, L-arginine, and tryptophan (Raber et al. 2012; Srivastava et al. 2010). MDSCs also inhibit T cell function through the production of nitric oxide (NO), reactive oxygen species (ROS), and immunosuppressive cytokines, such as IL-10 and TGF-β (Molon et al. 2011; Ostrand-Rosenberg and Sinha 2009). MDSCs could enhance angiogenesis and metastasis by producing MMP9, VEGF, and prokineticin 2.

1.3.3 The Controversial Immune Cell Type in Tumors: B Cells

The function of B cells in the TME is relatively unclear and controversial. B cells may concentrate at tumor margins or form tumor-associated immune aggregates of various complexity, ranging from small unorganized clusters to structured tertiary lymphoid structures (TLS) (Kroeger et al. 2016; Pitzalis et al. 2014). Antigen-specific interaction between T cells and B cells in the TLS is critical for the protective role of T cells. Co-transfer of ex vivo activated B cells and T cells in the 4T1 mouse model of breast cancer leads to better outcomes than the transfer of either population alone (Li et al. 2011). B cells can produce cytokines to coordinate with cytotoxic T lymphocyte (CTL) activity and serve as APCs to provide initial T cell activation in the lymph nodes (Rubtsov et al. 2015). High levels of expression of B cell signature genes result in increased overall survival in patients with melanoma, lung cancer, pancreatic cancer, and head and neck squamous cell carcinoma.

Although B cells possess antitumor functions, some studies indicate that B cells also have protumor effects. T cell infiltration into tumors and cytokine production are both increased in B cell-deficient mice (Shah et al. 2005). Moreover, B cells produce cytokines to recruit MDSCs, enhance angiogenesis, and inhibit the CTL functions (Tsou et al. 2016; de Visser et al. 2005). High levels of B cells are positively correlated with increased invasiveness in bladder cancer and reduced survival of patients with renal cell carcinoma (Woo et al. 2014; Ou et al. 2015). Further studies should be carried out to elucidate the mechanism by which B cells exert their functions.

1.3.4 The cGAS-STING Pathway

Genomic instability is a hallmark as well as a major driving force of cancer (Hanahan and Weinberg 2011; Wang et al. 2019; Wang et al. 2021a). However, genomic instability resulting from DNA damage also promotes antitumor immunity. DNA damage usually leads to cytosolic double-strand DNA (dsDNA) and micronuclei, which activate the innate immune sensor cyclic GMP-AMP synthase (cGAS) and its downstream effector stimulator of interferon genes (STING, also known as MITA, MPYS, and ERIS) (Zhang et al. 2020). In the cytoplasm, DNA fragments bind to cGAS in a sequence-independent but length-dependent manner. DNA ligands promote cGAS activation primarily by inducing conformational changes of the catalytic center of cGAS, thereby facilitating this enzyme to convert GTP and ATP into the second messenger cyclic GMP-AMP (cGAMP), which is a high-affinity ligand for STING (Wu et al. 2013) (Fig. 1.3). STING is a small endoplasmic reticulum (ER)-located protein with four putative transmembrane domains (Wu et al. 2013; Motwani et al. 2019). In the resting state, STING is tethered to the ER by binding to the Ca^{2+} sensor stromal interaction molecule 1 (STIM1) (Srikanth et al. 2019). Upon interaction with cyclic dinucleotides (CDNs), such as cGAMP, STING translocates from the ER to the Golgi apparatus due to the conformational changes (Ishikawa and Barber 2008). In the Golgi, STING is palmitoylated at cysteine 88 and cysteine 91 residues, a posttranslational modification necessary for STING activation (Mukai et al. 2016). Palmitoylated STING is phosphorylated by the kinase

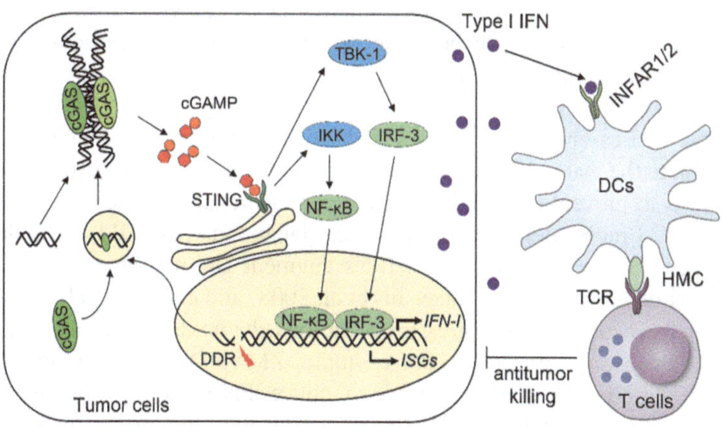

Fig. 1.3 The cGAS-STING pathway. Genomic instability generates micronucleus and cytoplasmic DNA fragments, which trigger the activation of cGAS. Active cGAS dimerizes to synthesize cGAMP from GTP and ATP. cGAMP functions as a second messenger to activate STING on the endoplasmic reticulum (ER). Subsequently, STING activates IRF-3 and NF-κB via TBK-1 and IKK, respectively. Activated IRF-3 and NF-κB translocate into the nucleus to induce the expression of type I interferon (IFN-1) and interferon-stimulated genes (ISGs). Type I IFN binds to the interferon α/β receptor 1 or 2 (IFNAR1/2) on dendritic cells (DCs) to promote the maturation of DCs, thereby priming T cells to kill tumor cells

TANK-binding kinase 1 (TBK1), and phosphorylated STING recruits interferon regulatory factor s (IRF3), which is also phosphorylated by TBK1 to form a homodimer. Dimerized IRF3 enters the nucleus to promote the transcription of type I interferons (IFNs) and interferon-stimulated genes (ISGs) (Tao et al. 2016) (Fig. 1.3). In parallel, STING can also associate with IκB kinase (IKK) and activates the transcription of NF-κB-mediated inflammatory genes (Fig. 1.3). When the stimulant is eliminated, STING will be degraded in the endolysosome (Ma et al. 2015b). The tumor hypoxic and acidic environment promotes the nuclear or mitochondrial DNA to be packaged into exosomes. Exosomal DNA stimulates cGAS-STING pathway once it is absorbed by tumor-infiltrating DCs (Diamond et al. 2018). Additionally, tumor cell-produced cGAMP can be transferred by the folate transporter SLC19A1 into tumor-infiltrating DCs where it can bind to and activate STING (Ritchie et al. 2019).

Activated cGAS-STING signaling exerts antitumor functions in tumor cells via inducing cellular senescence and releasing type I IFNs (Dou et al. 2017; Yang et al. 2017a). cGAS is required for the expression of senescence markers, including $p16^{INK4a}$ and senescence-associated β-galactosidase (SA-β-Gal). Besides, genotoxic stress and oncogene activation all induce cellular senescence in a cGAS-dependent way. However, cGAS-STING signaling is not an integral component of $p16^{INK4a}$ or SA-β-Gal pathways, suggesting that the cGAS-STING pathway induces senescence through other pathways. One possibility is that the cGAS-STING pathway promotes the production and release of some chemokines, cytokines, and proteases to modulate senescent cells and their microenvironment, a phenomenon that is known as senescence-associated secretory phenotype (SASP) (Li and Chen 2018). Another possibility is that senescence stimuli, including DNA damaging drugs, irradiation, and mitogenic stress, stimulate the production of cytokines and chemokines, such as IFN-β, IL-6, IL-8, and IL-1β. IL-8, and IL-6, enhance senescence signaling in the secreting cells (Dou et al. 2017; Acosta et al. 2008; Yang et al. 2017a). Type I IFNs are critical and essential for cGAS-STING-mediated antitumor functions. Type I IFNs promote the maturation of DCs and impede the endosome-lysosome acidification process to prevent engulfed tumor antigen clearance and increase the expression of MHC-I molecules on the cell surface (Diamond et al. 2011; Reboulet et al. 2010) (Fig. 1.3). In addition, type I IFNs augment DC migration toward lymph nodes to prime tumor T cells (Papewalis et al. 2008) and enhance CTL transfer and infiltration by increasing the expression of various chemokines, such as CXCL9 and CXCL10 (Padovan et al. 2002). For example, ZMYND8 depletion activates the cGAS-STING-NF-κB signaling cascade to induce IFN-β and ISGs in breast cancer cells and tumors (Wang et al. 2021c). Collectively, these findings provide convincing evidence for the feasibility of cGAS and STING inhibitors to suppress tumors.

Intriguingly, the cGAS-STING pathway also possesses pro-tumor functions by creating an immune-suppressive tumor microenvironment and enhancing tumor metastasis. STING depletion significantly reduces immune inhibitory cells and the expression of the immune checkpoint IDO, which is pivotal for immune evasion (Lemos et al. 2016). Besides, STING activation upregulates the expression of PD-L1 and CCR2 in a xenograft mouse model (Fu et al. 2015). The binding of

PD-L1 to its receptor PD-1 on activated T cells inhibits antitumor immunity by counteracting T cell-activated signals. High expression of CCR2 in MDMCs causes the aggregation of tumor-promoting monocytes (Liang et al. 2017). STING activation triggers non-canonical NF-κB signaling and inflammatory responses, consequently promoting the epithelial-to-mesenchymal (EMT) transition and metastasis (Bakhoum et al. 2018). The determinant that induces switching of the cGAS-STING pathway from pro-tumor functions to antitumor functions or vice versa needs to be elucidated.

1.4 Endoplasmic Reticulum Stress and Unfolded Protein Response in Tumor

The endoplasmic reticulum (ER) is a network of branching tubules and flatted sacs that govern protein synthesis, folding, and processing. Under stress conditions, such as hypoxia, ROS accumulation, ATP depletion, and nutrient deprivation, misfolded or unfolded proteins are accumulated in the ER of tumor cells and subsequently stimulate ER stress. To cope with this, cancer cells evoke an adaptive mechanism to restore ER proteostasis known as unfolded protein response (UPR). The three main UPR signaling branches are initiated by stress sensor inositol-requiring enzyme 1α (IRE1α), activating transcription factor 6α (ATF6), and PKR-like ER kinase (PERK), respectively (Hetz et al. 2020). UPR signaling regulates the expression of numerous genes that maintain homeostasis in the ER or induce apoptosis if ER stress remains unmitigated. ER stress and UPR have been documented in most major types of human cancer, such as breast, pancreatic, lung, skin, prostate, and brain cancers (Wang and Kaufman 2014).

1.4.1 The Three UPR Branches

Under normal conditions, IRE1α, PERK, and ATF6 are all bound by the chaperone protein BiP, which keeps them in monomeric and inactive states (Luo and Lee 2013; Lee 2005). ER-associated perturbations lead to dissociation of BiP from IRE1α, PERK, and ATF6. These perturbations include (I) ER protein overloading due to upregulated protein translation or inhibition of misfolded proteins; (II) hypoxia, glucose/nutrient deprivation; (III) disruption of the Ca^{2+} pool in ER; (IV) altered redox regulation of the ER lumen (Xu et al. 2005). Activated sensors subsequently stimulate the downstream signaling pathways to correct the protein-misfolding stress. If the burden can be reduced quickly, the cells can adapt to the insult. Otherwise, insufficient clearance leads to apoptotic cell death.

IRE1α, a type I ER transmembrane protein, possesses kinase and endoribonuclease activity (Fig. 1.4). Under ER stress, IRE1α dissociates from BiP, dimerizes, and

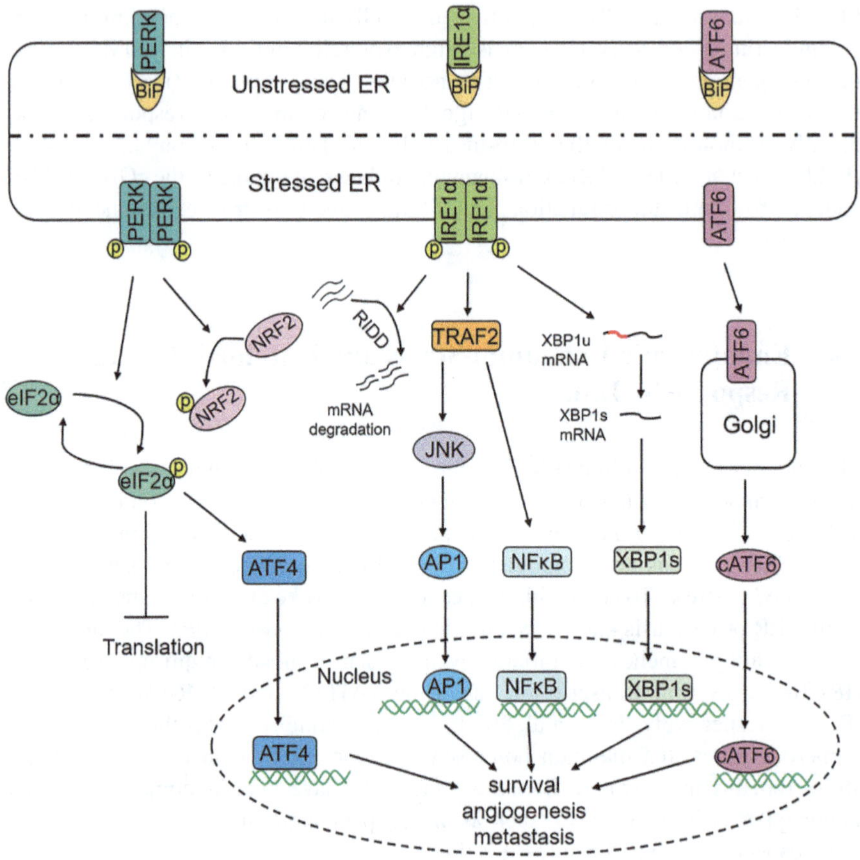

Fig. 1.4 Signaling through the unfolded protein response. Upon endoplasmic reticulum (ER) stress, BiP dissociates from PRK-like ER kinase (PERK), activating transcription factor 6 (ATF6), and inositol-requiring enzyme 1 (IRE1α). PERK homodimerizes and autophosphorylates to block general translation via eukaryotic translation initiation factor 2α (eIF2α) phosphorylation and promotes the expression of ATF4. Moreover, PERK increases the stability of nuclear factor erythroid 2-related factor (NRF2). Released IRE1α also homodimerizes and autophosphorylates. Active IRE1α enhances the expression of the transcription factor X-box binding protein-1 s (XBP1s) and promotes mRNA degradation via regulated URE1-1dependent decay (RIDD). Additionally, IRE1α activates NFκB and AP1 via TNF receptor-associated factor 2 (TRAF2). ATF6 is cleaved at the Golgi and generates active transcription factor cATF6. Ultimately, XBP1s, cATF6, and ATF4 regulate the expression of multiple genes involved in survival, angiogenesis, and metastasis

autophosphorylates to trigger its endoribonuclease activity (Credle et al. 2005; Zhou et al. 2006). The nuclease then excises a small 26-nucleotide intron from the mRNA encoding the transcriptional factor X-box-binding protein 1(XBP1) to generate an active XBP1 transcriptional factor known as XBP1s (Yoshida et al. 2001; Calfon et al. 2002). XBP1s upregulates the expression of foldases, oxidoreductases, intracellular trafficking components, ER-associated degradation machinery, and glycosylases to maintain homeostasis (Shoulders et al. 2013). Under severe ER stress, IRE1α evokes a pro-apoptotic process known as regulated IRE1α-dependent decay (RIDD), which leads to the degradation of multiple ER-localized mRNAs and microRNAs (Hollien et al. 2009; Hollien and Weissman 2006; Upton et al. 2012; Wang et al. 2017). Upon pharmacological ER stress, phosphorylated IRE1α recruits TRAF2 to activate JNK and NF-κB pathways (Tam et al. 2012; Urano et al. 2000). IRE1α is constitutively associated with STAT3 to augment STAT3 phosphorylation and subsequent liver regeneration (Liu et al. 2015). Additionally, IRE1α can form a complex with signal recognition particle RNAs, ribosomal RNAs, and transfer RNAs. However, the biological significance of the IRE1α-small RNA complexes remains unclear (Hassler et al. 2015).

Like IRE1α, the kinase PERK homodimerizes and autophosphorylates upon BiP dissociation (Fig. 1.4). Activated PERK phosphorylates eukaryotic translation factor 2 subunit-α (eIF2α), leading to the attenuation of 5′ cap-dependent translation and subsequent prevention of the nascent proteins flux into the ER (Harding et al. 1999; Vattem and Wek 2004). Paradoxically, phosphorylated eIF2α selectively increases the translation of the transcription factor ATF4, which directly upregulates the transcription factor C/EBP-homologous protein (CHOP). Subsequently, ATF4 and CHOP synergistically induce multiple genes involved in amino acid biosynthesis, amino acid transport, and the intracellular recycling system autophagy (B'Chir et al. 2013; Han et al. 2013). Accelerated amino acid biosynthesis usually generates reactive oxygen species (ROS). However, PERK limits ROS accumulation by phosphorylating NRF2 (Cullinan et al. 2003). Intriguingly, ATF4 participates in a feedback loop to dephosphorylate eIF2α to restore protein synthesis by inducing the expression of the protein phosphatase 1 (PP1) regulatory subunit GADD34 (Harding et al. 1999; Harding et al. 2003; Han et al. 2013).

Upon ER stress, full-length ATF6 (ATF6p90) translocates from the ER to the Golgi apparatus, where it is cleaved by site-1 protease (S1P) and site-2 protease (S2P) to generate an active ATF6 transcription factor ATF6 (ATF6p50) (Haze et al. 1999; Hetz and Papa 2018) (Fig. 1.4). Thereafter, ATF6p50 enters the nucleus to induce the expression of various genes. Unlike PERK and IRE1α, reduced ER membrane fluidity does not activate ATF6, perhaps because dimerization is unfavorable for ATF6 activation. ATF6p50 and XBP1s act coordinately or in parallel to regulate transcription of genes encoding ER chaperones and enzymes that promote ER protein translocation, folding, maturation, secretion, as well as the degradation of the unfolded or misfolded protein (Hassler et al. 2015; Wu et al. 2007; Shoulders et al. 2013).

1.4.2 The Role of ER Stress and UPR in Cell Survival

Severe ER stress in tumors leads to cell death mainly via the PERK- and IRE1α-dependent mechanisms. IRE1α-mediated JNK activation promotes cell death by suppressing the anti-apoptotic activity of BCL2 (Wang and Kaufman 2014). In parallel, RIDD cleaves miR-17, miR-96, and miR-34a, which target pro-apoptotic caspase-2 (Lerner et al. 2012). Excessive intratumor ER stress can also trigger cell death by enhancing immunosurveillance (Kepp et al. 2013). Tumor cells have multiple adaptive mechanisms to repress cell death. ER stress leads to decreased mRNA stability of pro-apoptotic gene *Ddit3* (Rutkowski et al. 2006). IRE1α and PERK activate STAT3 and NF-κB, which upregulate several anti-apoptotic proteins, such as BCL2, caspase-8 inhibitors c-FLIP and MCL-1, and inhibitor of apoptosis proteins (IAP) (Grivennikov and Karin 2010). PERK-mediated NRF2 phosphorylation, glutathione synthesis, and HO-1 upregulation coordinately reduce cytotoxic ROS levels to facilitate cancer cell growth (Dey et al. 2015; Rouschop et al. 2013).

1.4.3 The Role of ER Stress and UPR in Angiogenesis

Angiogenesis is essential for tumor growth and metastasis by supplying nutrients and oxygen. UPR can promote angiogenesis by promoting the transcription of multiple pro-angiogenic factors. Under ER stress, ATF4 directly binds to the *VEGF* promoter to upregulate its transcription, and UPR induction stabilizes the mRNA through the MAPK pathway (Pereira et al. 2010; Binet and Sapieha 2015). Under hypoxia and glucose deprivation, XBP1s, ATF4, and ATF6p50 transcriptionally upregulate VEGFA by directly binding to the *VEGFA* promoter (Ghosh et al. 2010).

Moreover, XBP1s interacts with HIF-1α to induce the transcription of VEGFA (Chen et al. 2014). Surprisingly, VEGF also activates UPR in endothelial cells via mTORC1 to promote endothelial cell survival and angiogenesis (Karali et al. 2014). Furthermore, IRE1α also triggers the expression of multiple pro-angiogenic factors, including fibroblast growth factor 2 (FGF2), IL-6, IL-8, and angiogenin. Thus, UPR and VEGF signaling may form a positive feedback loop to sustain angiogenesis.

1.4.4 The Role of ER Stress and UPR in Metastasis

Metastasis is the leading cause of tumor-associated death. The UPR contributes to multiple steps of the metastasis cascade. ATF4 target gene expression strongly correlates with an EMT gene signature in the breast, colon, gastric, and lung cancer. PERK inhibitors or ATF silencing can dramatically reduce ATF4-mediated EMT (Dey et al. 2015; Feng et al. 2014). ATF4-mediated activation of lysosome-associated membrane protein 3 (LAMP3) also contributes to migration and invasion of hypoxia breast cancer cells in vitro and in vivo (Mujcic et al. 2013). Furthermore, PERK-ATF4 is critical for the expression of EMT-associated gene CREB3L1, which is an ER-located transcription factor (Feng et al. 2017). CREB3L1 activation is stimulated by cleavage via S1P and S2P (Feng et al. 2017). Therefore, PERK is a potential therapeutic target to reduce metastasis. Moreover, IRE1α activity modulates the adhesion and migration properties of GMB cells, whereas XBP1s enhances TNBC metastasis to the lungs (Auf et al. 2010; Chen et al. 2014).

1.5 Tumor Hypoxia

Numerous biochemical reactions in cells require oxygen for electron transport and ATP generation. However, in the core of solid tumors is a hypoxic condition that is caused by excessive oxygen consumption and impaired oxygen supply due to dysfunctional vasculature (Chen et al. 2020; Wang et al. 2021b). Hypoxia significantly affects diverse aspects of tumor biology, including metabolic reprogramming, immune evasion, angiogenesis, metastasis, and drug resistance (Wang et al. 2020b; Chen et al. 2020; Lei et al. 2019). Hence, hypoxia has been considered a hallmark of solid tumors. Upon hypoxia, cancer cells utilize multiple mechanisms to cope with hypoxic stress, among which the hypoxia-inducible factor (HIF) pathway is the key player. HIF is a heterodimer comprising an oxygen-regulated HIF-α subunit and a constitutively expressed HIF-β subunit. In numerous tumors, HIF-α expression is significantly upregulated, and high levels of HIF-α are correlated with poor clinical outcomes. Thus, the HIF-α subunit is an attractive therapeutic target for cancer treatment. HIFs are members of the basic helix-loop-helix-Arn-Sim (bHLH-SAP) transcription factor superfamily. So far, three HIF-α subunits (HIF-1α, HIF-2α, and HIF-3α) and two HIF-β subunits (HIF-1β, also known as ARNT, and ARNT2) have been identified (Luo and Wang 2018; Wang et al. 2021b; Chen et al. 2020). HIF-1α and HIF-2α have been extensively studied, whereas HIF-3α is less investigated. Hereafter, HIF-1α and HIF-2α are collectively referred to as HIF-α.

1.5.1 Regulation of HIF

Under aerobic conditions, the prolyl hydroxylases (PHDs) catalyze the hydroxylation of two conserved proline residues within the oxygen-dependent degradation (ODD) domain of the HIF-α subunit (Fig. 1.5). The hydroxylated HIF-α subunit is recognized and bound by the Von Hipple-Lindau (VHL)/Cullin-2/Elongin-B/C ubiquitin E3 ligase complex, thus promoting the ubiquitylation and proteasomal degradation of HIF-α (Lee et al. 2020). Moreover, factor inhibiting HIF (FIH) is another hydroxylase that hydroxylates HIF-α at a specific asparagine residue, thereby blocking the interaction between HIF-α and its coactivator p300/CBP, and ultimately diminishing HIF transcriptional activity (Fig. 1.5) (Schodel and Ratcliffe 2019). Upon hypoxia, the activity of PHDs and FIH is dramatically reduced, resulting in HIF-α stabilization. The stabilized HIF-α subunit translocates into the nucleus to form a heterodimer with HIF-β. The heterodimer binds to the hypoxia response element (HRE) and recruits p300/CBP, thereby stimulating the expression of hundreds of downstream target genes involved in angiogenesis, metastasis, metabolic reprogramming, pluripotency maintenance, and immune escape (Fig. 1.5) (Lee et al. 2020).

In addition to oxygen-dependent regulation, HIF is also modulated through a myriad of oxygen-independent mechanisms, including but not limited to (1) HIF-α

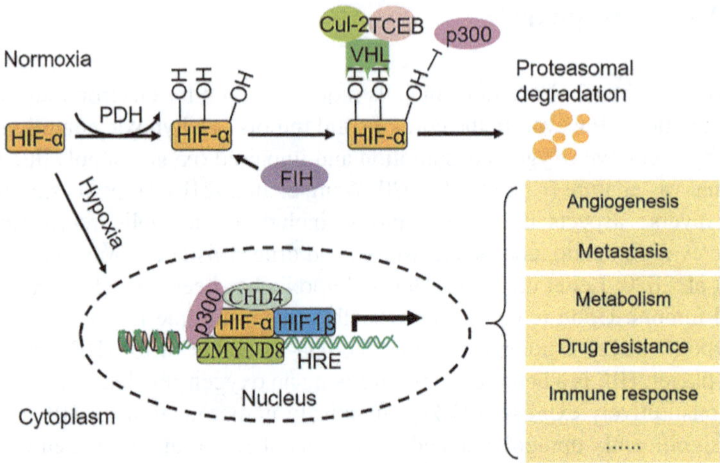

Fig. 1.5 Regulation of hypoxia-inducible factors. Under normoxic conditions, HIF-α subunit is hydroxylated by prolyl hydroxylase (PHD) and factor inhibiting HIF (FIH). PHD-mediated HIF-αhydroxylation leads to HIF-α polyubiquitylation and degradation via the Von Hipple-Lindau (VHL)/Cullin-2/Elongin-B/C ubiquitin E3 ligase complex, whereas FIH-mediated HIF-αhydroxylation impairs the interaction between HIF-α and its coactivator p300, thereby inhibiting HIF transcriptional activity. Both PHD and FIH require oxygen as co-substrate. Under hypoxic conditions, stabilized HIF-α enters the nucleus to interacts with HIF-1β to form a heterodimer which binds to the hypoxia response element (HRE) and recruits p300 to stimulate the expression of downstream genes that involved in multiple biological processes, including angiogenesis, metabolic reprogramming, and immune evasion. ZMYND8 and CHD4 are also HIF coactivators

stability: Numerous proteins and intermediates are involved in the regulation of HIF-α stability, such as USP20, TRAF6, CHIP, RACK1, WSB1lactate, succinate, pyruvate, and fumarate (Semenza 2017; Wang et al. 2020b; Cao et al. 2015; Kim et al. 2015); (2) DNA binding of HIF-α: Chromobox homolog 4 (Cbx4) and YAP enhance the HRE binding of HIF-1α, whereas SET7 has an opposite role (Li et al. 2014; Ma et al. 2015a); (3) formation of HIF heterodimer: Both COMMD1 and MGCRACGAP block the interaction between HIF-1α and HIF-1β; (4) formation of HIF transcriptional complex: ZMYND8, BRD4, CHD4, long and non-coding RNA (lncRNA) RAB11B-AS1 all promote the formation of HIF transcriptional complex and the expression of HIF target genes (Fig. 1.5) (Wang et al. 2020a; Chen et al. 2018a, b; Niu et al. 2020); (5) nuclear localization of HIF-α subunits: γH2AX sequesters HIF-1α in the nucleus (Rezaeian et al. 2017); (6) transcription of HIF-α: JAK/STAT3 and NF-κB signaling promote the transcription of HIF-α (Wang et al. 2020b); and (7) translation of HIF-α: mTOR pathway plays predominant roles in HIF-α translation (Wang et al. 2020b). All these mechanisms contribute to the complexity of HIF regulation.

1.5.2 Hypoxia and HIF in Tumor Angiogenesis

As mentioned above, abnormal structure and function of tumor blood vessels lead to tumor hypoxia and stimulated HIF signaling, which in turn promotes the expression of many genes involved in multiple steps of angiogenesis, such as VEGFA, ANGPTL4, SDF-1, CXCR4, PDGFB, and MMP1 (Luo and Wang 2019). Thus, HIF participates in vascular permeability, endothelial cell proliferation, sprouting, tube formation, and basement membrane degradation. Also, hypoxia activates HIF-independent pathways, such as mTOR and unfolded protein response (UPR), to promote angiogenesis in tumors (Schito 2018). Furthermore, vascularization in HIF-α and HIF-1β knockout mice is defective, suggesting a critical role for HIF in vascular remodeling (de Heer et al. 2020). The roles of HIFs in hypoxia-induced tumor angiogenesis are also well documented (Rapisarda and Melillo 2012).

1.5.3 Hypoxia and HIF in Tumor Metabolic Reprogramming

Hypoxia widely affects cellular metabolism, including glycolysis, glycogen synthesis, oxidative phosphorylation, amino acid metabolism, and lipid metabolism. The expression of most glycolytic genes is directly regulated by HIF, including GLUT1, GLUT3, HK1/2, GP1, PFK1, ALDOA, TP1, GAPDH, LDAHA, PGAM1, ENO1, PKM2, LDHA, and MCT-1/4 (Zhang et al. 2019). GLUT1 and GLUT3 are glucose transporters, LDHA enhances the conversion of pyruvate to lactate, and MCT4 mediates the export of lactate. Thus, hypoxia and HIF promote glucose uptake and lactate secretion by tumor cells. In contrast, glycolytic flux into the TCA cycle in cancer cells is suppressed. Instead, HIF enhances glutaminolysis to sustain the TCA cycle and lipogenesis and ultimately promotes cancer cell survival. Hypoxia augments the reductive activity of IDH1/2, which is essential for de novo lipogenesis. Moreover, FABP3, FABP7, and ADRP are all induced by hypoxia via HIF-1. Thus HIF-1 is required for fatty acids uptake and LD accumulation (Bensaad et al. 2014; Du et al. 2017; Lisec et al. 2019). Decreased OXPHOS and accumulated LD abolish the generation of ROS, thereby promoting the survival of cells under hypoxia-reoxygenation conditions.

1.5.4 Hypoxia and HIF in Tumor Immunity

Hypoxia and HIF profoundly influence most aspects of innate immune cells. HIF-1α depletion leads to dramatically decreased glycolysis and ATP levels in macrophages as well as attenuated aggregation, motility, and bacteria-killing (Cramer et al. 2003). HIF-1α also regulates macrophage survival and M1 macrophage polarization (Tannahill et al. 2013; Staples et al. 2011). The absence of HIF-2α in tumor-associated macrophages (TAMs) abolishes the expression of chemotactic receptors and the migration and infiltration of TAMs. HIF-1α expression levels are usually upregulated in dendritic cells (DCs), which are often exposed to the hypoxic microenvironment, leading to enhanced glycolysis, DC maturation, and T cell activation by DCs (Jantsch et al. 2008). In addition, HIF-1α promotes DC migration via CCR7 (Liu et al. 2019). Moreover, HIF-1α promotes neutrophil glycolysis, phagocytosis, survival, and neutrophil extracellular trap (NET) formation and release (McGettrick and O'Neill 2020).

HIF has been demonstrated to control the differentiation and function of different T cell subsets in both hypoxic and normoxic conditions. HIF-1α enhances Th17 cell differentiation via transcriptional activation of RORγt, which in turn forms a complex with HIF-1α and p300 to stimulate the expression of IL-17 (Dang et al. 2011). In CD8+ T cells, HIF-1α expression is upregulated by the mTOR pathway (Finlay et al. 2012). HIF-1α contributes to the differentiation and function of effector and memory CD8+ T cells by elevating the expression of cytokines IFNγ and TNFα (Palazon et al. 2017). In contrast to Th17 and CD8+ T cells, hypoxia and HIF-1α have opposite effects on Treg and Th1 cells since they are acting through different mechanisms (McGettrick and O'Neill 2020).

HIF-1α promotes tumor immune evasion by upregulating the expression of immune checkpoint molecules HLA-G, PD-L1, PD-L2, and soluble CD137 (sCD137). HLA-G suppresses IFNγ and TNFα and subsequently blocks tumor cell recognition by T cells. PD-L1 and PD-L2 on the tumor cell surface interact with PD-1 on the T cell surface to inactivate T cells, thus leading to immune escape. CD137 of tumor cells binds to CD137L that is expressed by DCs, preventing the interaction between CD137L and T cell-expressed CD137, thereby promoting adaptive evasion (You et al. 2020).

1.6 Conclusions

The rapid and unlimited tumor growth dramatically changes the tumor microenvironment, including blood vessels, oxygen concentration, metabolism, ER homeostasis, and immune response, which in turn profoundly influence the tumor growth and progression. Moreover, tumor angiogenesis, metabolism/metabolites, hypoxia, ER stress-induced UPR, and the immune system can mutually regulate each other to form cross-linked networks, all of which contribute to the complexity of tumor biology and chemotherapy resistance. On the other hand, the crosstalk between these pathways provides potential therapeutic targets and rationality for combination therapies. Thus, elucidation of tumor biology will provide novel strategies for drug delivery and improve the efficacy of chemotherapy.

Acknowledgments This work was supported by grants from Taishan Scholar Young Expert Program of Shandong Province (tsqn202306155), Shandong Provincial Natural Science Foundation (ZR2023MH252), the National Natural Science Foundation of China (81902691),

Guangzhou Science and Technology Program (202102080147), and Guangdong Basic and Applied Basic Research Foundation (2021A1515011224, 2022A1515011738) to Y.C.; National Natural Science Foundation of China (82203297), Shandong Excellent Young Scientists Fund Program (Overseas) (2022HWYQ-077), and Taishan Scholar Young Expert Program of Shandong Province (tsqn202306154) to Y.W., and National Natural Science Foundation of China (81773758) and Guangdong Natural Science Foundation (2019A1515011934) to T.L.

References

Acosta JC, O'Loghlen A, Banito A et al (2008) Chemokine signaling via the CXCR2 receptor reinforces senescence. Cell 133(6):1006–1018. https://doi.org/10.1016/j.cell.2008.03.038

Adams CM, Reitz J, De Brabander JK et al (2004) Cholesterol and 25-hydroxycholesterol inhibit activation of SREBPs by different mechanisms, both involving SCAP and Insigs. J Biol Chem 279(50):52772–52780. https://doi.org/10.1074/jbc.M410302200

Amelio I, Cutruzzola F, Antonov A et al (2014) Serine and glycine metabolism in cancer. Trends Biochem Sci 39(4):191–198. https://doi.org/10.1016/j.tibs.2014.02.004

Angara K, Borin TF, Arbab AS (2017) Vascular Mimicry: a novel neovascularization mechanism driving anti-angiogenic therapy (AAT) resistance in glioblastoma. Transl Oncol 10(4):650–660. https://doi.org/10.1016/j.tranon.2017.04.007

Auf G, Jabouille A, Guerit S et al (2010) Inositol-requiring enzyme 1alpha is a key regulator of angiogenesis and invasion in malignant glioma. Proc Natl Acad Sci USA 107(35):15553–15558. https://doi.org/10.1073/pnas.0914072107

B'Chir W, Maurin AC, Carraro V et al (2013) The eIF2alpha/ATF4 pathway is essential for stress-induced autophagy gene expression. Nucleic Acids Res 41(16):7683–7699. https://doi.org/10.1093/nar/gkt563

Bakhoum SF, Ngo B, Laughney AM et al (2018) Chromosomal instability drives metastasis through a cytosolic DNA response. Nature 553(7689):467–472. https://doi.org/10.1038/nature25432

Barthel A, Okino ST, Liao J et al (1999) Regulation of GLUT1 gene transcription by the serine/threonine kinase Akt1. J Biol Chem 274(29):20281–20286. https://doi.org/10.1074/jbc.274.29.20281

Bennett MK, Lopez JM, Sanchez HB et al (1995) Sterol regulation of fatty acid synthase promoter. Coordinate feedback regulation of two major lipid pathways. J Biol Chem 270(43):25578–25583. https://doi.org/10.1074/jbc.270.43.25578

Bensaad K, Favaro E, Lewis CA et al (2014) Fatty acid uptake and lipid storage induced by HIF-1alpha contribute to cell growth and survival after hypoxia-reoxygenation. Cell Rep 9(1):349–365. https://doi.org/10.1016/j.celrep.2014.08.056

Binet F, Sapieha P (2015) ER stress and angiogenesis. Cell Metab 22(4):560–575. https://doi.org/10.1016/j.cmet.2015.07.010

Bottcher JP, Bonavita E, Chakravarty P et al (2018) NK Cells Stimulate Recruitment of cDC1 into the tumor microenvironment promoting cancer immune control. Cell 172(5):1022–1037.e1014. https://doi.org/10.1016/j.cell.2018.01.004

Brown MS, Goldstein JL (1997) The SREBP pathway: regulation of cholesterol metabolism by proteolysis of a membrane-bound transcription factor. Cell 89(3):331–340. https://doi.org/10.1016/s0092-8674(00)80213-5

Calfon M, Zeng H, Urano F et al (2002) IRE1 couples endoplasmic reticulum load to secretory capacity by processing the XBP-1 mRNA. Nature 415(6867):92–96. https://doi.org/10.1038/415092a

Cantor JR, Sabatini DM (2012) Cancer cell metabolism: one hallmark, many faces. Cancer Discov 2(10):881–898. https://doi.org/10.1158/2159-8290.CD-12-0345

Cao J, Wang Y, Dong R et al (2015) Hypoxia-induced WSB1 promotes the metastatic potential of osteosarcoma cells. Cancer Res 75(22):4839–4851. https://doi.org/10.1158/0008-5472.CAN-15-0711

Carracedo A, Cantley LC, Pandolfi PP (2013) Cancer metabolism: fatty acid oxidation in the limelight. Nat Rev Cancer 13(4):227–232. https://doi.org/10.1038/nrc3483

Chang EI, Chang EI, Thangarajah H et al (2007) Hypoxia, hormones, and endothelial progenitor cells in hemangioma. Lymphat Res Biol 5(4):237–243. https://doi.org/10.1089/lrb.2007.1014

Chen X, Iliopoulos D, Zhang Q et al (2014) XBP1 promotes triple-negative breast cancer by controlling the HIF1alpha pathway. Nature 508(7494):103–107. https://doi.org/10.1038/nature13119

Chen Y, Wang Y, Luo W (2018a) ZMYND8 is a primary HIF coactivator that mediates breast cancer progression. Mol Cell Oncol 5(4):e1479619. https://doi.org/10.1080/23723556.2018.1479619

Chen Y, Zhang B, Bao L et al (2018b) ZMYND8 acetylation mediates HIF-dependent breast cancer progression and metastasis. J Clin Invest 128(5):1937–1955. https://doi.org/10.1172/JCI95089

Chen Y, Liu X, Yuan H et al (2019) Therapeutic remodeling of the tumor microenvironment enhances nanoparticle delivery. Adv Sci (Weinh) 6(5):1802070. https://doi.org/10.1002/advs.201802070

Chen Y, Liu M, Niu Y et al (2020) Romance of the three kingdoms in hypoxia: HIFs, epigenetic regulators, and chromatin reprogramming. Cancer Lett 495:211–223. https://doi.org/10.1016/j.canlet.2020.09.009

Cheng JT, Deng YN, Yi HM et al (2016) Hepatic carcinoma-associated fibroblasts induce IDO-producing regulatory dendritic cells through IL-6-mediated STAT3 activation. Oncogenesis 5:e198. https://doi.org/10.1038/oncsis.2016.7

Chopra H, Hung MK, Kwong DL et al (2018) Insights into endothelial progenitor cells: origin, classification, potentials, and prospects. Stem Cells Int 2018:9847015. https://doi.org/10.1155/2018/9847015

Chou TY, Hart GW, Dang CV (1995) c-Myc is glycosylated at threonine 58, a known phosphorylation site and a mutational hot spot in lymphomas. J Biol Chem 270(32):18961–18965. https://doi.org/10.1074/jbc.270.32.18961

Cortes-Cros M, Hemmerlin C, Ferretti S et al (2013) M2 isoform of pyruvate kinase is dispensable for tumor maintenance and growth. Proc Natl Acad Sci USA 110(2):489–494. https://doi.org/10.1073/pnas.1212780110

Cox AG, Tsomides A, Yimlamai D et al (2018) Yap regulates glucose utilization and sustains nucleotide synthesis to enable organ growth. EMBO J 37(22):10.15252/embj.2018100294

Cramer T, Yamanishi Y, Clausen BE et al (2003) HIF-1alpha is essential for myeloid cell-mediated inflammation. Cell 112(5):645–657. https://doi.org/10.1016/s0092-8674(03)00154-5

Credle JJ, Finer-Moore JS, Papa FR et al (2005) On the mechanism of sensing unfolded protein in the endoplasmic reticulum. Proc Natl Acad Sci USA 102(52):18773–18784. https://doi.org/10.1073/pnas.0509487102

Cullinan SB, Zhang D, Hannink M et al (2003) Nrf2 is a direct PERK substrate and effector of PERK-dependent cell survival. Mol Cell Biol 23(20):7198–7209. https://doi.org/10.1128/mcb.23.20.7198-7209.2003

Currie E, Schulze A, Zechner R et al (2013) Cellular fatty acid metabolism and cancer. Cell Metab 18(2):153–161. https://doi.org/10.1016/j.cmet.2013.05.017

Dang EV, Barbi J, Yang HY et al (2011) Control of T(H)17/T(reg) balance by hypoxia-inducible factor 1. Cell 146(5):772–784. https://doi.org/10.1016/j.cell.2011.07.033

Dasgupta S, Rajapakshe K, Zhu B et al (2018) Metabolic enzyme PFKFB4 activates transcriptional coactivator SRC-3 to drive breast cancer. Nature 556(7700):249–254. https://doi.org/10.1038/s41586-018-0018-1

de Heer EC, Jalving M, Harris AL (2020) HIFs, angiogenesis, and metabolism: elusive enemies in breast cancer. J Clin Invest 130(10):5074–5087. https://doi.org/10.1172/JCI137552

de Visser KE, Korets LV, Coussens LM (2005) De novo carcinogenesis promoted by chronic inflammation is B lymphocyte dependent. Cancer Cell 7(5):411–423. https://doi.org/10.1016/j.ccr.2005.04.014

Dey S, Sayers CM, Verginadis II et al (2015) ATF4-dependent induction of heme oxygenase 1 prevents anoikis and promotes metastasis. J Clin Invest 125(7):2592–2608. https://doi.org/10.1172/JCI78031

Diamond MS, Kinder M, Matsushita H et al (2011) Type I interferon is selectively required by dendritic cells for immune rejection of tumors. J Exp Med 208(10):1989–2003. https://doi.org/10.1084/jem.20101158

Diamond JM, Vanpouille-Box C, Spada S et al (2018) Exosomes shuttle TREX1-sensitive IFN-stimulatory dsDNA from irradiated cancer cells to DCs. Cancer Immunol Res 6(8):910–920. https://doi.org/10.1158/2326-6066.CIR-17-0581

Dou Z, Ghosh K, Vizioli MG et al (2017) Cytoplasmic chromatin triggers inflammation in senescence and cancer. Nature 550(7676):402–406. https://doi.org/10.1038/nature24050

Du W, Zhang L, Brett-Morris A et al (2017) HIF drives lipid deposition and cancer in ccRCC via repression of fatty acid metabolism. Nat Commun 8(1):1769. https://doi.org/10.1038/s41467-017-01965-8

Feng YX, Sokol ES, Del Vecchio CA et al (2014) Epithelial-to-mesenchymal transition activates PERK-eIF2alpha and sensitizes cells to endoplasmic reticulum stress. Cancer Discov 4(6):702–715. https://doi.org/10.1158/2159-8290.CD-13-0945

Feng YX, Jin DX, Sokol ES et al (2017) Cancer-specific PERK signaling drives invasion and metastasis through CREB3L1. Nat Commun 8(1):1079. https://doi.org/10.1038/s41467-017-01052-y

Ferrer CM, Lynch TP, Sodi VL et al (2014) O-GlcNAcylation regulates cancer metabolism and survival stress signaling via regulation of the HIF-1 pathway. Mol Cell 54(5):820–831. https://doi.org/10.1016/j.molcel.2014.04.026

Finlay DK, Rosenzweig E, Sinclair LV et al (2012) PDK1 regulation of mTOR and hypoxia-inducible factor 1 integrate metabolism and migration of CD8+ T cells. J Exp Med 209(13):2441–2453. https://doi.org/10.1084/jem.20112607

Folkman J (2007) Angiogenesis: an organizing principle for drug discovery? Nat Rev Drug Discov 6(4):273–286. https://doi.org/10.1038/nrd2115

Fong L, Small EJ (2008) Anti-cytotoxic T-lymphocyte antigen-4 antibody: the first in an emerging class of immunomodulatory antibodies for cancer treatment. J Clin Oncol 26(32):5275–5283. https://doi.org/10.1200/JCO.2008.17.8954

Fu J, Kanne DB, Leong M et al (2015) STING agonist formulated cancer vaccines can cure established tumors resistant to PD-1 blockade. Sci Transl Med 7(283):283ra252. https://doi.org/10.1126/scitranslmed.aaa4306

Fu V, Moroishi T, Guan KL (2016) Glycoholics anonymous: cancer sobers up with mTORC1. Cancer Cell 29(4):432–434. https://doi.org/10.1016/j.ccell.2016.03.016

Furuta E, Pai SK, Zhan R et al (2008) Fatty acid synthase gene is up-regulated by hypoxia via activation of Akt and sterol regulatory element binding protein-1. Cancer Res 68(4):1003–1011. https://doi.org/10.1158/0008-5472.CAN-07-2489

Gao P, Tchernyshyov I, Chang TC et al (2009) c-Myc suppression of miR-23a/b enhances mitochondrial glutaminase expression and glutamine metabolism. Nature 458(7239):762–765. https://doi.org/10.1038/nature07823

Gerhardt H, Golding M, Fruttiger M et al (2003) VEGF guides angiogenic sprouting utilizing endothelial tip cell filopodia. J Cell Biol 161(6):1163–1177. https://doi.org/10.1083/jcb.200302047

Ghosh R, Lipson KL, Sargent KE et al (2010) Transcriptional regulation of VEGF-A by the unfolded protein response pathway. PLoS One 5(3):e9575. https://doi.org/10.1371/journal.pone.0009575

Gottlob K, Majewski N, Kennedy S et al (2001) Inhibition of early apoptotic events by Akt/PKB is dependent on the first committed step of glycolysis and mitochondrial hexokinase. Genes Dev 15(11):1406–1418. https://doi.org/10.1101/gad.889901

Greenfield JP, Cobb WS, Lyden D (2010) Resisting arrest: a switch from angiogenesis to vasculogenesis in recurrent malignant gliomas. J Clin Invest 120(3):663–667. https://doi.org/10.1172/JCI42345

Grivennikov SI, Karin M (2010) Dangerous liaisons: STAT3 and NF-kappaB collaboration and crosstalk in cancer. Cytokine Growth Factor Rev 21(1):11–19. https://doi.org/10.1016/j.cytogfr.2009.11.005

Guillerey C, Huntington ND, Smyth MJ (2016) Targeting natural killer cells in cancer immunotherapy. Nat Immunol 17(9):1025–1036. https://doi.org/10.1038/ni.3518

Han J, Back SH, Hur J et al (2013) ER-stress-induced transcriptional regulation increases protein synthesis leading to cell death. Nat Cell Biol 15(5):481–490. https://doi.org/10.1038/ncb2738

Hanahan D, Weinberg RA (2011) Hallmarks of cancer: the next generation. Cell 144(5):646–674. https://doi.org/10.1016/j.cell.2011.02.013

Harding HP, Zhang Y, Ron D (1999) Protein translation and folding are coupled by an endoplasmic-reticulum-resident kinase. Nature 397(6716):271–274. https://doi.org/10.1038/16729

Harding HP, Zhang Y, Zeng H et al (2003) An integrated stress response regulates amino acid metabolism and resistance to oxidative stress. Mol Cell 11(3):619–633. https://doi.org/10.1016/s1097-2765(03)00105-9

Hassanein M, Qian J, Hoeksema MD et al (2015) Targeting SLC1a5-mediated glutamine dependence in non-small cell lung cancer. Int J Cancer 137(7):1587–1597. https://doi.org/10.1002/ijc.29535

Hassler JR, Scheuner DL, Wang S et al (2015) The IRE1alpha/XBP1s pathway is essential for the glucose response and protection of beta cells. PLoS Biol 13(10):e1002277. https://doi.org/10.1371/journal.pbio.1002277

Hay N (2016) Reprogramming glucose metabolism in cancer: can it be exploited for cancer therapy? Nat Rev Cancer 16(10):635–649. https://doi.org/10.1038/nrc.2016.77

Haze K, Yoshida H, Yanagi H et al (1999) Mammalian transcription factor ATF6 is synthesized as a transmembrane protein and activated by proteolysis in response to endoplasmic reticulum stress. Mol Biol Cell 10(11):3787–3799. https://doi.org/10.1091/mbc.10.11.3787

He S, Li X, Li R et al (2016) Annexin A2 modulates ROS and impacts inflammatory response via IL-17 signaling in polymicrobial sepsis mice. PLoS Pathog 12(7):e1005743. https://doi.org/10.1371/journal.ppat.1005743

Heissig B, Hattori K, Dias S et al (2002) Recruitment of stem and progenitor cells from the bone marrow niche requires MMP-9 mediated release of kit-ligand. Cell 109(5):625–637. https://doi.org/10.1016/s0092-8674(02)00754-7

Hetz C, Papa FR (2018) The unfolded protein response and cell fate control. Mol Cell 69(2):169–181. https://doi.org/10.1016/j.molcel.2017.06.017

Hetz C, Zhang K, Kaufman RJ (2020) Mechanisms, regulation and functions of the unfolded protein response. Nat Rev Mol Cell Biol 21(8):421–438. https://doi.org/10.1038/s41580-020-0250-z

Hirschey MD, DeBerardinis RJ, Diehl AME et al (2015) Dysregulated metabolism contributes to oncogenesis. Semin Cancer Biol 35(Suppl):S129–S150. https://doi.org/10.1016/j.semcancer.2015.10.002

Hollien J, Weissman JS (2006) Decay of endoplasmic reticulum-localized mRNAs during the unfolded protein response. Science 313(5783):104–107. https://doi.org/10.1126/science.1129631

Hollien J, Lin JH, Li H et al (2009) Regulated Ire1-dependent decay of messenger RNAs in mammalian cells. J Cell Biol 186(3):323–331. https://doi.org/10.1083/jcb.200903014

Huang LT, Li X, Zhang L et al (2014) HIF-1-mediated suppression of acyl-CoA dehydrogenases and fatty acid oxidation is critical for cancer progression. Cell Rep 8(6):1930–1942. https://doi.org/10.1016/j.celrep.2014.08.028

Ikeda S, Kawahara-Miki R, Iwata H et al (2017) Role of methionine adenosyltransferase 2A in bovine preimplantation development and its associated genomic regions. Sci Rep 7(1):3800. https://doi.org/10.1038/s41598-017-04003-1

Ingram DA, Mead LE, Tanaka H et al (2004) Identification of a novel hierarchy of endothelial progenitor cells using human peripheral and umbilical cord blood. Blood 104(9):2752–2760. https://doi.org/10.1182/blood-2004-04-1396

Ishikawa H, Barber GN (2008) STING is an endoplasmic reticulum adaptor that facilitates innate immune signalling. Nature 455(7213):674–678. https://doi.org/10.1038/nature07317

Israelsen WJ, Dayton TL, Davidson SM et al (2013) PKM2 isoform-specific deletion reveals a differential requirement for pyruvate kinase in tumor cells. Cell 155(2):397–409. https://doi.org/10.1016/j.cell.2013.09.025

Jakobsson L, Bentley K, Gerhardt H (2009) VEGFRs and notch: a dynamic collaboration in vascular patterning. Biochem Soc Trans 37(Pt 6):1233–1236. https://doi.org/10.1042/BST0371233

Jantsch J, Chakravortty D, Turza N et al (2008) Hypoxia and hypoxia-inducible factor-1 alpha modulate lipopolysaccharide-induced dendritic cell activation and function. J Immunol 180(7):4697–4705. https://doi.org/10.4049/jimmunol.180.7.4697

Jiang L, Shestov AA, Swain P et al (2016) Reductive carboxylation supports redox homeostasis during anchorage-independent growth. Nature 532(7598):255–258. https://doi.org/10.1038/nature17393

Jin F, Brockmeier U, Otterbach F et al (2012) New insight into the SDF-1/CXCR4 axis in a breast carcinoma model: hypoxia-induced endothelial SDF-1 and tumor cell CXCR4 are required for tumor cell intravasation. Mol Cancer Res 10(8):1021–1031. https://doi.org/10.1158/1541-7786.MCR-11-0498

Jung SY, Jeon HK, Choi JS et al (2012) Reduced expression of FASN through SREBP-1 down-regulation is responsible for hypoxic cell death in HepG2 cells. J Cell Biochem 113(12):3730–3739. https://doi.org/10.1002/jcb.24247

Karali E, Bellou S, Stellas D et al (2014) VEGF signals through ATF6 and PERK to promote endothelial cell survival and angiogenesis in the absence of ER stress. Mol Cell 54(4):559–572. https://doi.org/10.1016/j.molcel.2014.03.022

Kawauchi K, Araki K, Tobiume K et al (2008) p53 regulates glucose metabolism through an IKK-NF-kappaB pathway and inhibits cell transformation. Nat Cell Biol 10(5):611–618. https://doi.org/10.1038/ncb1724

Kepp O, Menger L, Vacchelli E et al (2013) Crosstalk between ER stress and immunogenic cell death. Cytokine Growth Factor Rev 24(4):311–318. https://doi.org/10.1016/j.cytogfr.2013.05.001

Kim JJ, Lee SB, Jang J et al (2015) WSB1 promotes tumor metastasis by inducing pVHL degradation. Genes Dev 29(21):2244–2257. https://doi.org/10.1101/gad.268128.115

Kopp HG, Hooper AT, Broekman MJ et al (2006) Thrombospondins deployed by thrombopoietic cells determine angiogenic switch and extent of revascularization. J Clin Invest 116(12):3277–3291. https://doi.org/10.1172/JCI29314

Kroeger DR, Milne K, Nelson BH (2016) Tumor-infiltrating plasma cells are associated with tertiary lymphoid structures, cytolytic T-Cell responses, and superior prognosis in ovarian cancer. Clin Cancer Res 22(12):3005–3015. https://doi.org/10.1158/1078-0432.CCR-15-2762

Ladanyi A, Mukherjee A, Kenny HA et al (2018) Adipocyte-induced CD36 expression drives ovarian cancer progression and metastasis. Oncogene 37(17):2285–2301. https://doi.org/10.1038/s41388-017-0093-z

Lee AS (2005) The ER chaperone and signaling regulator GRP78/BiP as a monitor of endoplasmic reticulum stress. Methods 35(4):373–381. https://doi.org/10.1016/j.ymeth.2004.10.010

Lee P, Chandel NS, Simon MC (2020) Cellular adaptation to hypoxia through hypoxia inducible factors and beyond. Nat Rev Mol Cell Biol 21(5):268–283. https://doi.org/10.1038/s41580-020-0227-y

Lei Y, Wang K, Li X et al (2019) Cell-surface translocation of annexin A2 contributes to bleomycin-induced pulmonary fibrosis by mediating inflammatory response in mice. Clin Sci (Lond) 133(7):789–804. https://doi.org/10.1042/CS20180687

Lemos H, Mohamed E, Huang L et al (2016) STING promotes the growth of tumors characterized by low antigenicity via IDO activation. Cancer Res 76(8):2076–2081. https://doi.org/10.1158/0008-5472.CAN-15-1456

Lerner AG, Upton JP, Praveen PV et al (2012) IRE1alpha induces thioredoxin-interacting protein to activate the NLRP3 inflammasome and promote programmed cell death under irremediable ER stress. Cell Metab 16(2):250–264. https://doi.org/10.1016/j.cmet.2012.07.007

Li T, Chen ZJ (2018) The cGAS-cGAMP-STING pathway connects DNA damage to inflammation, senescence, and cancer. J Exp Med 215(5):1287–1299. https://doi.org/10.1084/jem.20180139

Li Q, Lao X, Pan Q et al (2011) Adoptive transfer of tumor reactive B cells confers host T-cell immunity and tumor regression. Clin Cancer Res 17(15):4987–4995. https://doi.org/10.1158/1078-0432.CCR-11-0207

Li J, Xu Y, Long XD et al (2014) Cbx4 governs HIF-1alpha to potentiate angiogenesis of hepatocellular carcinoma by its SUMO E3 ligase activity. Cancer Cell 25(1):118–131. https://doi.org/10.1016/j.ccr.2013.12.008

Liang H, Deng L, Hou Y et al (2017) Host STING-dependent MDSC mobilization drives extrinsic radiation resistance. Nat Commun 8(1):1736. https://doi.org/10.1038/s41467-017-01566-5

Lisec J, Jaeger C, Rashid R et al (2019) Cancer cell lipid class homeostasis is altered under nutrient-deprivation but stable under hypoxia. BMC Cancer 19(1):501. https://doi.org/10.1186/s12885-019-5733-y

Liu Q, Wu K, Zhu Y et al (2007) Silencing MAT2A gene by RNA interference inhibited cell growth and induced apoptosis in human hepatoma cells. Hepatol Res 37(5):376–388. https://doi.org/10.1111/j.1872-034X.2007.00041.x

Liu Y, Shao M, Wu Y et al (2015) Role for the endoplasmic reticulum stress sensor IRE1alpha in liver regenerative responses. J Hepatol 62(3):590–598. https://doi.org/10.1016/j.jhep.2014.10.022

Liu J, Zhang X, Chen K et al (2019) CCR7 chemokine receptor-inducible lnc-Dpf3 restrains dendritic cell migration by inhibiting HIF-1alpha-mediated glycolysis. Immunity 50(3):600–615 e615. https://doi.org/10.1016/j.immuni.2019.01.021

Locasale JW, Grassian AR, Melman T et al (2011) Phosphoglycerate dehydrogenase diverts glycolytic flux and contributes to oncogenesis. Nat Genet 43(9):869–874. https://doi.org/10.1038/ng.890

Lugano R, Ramachandran M, Dimberg A (2020) Tumor angiogenesis: causes, consequences, challenges and opportunities. Cell Mol Life Sci 77(9):1745–1770. https://doi.org/10.1007/s00018-019-03351-7

Luo B, Lee AS (2013) The critical roles of endoplasmic reticulum chaperones and unfolded protein response in tumorigenesis and anticancer therapies. Oncogene 32(7):805–818. https://doi.org/10.1038/onc.2012.130

Luo W, Semenza GL (2012) Emerging roles of PKM2 in cell metabolism and cancer progression. Trends Endocrinol Metab 23(11):560–566. https://doi.org/10.1016/j.tem.2012.06.010

Luo W, Wang Y (2018) Epigenetic regulators: multifunctional proteins modulating hypoxia-inducible factor-alpha protein stability and activity. Cell Mol Life Sci 75(6):1043–1056. https://doi.org/10.1007/s00018-017-2684-9

Luo W, Wang Y (2019) Hypoxia mediates tumor malignancy and therapy resistance. Adv Exp Med Biol 1136:1–18. https://doi.org/10.1007/978-3-030-12734-3_1

Luo W, Hu H, Chang R et al (2011) Pyruvate kinase M2 is a PHD3-stimulated coactivator for hypoxia-inducible factor 1. Cell 145(5):732–744. https://doi.org/10.1016/j.cell.2011.03.054

Ma Z, Vocadlo DJ, Vosseller K (2013) Hyper-O-GlcNAcylation is anti-apoptotic and maintains constitutive NF-kappaB activity in pancreatic cancer cells. J Biol Chem 288(21):15121–15130. https://doi.org/10.1074/jbc.M113.470047

Ma B, Chen Y, Chen L et al (2015a) Hypoxia regulates Hippo signalling through the SIAH2 ubiquitin E3 ligase. Nat Cell Biol 17(1):95–103. https://doi.org/10.1038/ncb3073

Ma Z, Jacobs SR, West JA et al (2015b) Modulation of the cGAS-STING DNA sensing pathway by gammaherpesviruses. Proc Natl Acad Sci USA 112(31):E4306–E4315. https://doi.org/10.1073/pnas.1503831112

Mahmood S, Upreti D, Sow I et al (2015) Bidirectional interactions of NK cells and dendritic cells in immunotherapy: current and future perspective. Immunotherapy 7(3):301–308. https://doi.org/10.2217/imt.14.122

Maniotis AJ, Folberg R, Hess A et al (1999) Vascular channel formation by human melanoma cells in vivo and in vitro: vasculogenic mimicry. Am J Pathol 155(3):739–752. https://doi.org/10.1016/S0002-9440(10)65173-5

Mattaini KR, Sullivan MR, Vander Heiden MG (2016) The importance of serine metabolism in cancer. J Cell Biol 214(3):249–257. https://doi.org/10.1083/jcb.201604085

McGettrick AF, O'Neill LAJ (2020) The role of HIF in immunity and inflammation. Cell Metab 32(4):524–536. https://doi.org/10.1016/j.cmet.2020.08.002

Metallo CM, Gameiro PA, Bell EL et al (2011) Reductive glutamine metabolism by IDH1 mediates lipogenesis under hypoxia. Nature 481(7381):380–384. https://doi.org/10.1038/nature10602

Michea P, Noel F, Zakine E et al (2018) Adjustment of dendritic cells to the breast-cancer microenvironment is subset specific. Nat Immunol 19(8):885–897. https://doi.org/10.1038/s41590-018-0145-8

Michielsen AJ, Hogan AE, Marry J et al (2011) Tumour tissue microenvironment can inhibit dendritic cell maturation in colorectal cancer. PLoS One 6(11):e27944. https://doi.org/10.1371/journal.pone.0027944

Molon B, Ugel S, Del Pozzo F et al (2011) Chemokine nitration prevents intratumoral infiltration of antigen-specific T cells. J Exp Med 208(10):1949–1962. https://doi.org/10.1084/jem.20101956

Motwani M, Pesiridis S, Fitzgerald KA (2019) DNA sensing by the cGAS-STING pathway in health and disease. Nat Rev Genet 20(11):657–674. https://doi.org/10.1038/s41576-019-0151-1

Mujcic H, Nagelkerke A, Rouschop KM et al (2013) Hypoxic activation of the PERK/eIF2alpha arm of the unfolded protein response promotes metastasis through induction of LAMP3. Clin Cancer Res 19(22):6126–6137. https://doi.org/10.1158/1078-0432.CCR-13-0526

Mukai K, Konno H, Akiba T et al (2016) Activation of STING requires palmitoylation at the Golgi. Nat Commun 7:11932. https://doi.org/10.1038/ncomms11932

Mullarky E, Lucki NC, Beheshti Zavareh R et al (2016) Identification of a small molecule inhibitor of 3-phosphoglycerate dehydrogenase to target serine biosynthesis in cancers. Proc Natl Acad Sci USA 113(7):1778–1783. https://doi.org/10.1073/pnas.1521548113

Mullen AR, Wheaton WW, Jin ES et al (2011) Reductive carboxylation supports growth in tumour cells with defective mitochondria. Nature 481(7381):385–388. https://doi.org/10.1038/nature10642

Nakakuki M, Kawano H, Notsu T et al (2014) A novel processing system of sterol regulatory element-binding protein-1c regulated by polyunsaturated fatty acid. J Biochem 155(5):301–313. https://doi.org/10.1093/jb/mvu019

Nicklin P, Bergman P, Zhang B et al (2009) Bidirectional transport of amino acids regulates mTOR and autophagy. Cell 136(3):521–534. https://doi.org/10.1016/j.cell.2008.11.044

Niu Y, Bao L, Chen Y et al (2020) HIF2-Induced long noncoding RNA RAB11B-AS1 promotes hypoxia-mediated angiogenesis and breast cancer metastasis. Cancer Res 80(5):964–975. https://doi.org/10.1158/0008-5472.CAN-19-1532

Nokin MJ, Durieux F, Peixoto P et al (2016) Methylglyoxal, a glycolysis side-product, induces Hsp90 glycation and YAP-mediated tumor growth and metastasis. elife 5. https://doi.org/10.7554/eLife.19375

Nowotarski SL, Woster PM, Casero RA et al (2013) Polyamines and cancer: implications for chemotherapy and chemoprevention. Expert Rev Mol Med 15:e3. https://doi.org/10.1017/erm.2013.3

Nusse R, Clevers H (2017) Wnt/beta-catenin signaling, disease, and emerging therapeutic modalities. Cell 169(6):985–999. https://doi.org/10.1016/j.cell.2017.05.016

Olzmann JA, Carvalho P (2019) Dynamics and functions of lipid droplets. Nat Rev Mol Cell Biol 20(3):137–155. https://doi.org/10.1038/s41580-018-0085-z

Osthus RC, Shim H, Kim S et al (2000) Deregulation of glucose transporter 1 and glycolytic gene expression by c-Myc. J Biol Chem 275(29):21797–21800. https://doi.org/10.1074/jbc.C000023200

Ostrand-Rosenberg S, Sinha P (2009) Myeloid-derived suppressor cells: linking inflammation and cancer. J Immunol 182(8):4499–4506. https://doi.org/10.4049/jimmunol.0802740

Ou Z, Wang Y, Liu L et al (2015) Tumor microenvironment B cells increase bladder cancer metastasis via modulation of the IL-8/androgen receptor (AR)/MMPs signals. Oncotarget 6(28):26065–26078. https://doi.org/10.18632/oncotarget.4569

Pacold ME, Brimacombe KR, Chan SH et al (2016) A PHGDH inhibitor reveals coordination of serine synthesis and one-carbon unit fate. Nat Chem Biol 12(6):452–458. https://doi.org/10.1038/nchembio.2070

Padovan E, Spagnoli GC, Ferrantini M et al (2002) IFN-alpha2a induces IP-10/CXCL10 and MIG/CXCL9 production in monocyte-derived dendritic cells and enhances their capacity to attract and stimulate CD8+ effector T cells. J Leukoc Biol 71(4):669–676

Palazon A, Tyrakis PA, Macias D et al (2017) An HIF-1alpha/VEGF-A Axis in cytotoxic T cells regulates tumor progression. Cancer Cell 32(5):669–683 e665. https://doi.org/10.1016/j.ccell.2017.10.003

Paluskievicz CM, Cao X, Abdi R et al (2019) T regulatory cells and priming the suppressive tumor microenvironment. Front Immunol 10:2453. https://doi.org/10.3389/fimmu.2019.02453

Papewalis C, Jacobs B, Wuttke M et al (2008) IFN-alpha skews monocytes into CD56+-expressing dendritic cells with potent functional activities in vitro and in vivo. J Immunol 180(3):1462–1470. https://doi.org/10.4049/jimmunol.180.3.1462

Park JH, Pyun WY, Park HW (2020) Cancer metabolism: phenotype, signaling and therapeutic targets. Cells 9(10). https://doi.org/10.3390/cells9102308

Patra KC, Hay N (2014) The pentose phosphate pathway and cancer. Trends Biochem Sci 39(8):347–354. https://doi.org/10.1016/j.tibs.2014.06.005

Peng T, Golub TR, Sabatini DM (2002) The immunosuppressant rapamycin mimics a starvation-like signal distinct from amino acid and glucose deprivation. Mol Cell Biol 22(15):5575–5584. https://doi.org/10.1128/mcb.22.15.5575-5584.2002

Peng C, Zhu Y, Zhang W et al (2017) Regulation of the Hippo-YAP pathway by glucose sensor O-GlcNAcylation. Mol Cell 68(3):591–604 e595. https://doi.org/10.1016/j.molcel.2017.10.010

Pereira ER, Liao N, Neale GA et al (2010) Transcriptional and post-transcriptional regulation of proangiogenic factors by the unfolded protein response. PLoS One 5(9). https://doi.org/10.1371/journal.pone.0012521

Petan T, Jarc E, Jusovic M (2018) Lipid droplets in cancer: guardians of fat in a stressful world. Molecules 23(8). https://doi.org/10.3390/molecules23081941

Pitzalis C, Jones GW, Bombardieri M et al (2014) Ectopic lymphoid-like structures in infection, cancer and autoimmunity. Nat Rev Immunol 14(7):447–462. https://doi.org/10.1038/nri3700

Possemato R, Marks KM, Shaul YD et al (2011) Functional genomics reveal that the serine synthesis pathway is essential in breast cancer. Nature 476(7360):346–350. https://doi.org/10.1038/nature10350

Qing G, Li B, Vu A et al (2012) ATF4 regulates MYC-mediated neuroblastoma cell death upon glutamine deprivation. Cancer Cell 22(5):631–644. https://doi.org/10.1016/j.ccr.2012.09.021

Quinlan CL, Kaiser SE, Bolanos B et al (2017) Targeting S-adenosylmethionine biosynthesis with a novel allosteric inhibitor of Mat2A. Nat Chem Biol 13(7):785–792. https://doi.org/10.1038/nchembio.2384

Qureshi OS, Zheng Y, Nakamura K et al (2011) Trans-endocytosis of CD80 and CD86: a molecular basis for the cell-extrinsic function of CTLA-4. Science 332(6029):600–603. https://doi.org/10.1126/science.1202947

Raber P, Ochoa AC, Rodriguez PC (2012) Metabolism of L-arginine by myeloid-derived suppressor cells in cancer: mechanisms of T cell suppression and therapeutic perspectives. Immunol Investig 41(6–7):614–634. https://doi.org/10.3109/08820139.2012.680634

Ramani K, Lu SC (2017) Methionine adenosyltransferases in liver health and diseases. Liver Res 1(2):103–111. https://doi.org/10.1016/j.livres.2017.07.002

Rapisarda A, Melillo G (2012) Overcoming disappointing results with antiangiogenic therapy by targeting hypoxia. Nat Rev Clin Oncol 9(7):378–390. https://doi.org/10.1038/nrclinonc.2012.64

Reboulet RA, Hennies CM, Garcia Z et al (2010) Prolonged antigen storage endows merocytic dendritic cells with enhanced capacity to prime anti-tumor responses in tumor-bearing mice. J Immunol 185(6):3337–3347. https://doi.org/10.4049/jimmunol.1001619

Ren K, Yao N, Wang G et al (2014) Vasculogenic mimicry: a new prognostic sign of human osteosarcoma. Hum Pathol 45(10):2120–2129. https://doi.org/10.1016/j.humpath.2014.06.013

Ren P, Yue M, Xiao D et al (2015) ATF4 and N-Myc coordinate glutamine metabolism in MYCN-amplified neuroblastoma cells through ASCT2 activation. J Pathol 235(1):90–100. https://doi.org/10.1002/path.4429

Renoux VM, Zriwil A, Peitzsch C et al (2015) Identification of a human natural killer cell lineage-restricted progenitor in fetal and adult tissues. Immunity 43(2):394–407. https://doi.org/10.1016/j.immuni.2015.07.011

Rey S, Schito L, Wouters BG et al (2017) Targeting hypoxia-inducible factors for antiangiogenic cancer therapy. Trends Cancer 3(7):529–541. https://doi.org/10.1016/j.trecan.2017.05.002

Rezaeian AH, Li CF, Wu CY et al (2017) A hypoxia-responsive TRAF6-ATM-H2AX signalling axis promotes HIF1alpha activation, tumorigenesis and metastasis. Nat Cell Biol 19(1):38–51. https://doi.org/10.1038/ncb3445

Ritchie C, Cordova AF, Hess GT et al (2019) SLC19A1 is an importer of the immunotransmitter cGAMP. Mol Cell 75(2):372–381 e375. https://doi.org/10.1016/j.molcel.2019.05.006

Roberts DJ, Tan-Sah VP, Smith JM et al (2013) Akt phosphorylates HK-II at Thr-473 and increases mitochondrial HK-II association to protect cardiomyocytes. J Biol Chem 288(33):23798–23806. https://doi.org/10.1074/jbc.M113.482026

Rouschop KM, Dubois LJ, Keulers TG et al (2013) PERK/eIF2alpha signaling protects therapy resistant hypoxic cells through induction of glutathione synthesis and protection against ROS. Proc Natl Acad Sci USA 110(12):4622–4627. https://doi.org/10.1073/pnas.1210633110

Rubtsov AV, Rubtsova K, Kappler JW et al (2015) CD11c-expressing B cells are located at the T cell/B cell border in spleen and are potent APCs. J Immunol 195(1):71–79. https://doi.org/10.4049/jimmunol.1500055

Rutkowski DT, Arnold SM, Miller CN et al (2006) Adaptation to ER stress is mediated by differential stabilities of pro-survival and pro-apoptotic mRNAs and proteins. PLoS Biol 4(11):e374. https://doi.org/10.1371/journal.pbio.0040374

Samstein RM, Arvey A, Josefowicz SZ et al (2012) Foxp3 exploits a pre-existent enhancer landscape for regulatory T cell lineage specification. Cell 151(1):153–166. https://doi.org/10.1016/j.cell.2012.06.053

Scatena R, Bottoni P, Pontoglio A et al (2008) Glycolytic enzyme inhibitors in cancer treatment. Expert Opin Investig Drugs 17(10):1533–1545. https://doi.org/10.1517/13543784.17.10.1533

Schafer ZT, Grassian AR, Song L et al (2009) Antioxidant and oncogene rescue of metabolic defects caused by loss of matrix attachment. Nature 461(7260):109–113. https://doi.org/10.1038/nature08268

Schito L (2018) Bridging angiogenesis and immune evasion in the hypoxic tumor microenvironment. Am J Physiol Regul Integr Comp Physiol 315(6):R1072–R1084. https://doi.org/10.1152/ajpregu.00209.2018

Schodel J, Ratcliffe PJ (2019) Mechanisms of hypoxia signalling: new implications for nephrology. Nat Rev Nephrol 15(10):641–659. https://doi.org/10.1038/s41581-019-0182-z

Seftor EA, Brown KM, Chin L et al (2005) Epigenetic transdifferentiation of normal melanocytes by a metastatic melanoma microenvironment. Cancer Res 65(22):10164–10169. https://doi.org/10.1158/0008-5472.CAN-05-2497

Semenza GL (2017) A compendium of proteins that interact with HIF-1alpha. Exp Cell Res 356(2):128–135. https://doi.org/10.1016/j.yexcr.2017.03.041

Shah S, Divekar AA, Hilchey SP et al (2005) Increased rejection of primary tumors in mice lacking B cells: inhibition of anti-tumor CTL and TH1 cytokine responses by B cells. Int J Cancer 117(4):574–586. https://doi.org/10.1002/ijc.21177

Shan C, Elf S, Ji Q et al (2014) Lysine acetylation activates 6-phosphogluconate dehydrogenase to promote tumor growth. Mol Cell 55(4):552–565. https://doi.org/10.1016/j.molcel.2014.06.020

Shawber CJ, Funahashi Y, Francisco E et al (2007) Notch alters VEGF responsiveness in human and murine endothelial cells by direct regulation of VEGFR-3 expression. J Clin Invest 117(11):3369–3382. https://doi.org/10.1172/JCI24311

Shimasaki N, Jain A, Campana D (2020) NK cells for cancer immunotherapy. Nat Rev Drug Discov 19(3):200–218. https://doi.org/10.1038/s41573-019-0052-1

Shin JW, Lee DW, Kim MJ et al (2005) Isolation of endothelial progenitor cells from cord blood and induction of differentiation by ex vivo expansion. Yonsei Med J 46(2):260–267. https://doi.org/10.3349/ymj.2005.46.2.260

Shoulders MD, Ryno LM, Genereux JC et al (2013) Stress-independent activation of XBP1s and/or ATF6 reveals three functionally diverse ER proteostasis environments. Cell Rep 3(4):1279–1292. https://doi.org/10.1016/j.celrep.2013.03.024

Smyth MJ, Thia KY, Street SE et al (2000) Perforin-mediated cytotoxicity is critical for surveillance of spontaneous lymphoma. J Exp Med 192(5):755–760. https://doi.org/10.1084/jem.192.5.755

Spring H, Schuler T, Arnold B et al (2005) Chemokines direct endothelial progenitors into tumor neovessels. Proc Natl Acad Sci USA 102(50):18111–18116. https://doi.org/10.1073/pnas.0507158102

Srikanth S, Woo JS, Wu B et al (2019) The Ca(2+) sensor STIM1 regulates the type I interferon response by retaining the signaling adaptor STING at the endoplasmic reticulum. Nat Immunol 20(2):152–162. https://doi.org/10.1038/s41590-018-0287-8

Srivastava MK, Sinha P, Clements VK et al (2010) Myeloid-derived suppressor cells inhibit T-cell activation by depleting cystine and cysteine. Cancer Res 70(1):68–77. https://doi.org/10.1158/0008-5472.CAN-09-2587

Staples KJ, Sotoodehnejadnematalahi F, Pearson H et al (2011) Monocyte-derived macrophages matured under prolonged hypoxia transcriptionally up-regulate HIF-1alpha mRNA. Immunobiology 216(7):832–839. https://doi.org/10.1016/j.imbio.2010.12.005

Su X, Abumrad NA (2009) Cellular fatty acid uptake: a pathway under construction. Trends Endocrinol Metab 20(2):72–77. https://doi.org/10.1016/j.tem.2008.11.001

Sun LP, Li L, Goldstein JL et al (2005) Insig required for sterol-mediated inhibition of Scap/SREBP binding to COPII proteins in vitro. J Biol Chem 280(28):26483–26490. https://doi.org/10.1074/jbc.M504041200

Sun L, Song L, Wan Q et al (2015) cMyc-mediated activation of serine biosynthesis pathway is critical for cancer progression under nutrient deprivation conditions. Cell Res 25(4):429–444. https://doi.org/10.1038/cr.2015.33

Takahashi T, Tagami T, Yamazaki S et al (2000) Immunologic self-tolerance maintained by CD25(+) CD4(+) regulatory T cells constitutively expressing cytotoxic T lymphocyte-associated antigen 4. J Exp Med 192(2):303–310. https://doi.org/10.1084/jem.192.2.303

Tam AB, Mercado EL, Hoffmann A et al (2012) ER stress activates NF-kappaB by integrating functions of basal IKK activity, IRE1 and PERK. PLoS One 7(10):e45078. https://doi.org/10.1371/journal.pone.0045078

Tammela T, Zarkada G, Wallgard E et al (2008) Blocking VEGFR-3 suppresses angiogenic sprouting and vascular network formation. Nature 454(7204):656–660. https://doi.org/10.1038/nature07083

Tannahill GM, Curtis AM, Adamik J et al (2013) Succinate is an inflammatory signal that induces IL-1beta through HIF-1alpha. Nature 496(7444):238–242. https://doi.org/10.1038/nature11986

Tao J, Zhou X, Jiang Z (2016) cGAS-cGAMP-STING: the three musketeers of cytosolic DNA sensing and signaling. IUBMB Life 68(11):858–870. https://doi.org/10.1002/iub.1566

Togashi Y, Shitara K, Nishikawa H (2019) Regulatory T cells in cancer immunosuppression – implications for anticancer therapy. Nat Rev Clin Oncol 16(6):356–371. https://doi.org/10.1038/s41571-019-0175-7

Truxova I, Kasikova L, Hensler M et al (2018) Mature dendritic cells correlate with favorable immune infiltrate and improved prognosis in ovarian carcinoma patients. J Immunother Cancer 6(1):139. https://doi.org/10.1186/s40425-018-0446-3

Tsou P, Katayama H, Ostrin EJ et al (2016) The emerging role of B cells in tumor immunity. Cancer Res 76(19):5597–5601. https://doi.org/10.1158/0008-5472.CAN-16-0431

Upton JP, Wang L, Han D et al (2012) IRE1alpha cleaves select microRNAs during ER stress to derepress translation of proapoptotic Caspase-2. Science 338(6108):818–822. https://doi.org/10.1126/science.1226191

Urano F, Wang X, Bertolotti A et al (2000) Coupling of stress in the ER to activation of JNK protein kinases by transmembrane protein kinase IRE1. Science 287(5453):664–666. https://doi.org/10.1126/science.287.5453.664

Vasconcelos-Dos-Santos A, Oliveira IA, Lucena MC et al (2015) Biosynthetic machinery involved in aberrant glycosylation: promising targets for developing of drugs against cancer. Front Oncol 5:138. https://doi.org/10.3389/fonc.2015.00138

Vattem KM, Wek RC (2004) Reinitiation involving upstream ORFs regulates ATF4 mRNA translation in mammalian cells. Proc Natl Acad Sci USA 101(31):11269–11274. https://doi.org/10.1073/pnas.0400541101

Veglia F, Perego M, Gabrilovich D (2018) Myeloid-derived suppressor cells coming of age. Nat Immunol 19(2):108–119. https://doi.org/10.1038/s41590-017-0022-x

Wamelink MM, Struys EA, Jakobs C (2008) The biochemistry, metabolism and inherited defects of the pentose phosphate pathway: a review. J Inherit Metab Dis 31(6):703–717. https://doi.org/10.1007/s10545-008-1015-6

Wang M, Kaufman RJ (2014) The impact of the endoplasmic reticulum protein-folding environment on cancer development. Nat Rev Cancer 14(9):581–597. https://doi.org/10.1038/nrc3800

Wang Q, Beaumont KA, Otte NJ et al (2014) Targeting glutamine transport to suppress melanoma cell growth. Int J Cancer 135(5):1060–1071. https://doi.org/10.1002/ijc.28749

Wang W, Xiao ZD, Li X et al (2015) AMPK modulates Hippo pathway activity to regulate energy homeostasis. Nat Cell Biol 17(4):490–499. https://doi.org/10.1038/ncb3113

Wang J, Duan Z, Nugent Z et al (2016) Reprogramming metabolism by histone methyltransferase NSD2 drives endocrine resistance via coordinated activation of pentose phosphate pathway enzymes. Cancer Lett 378(2):69–79. https://doi.org/10.1016/j.canlet.2016.05.004

Wang JM, Qiu Y, Yang ZQ et al (2017) Inositol-requiring enzyme 1 facilitates diabetic wound healing through modulating MicroRNAs. Diabetes 66(1):177–192. https://doi.org/10.2337/db16-0052

Wang Y, Luo W, Wang Y (2019) PARP-1 and its associated nucleases in DNA damage response. DNA Repair (Amst) 81:102651. https://doi.org/10.1016/j.dnarep.2019.102651

Wang Y, Chen Y, Bao L et al (2020a) CHD4 promotes breast cancer progression as a coactivator of hypoxia-inducible factors. Cancer Res 80(18):3880–3891. https://doi.org/10.1158/0008-5472.CAN-20-1049

Wang Y, Li G, Deng M et al (2020b) The multifaceted functions of RNA helicases in the adaptive cellular response to hypoxia: from mechanisms to therapeutics. Pharmacol Ther 221:107783. https://doi.org/10.1016/j.pharmthera.2020.107783

Wang Y, Chen Y, Wang C et al (2021a) MIF is a 3' flap nuclease that facilitates DNA replication and promotes tumor growth. Nat Commun 12(1):2954. https://doi.org/10.1038/s41467-021-23264-z

Wang Y, Li G, Deng M et al (2021b) The multifaceted functions of RNA helicases in the adaptive cellular response to hypoxia: from mechanisms to therapeutics. Pharmacol Ther 221:107783. https://doi.org/10.1016/j.pharmthera.2020.107783

Wang Y, Luo M, Chen Y et al (2021c) ZMYND8 expression in breast cancer cells blocks T-lymphocyte surveillance to promote tumor growth. Cancer Res 81(1):174–186. https://doi.org/10.1158/0008-5472.CAN-20-1710

Watt MJ, Clark AK, Selth LA et al (2019) Suppressing fatty acid uptake has therapeutic effects in preclinical models of prostate cancer. Sci Transl Med 11(478). https://doi.org/10.1126/scitranslmed.aau5758

Wilson WR, Hay MP (2011) Targeting hypoxia in cancer therapy. Nat Rev Cancer 11(6):393–410. https://doi.org/10.1038/nrc3064

Woo JR, Liss MA, Muldong MT et al (2014) Tumor infiltrating B-cells are increased in prostate cancer tissue. J Transl Med 12:30. https://doi.org/10.1186/1479-5876-12-30

Wu J, Rutkowski DT, Dubois M et al (2007) ATF6alpha optimizes long-term endoplasmic reticulum function to protect cells from chronic stress. Dev Cell 13(3):351–364. https://doi.org/10.1016/j.devcel.2007.07.005

Wu J, Sun L, Chen X et al (2013) Cyclic GMP-AMP is an endogenous second messenger in innate immune signaling by cytosolic DNA. Science 339(6121):826–830. https://doi.org/10.1126/science.1229963

Xie Q, Liu Y, Li X (2020) The interaction mechanism between autophagy and apoptosis in colon cancer. Transl Oncol 13(12):100871. https://doi.org/10.1016/j.tranon.2020.100871

Xu C, Bailly-Maitre B, Reed JC (2005) Endoplasmic reticulum stress: cell life and death decisions. J Clin Invest 115(10):2656–2664. https://doi.org/10.1172/JCI26373

Yabe D, Brown MS, Goldstein JL (2002) Insig-2, a second endoplasmic reticulum protein that binds SCAP and blocks export of sterol regulatory element-binding proteins. Proc Natl Acad Sci USA 99(20):12753–12758. https://doi.org/10.1073/pnas.162488899

Yang M, Vousden KH (2016) Serine and one-carbon metabolism in cancer. Nat Rev Cancer 16(10):650–662. https://doi.org/10.1038/nrc.2016.81

Yang J, Goldstein JL, Hammer RE et al (2001) Decreased lipid synthesis in livers of mice with disrupted Site-1 protease gene. Proc Natl Acad Sci USA 98(24):13607–13612. https://doi.org/10.1073/pnas.201524598

Yang H, Wang H, Ren J et al (2017a) cGAS is essential for cellular senescence. Proc Natl Acad Sci USA 114(23):E4612–E4620. https://doi.org/10.1073/pnas.1705499114

Yang L, Venneti S, Nagrath D (2017b) Glutaminolysis: a hallmark of cancer metabolism. Annu Rev Biomed Eng 19:163–194. https://doi.org/10.1146/annurev-bioeng-071516-044546

Ying H, Kimmelman AC, Lyssiotis CA et al (2012) Oncogenic Kras maintains pancreatic tumors through regulation of anabolic glucose metabolism. Cell 149(3):656–670. https://doi.org/10.1016/j.cell.2012.01.058

Yokoyama C, Wang X, Briggs MR et al (1993) SREBP-1, a basic-helix-loop-helix-leucine zipper protein that controls transcription of the low density lipoprotein receptor gene. Cell 75(1):187–197

Yoshida H, Matsui T, Yamamoto A et al (2001) XBP1 mRNA is induced by ATF6 and spliced by IRE1 in response to ER stress to produce a highly active transcription factor. Cell 107(7):881–891. https://doi.org/10.1016/s0092-8674(01)00611-0

You L, Wu W, Wang X et al (2020) The role of hypoxia-inducible factor 1 in tumor immune evasion. Med Res Rev. https://doi.org/10.1002/med.21771

Yu J, Freud AG, Caligiuri MA (2013) Location and cellular stages of natural killer cell development. Trends Immunol 34(12):573–582. https://doi.org/10.1016/j.it.2013.07.005

Yun J, Rago C, Cheong I et al (2009) Glucose deprivation contributes to the development of KRAS pathway mutations in tumor cells. Science 325(5947):1555–1559. https://doi.org/10.1126/science.1174229

Zhang T, Suo C, Zheng C et al (2019) Hypoxia and metabolism in metastasis. Adv Exp Med Biol 1136:87–95. https://doi.org/10.1007/978-3-030-12734-3_6

Zhang X, Bai XC, Chen ZJ (2020) Structures and mechanisms in the cGAS-STING innate immunity pathway. Immunity 53(1):43–53. https://doi.org/10.1016/j.immuni.2020.05.013

Zhou J, Liu CY, Back SH et al (2006) The crystal structure of human IRE1 luminal domain reveals a conserved dimerization interface required for activation of the unfolded protein response. Proc Natl Acad Sci USA 103(39):14343–14348. https://doi.org/10.1073/pnas.0606480103

Zuazo-Gaztelu I, Casanovas O (2018) Unraveling the role of angiogenesis in cancer ecosystems. Front Oncol 8:248. https://doi.org/10.3389/fonc.2018.00248

Chapter 2
Development and Advantages of Drug Delivery Systems

Yuelan Zhang, Xiaoli Liu, and Xuefeng Li

Abstract Nowadays, formulation research utilizes multidisciplinary approaches to effectively deliver drugs to targeted site, regulating metabolomics, pharmacokinetics, toxicity, immunogenicity, and biometrics. Drug delivery technology was originated in 1952, and the first generation of drug delivery systems (DDSs) significantly promoted the development of oral and transdermal drugs, while the second generation attempted to achieve the smart DDSs. However, the third generation of DDSs is expected to overcome physicochemical and biological obstacles to achieve targeted drug delivery using various novel technologies, such as the pulsatile release DDSs and feedback-regulated DDSs. Subsequently, the integration of biology and biomedical materials science led to the development of biomaterials. Besides, advanced knowledge about the human body and its properties, as well as the occurring of the concept of biocompatibility, provides a theoretical basis to design and take advantage of biomaterials as DDSs. Thus, the development of new DDSs plays a critical role in the treatment of various diseases.

Keywords Drug delivery system · Target · Biomaterial · Biocompatibility

Y. Zhang
State Key Laboratory of Respiratory Disease, Sino-French Hoffmann Institute, School of Basic Medical Sciences, Guangzhou Medical University, Guangzhou, People's Republic of China

Department of Laboratory Medicine, School of Medicine, Foshan University, Foshan, People's Republic of China

X. Liu
Department of Health Technology, Centre for Nanomedicine and Theranostics, Technical University of Denmark, Lyngby, Denmark

X. Li (✉)
State Key Laboratory of Respiratory Disease, Sino-French Hoffmann Institute, School of Basic Medical Sciences, Guangzhou Medical University, Guangzhou, People's Republic of China

Cancer Center, The First Hospital of Jilin University, Changchun, People's Republic of China
e-mail: xuefeng_li@jlu.edu.cn

L. Teng et al. (eds.), *Drug Delivery to Tumors*,
https://doi.org/10.1007/978-981-19-8930-8_2

2.1 Introduction

The development of drugs or active pharmaceutical ingredients plays an important role in the treatment of specific disease pathologies (Zhao et al. 2020). However, the development of drugs often fails because of low safety and effectiveness. Generally, pharmacological treatments require the drug substances to reach to the pharmacological target with a sufficient amount that could maximize therapeutic effect but minimize collateral side effects (Chen et al. 2014a; Sun et al. 2019). However, a drug alone has to overcome numerous challenges, such as poor permeability of some tissues in vivo (Chen et al. 2020a, b; Gong et al. 2020; Yang et al. 2018). Actually, many new drug candidates suffer from poor stability (easy to degrade) in vivo and in vitro, while some drug candidates have limited transport capacity due to biological obstacles, including poor intestinal absorption and blocked diffusion through the blood-brain barrier. As a consequence, it is necessary to design intelligent drug delivery systems (DDSs) to solve these problems and enhance the therapeutic effect with reduced drug consumption (Jain 2014).

Drug delivery technologies have the ability to deal with these limitations mentioned above. DDSs are platforms of formulations or devices to deliver drugs into patient body efficiently and safely via controlling the place, time, and rate of drug release. There are more than 10 drug delivery routes in human body, such as digestive tract, respiratory tract, other lumens, blood vessel, tissue, subcutaneous, and muscle With unique biological and physicochemical properties, different drugs require suitable delivery systems to enhance their therapeutic effect (Gao and Jiang 2017). For instance, drugs for asthma or angina require quick released DDSs; while some drugs prefer long-term activity, such as the transdermal delivery system which releases drugs in a relatively slow and sustained manner (DeWeerdt 2016; Zhou et al. 2014a). In addition, drugs with different molecular structures as well as different targets in human body require specific DDSs design. For example, small-molecule drugs are relatively easy to cross the blood-brain barrier and get into the central nervous system compared to larger biologics such as proteins. Therefore, the utilized DDSs should be systematically evaluated and analyzed based on pharmacokinetic analysis (Hashida 2020).

To optimize the efficacy of drugs, achieve the accumulation in the target site, and minimize the side effects for other organs, the development of DDSs plays a pivotal role. When developing a new drug, three major aspects, including the choice of delivery route, the drug release behavior, and the targeting sites of action should be taken into consideration to optimize the function of the drug (Fan et al. 2021). At present, pharmaceutical research has entered a new era of controlled drug release, leading to the rapid development of various new drug release systems. The highly controllable DDSs allow active drug to be released at a desirable amount in proximity to a specific site (Youan 2010). However, even with properly prescribed medications, approximately 100,000 deaths are caused by adverse drug reactions in the United States (Helfand and Cowen 1983). Thus, in order to improve their efficacy and safety, it is necessary to revisit our current drug discovery and formulation approaches to explore more effective, safe, and robust DDSs to implement a clinical or therapeutic advantage.

2.2 The History of Drug Delivery Technologies

The concept of DDS appeared in the early 1950s and became a hot topic in formulation research in the 1980s. In the 1960s, the rise of bio-pharmaceutics and pharmacokinetics can determine the quantitative relationship between the absorption, distribution, metabolism, and excretion of drugs in the body, as well as the bioavailability of drugs. However, drugs in high concentrations can cause toxic effects, while they demonstrate minimal therapeutic effect when in low concentrations. Thus, in the initial development stage of DDSs, the corresponding products are controlled-release formulations which can maintain the drug levels in blood.

During the past seven decades, drug delivery technology has improved remarkably (Table 2.1). Before the 1950s, the drug release kinetics showed uncontrollable release profiles for pills and capsules where drugs underwent immediate release upon contacting with water. Fortunately, the first sustained-release formulation was introduced by Smithkline Beecham and French in 1952, which was able to control the release kinetics of dextroamphetamine (Dexedrine) for 12-h efficacy (Prausnitz et al. 2004). Then the mechanisms for the first generation (1950–1980) of controlled DDSs were established and became more sophisticated, including the demonstration of dissolution-, diffusion-, osmosis-, and ion exchange-based release behaviors, and all these significantly promoted the development of oral and transdermal drugs in controlled-release formulations.

In 1979, the US FDA approved the first transdermal patch which was a three-day patch delivering scopolamine to treat motion sickness (Fan et al. 2020). In comparison with the oral administration route, transdermal delivery system has several advantages, especially for the drugs that can be prematurely metabolized by the liver due to first-pass effect. Consequently, the transdermal delivery system became an attractive alternative to oral administration as well as hypodermic injection delivery of drugs, making considerable contribution to medical drug delivery applications. Subsequently, the transdermal DDSs have been widely used in many

Table 2.1 Features of three generations of DDSs (Cohen and Ball 1998)

First generation (1950–1980)	Second generation (1980–2010)	Third generation (from 2010)
Controlled DDSs	Smart DDSs	Regulated DDSs
Basically control the release kinetics of drugs	Attempt to overcome biological barriers	Aim to achieve targeted drug delivery
Oral delivery Transdermal delivery	Zero-order release Smart polymers & hydrogels Nanoparticles	On-off insulin release Targeted delivery Long-term DDSs In vitro-In vivo correlation

treatments, such as hormonal disorders, cancer, diabetes, vaccines, and contraception (Li et al. 2014a, b; Wang et al. 2014a; Zhou et al. 2014c).

As is well known, the accumulation of drugs in targeting sites will be effective for disease treatment, while no therapeutic effect, or even toxic side effects, will be caused in other parts of the body. Therefore, to achieve therapeutic effects and reduce the toxic side effects, approaches have been explored to accumulate drugs on the lesion site and minimize them in other sites (Yuan et al. 2020; Ye et al. 2014). This is of great significance for the treatment of local diseases, such as cancer and inflammation (Liang et al. 2014; Wu and Li 2015; Li et al. 2015; Zhou et al. 2015; Zhang et al. 2015; Cochrane et al. 2015). The lesion may be a diseased organ, or it may be cells or bacteria (Yun et al. 2015).

The uncontrollable initial burst release is a great challenge for the first generation (1G) of DDSs, while most formulations release 50% of the total drug in the first or second day (Khafagy el et al. 2007). Hence, the second generation (2G) (1980–2010) of DDSs has developed the ability to smartly deliver drugs. Unlike the first generation of drug delivery technologies, the second generation was aimed at dealing with biological barriers. However, the development of DDSs was going through a tough time in the first 20 years of the 2G period. For example, needle phobia of patients has prompted to exploit novel devices and delivery systems for delivering insulin. All promising routes with long-term possibilities have tried to overcome the inherent barriers of insulin uptake (Heinemann 2008). Nevertheless, several routes such as pulmonary delivery systems for insulin with low bioavailability ended in failure and led to the withdrawal of the product from the market. This is because insulin needs large dose than parenteral injection, and this can cause unexpected side effects (Yun et al. 2015; Stayton et al. 2005).

The drug delivery system developed during the 2G era needs to deal with more difficult problems, including the biological barrier as well as the drug release kinetics. The release of the drug in vivo depends not only on the nature of the drug itself, but also on the biological environment, which makes it more difficult to predict the release kinetics and bioavailability of drugs. This might due to no standard in vitro drug release test method to predict pharmacokinetic characteristics. Hence, various drug delivery strategies have been investigated and exploited during the 2G period, including peptide and protein delivery, smart polymers and hydrogels, and nanoparticles (Soppimath et al. 2002). A key feature of these biological delivery systems is that they can change the structure and functional characteristics according to the biological environment (such as pH, temperature), thereby showing excellent "smart" material properties (Yun et al. 2015). In short, these technologies laid foundation for the development of the third-generation drug delivery system.

2.3 The Third-Generation Drug Delivery Technologies

With the development of biotechnologies, the research on polypeptide and protein drug biologic formulations has become an important field of pharmacy research, which in parallel has brought new challenges to drug delivery methods. Since these molecules have short half-life in vivo, it is difficult for macromolecules to diffuse through the polymer materials via traditional drug delivery (controlled release) systems (Sun et al. 2021). To solve this problem, these molecules were physically embedded in the polymer to form a complex interconnected network of pores, through which drugs can subsequently diffuse during the 2G period. The 2G DDSs have partially achieved "smartly" delivering drugs. Because of the enormous toxicity of many drugs, especially anticancer drugs, the 3G (from 2010) of DDSs has been expected to overcome physicochemical and biological obstacles to achieve targeted-regulated drug delivery. Currently, the 3G targeted-regulated DDSs can be classified into two major categories: (1) closed-loop systems or self-regulating systems, which can activate or regulate the release rate of a drug via recognizing changes in biological media, such as temperature, pH, or concentration of certain substances (Alvarez-Lorenzo and Concheiro 2014; Sun et al. 2018; Pan et al. 2018); (2) systems that can switch drug release on/off to provide pulsed drug release with specific external stimuli, such as light, electricity, or magnetic field (Ma et al. 2020; Xie et al. 2020; Wang et al. 2020; Yang et al. 2020b). In the past decades, novel drug delivery technologies developed very rapidly, including sustained and controlled DDSs, target DDSs, NP DDSs, impulse systems, and feedback-regulated drug delivery systems (Lu et al. 2020). Actually, these novel DDSs are still in the developing process or early clinical phases.

2.3.1 Target DDS

In the last century, Paul Ehrlich proposed the idea of a drug working as a "magic bullet" that selectively kill pathogens without affecting normal cells. At the end of the 1970s, targeted drugs began to be studied comprehensively. Since then, the target drug system or targeted DDS has always been a hot research topic, which could make a drug to directly accumulate in the target site with high curative effects and low side effects.

Targeted drugs can be divided into three types: passive targeted agents, active targeted agents, and physicochemical agents. Passive targeted drugs can be ingested by macrophages in the mononuclear macrophage system, which can be selectively concentrated at specific sites. Passive targeting agents include liposomes, microemulsions, microcapsules, microspheres, and NPs that can be absorbed by host cells (Wu et al. 2019a, c). In contrast, active targeted agents utilize modified drug carriers as the "missile," which assists in delivering drugs to the target sites in terms of the carrier structure modification or antibody recognition, receptor recognition, and immune recognition (Langer 2001). Another strategy of active targeted agents is to synthesize precursor drugs that are released as active pharmacological drugs after enzymatic reactions or non-enzymatic reactions. Currently, the commonly used precursors include anticancer precursors, brain-targeted precursors, and colon-targeted precursors. However, physicochemical agents, such as magnetic targeting agents, heat sensitive targeting agents, pH targeting agents, and embolic agents, can release drugs via changing external environmental conditions.

With the development of advanced target DDSs, several novel target drugs have been explored in the past few years, such as the anticancer drug "Gliadel" and the systemic delivery of a protein, Nutropin Depot (Zhou et al. 2014b). In addition, a study has shown that triptolide lysozyme conjugates can significantly target proximal tubular epithelial cells (Xue et al. 2019). However, the biggest challenge for targeted drug delivery is overcoming difficult-to-cross cellular barriers, such as the barriers in the nose, intestine, and skin. Thus, materials for condensing or packaging drugs are needed to overcome these barriers, which can be taken up by cells, bypass or escape the cell's endocytic pathways, and deliver drugs to the target sites by releasing them in active forms.

In this respect, nanoparticles have proved to be suitable drug delivery vehicles, which can carry drugs or therapeutic agents to the desired sites and reduce their associated toxicity (Zhao et al. 2020). The small size of the nanomaterials allows the particles to be internalized by the cells, thus allowing them to interact with biomolecules and subsequently induce selective cellular responses. Nanomaterials are usually hydrophilic or hydrophobic, capable of forming a series of NP/biological interfaces relying on glue force and dynamic biophysical chemistry interactions. They can efficiently bind, absorb, and deliver compounds including DNA, RNA, and proteins (Lan et al. 2020; Xie et al. 2016; Chen et al. 2016; Li et al. 2014b). It can effectively co-deliver a series of cargoes, including medications, medicinal peptides, and organic compounds (Kumar et al. 2020).

At present, nano-DDSs have been studied in cancer chemotherapy chiefly. Some nanomaterials, including lipids/liposomes, polymeric NPs, inorganic NPs, have been designed and developed for anticancer drug delivery applications in the past few years (Yang et al. 2020a). For instance, liposome-clodronate NPs (LCL-PLP); PEG-histidine-modified alginate-galactosylated-dextran-oligodeoxynucleotide (PHA-Gal-dextran-ODN) NPs; mannan (Man)-targeted, hyaluronic acid (HA)-coated manganese dioxide (MnO_2) NPs, membrane-derived microparticle (MP)-coated codelivery Fe_3O_4; and liposome-CpG NPs (MP-Fe_3O_4-Lipo/CpG) have been developed for nano-DDSs to display antitumor efficacy (Gao et al. 2013; Kang et al. 2016; Yang et al. 2021). In addition, nano-DDSs can be modified to improve the pharmacokinetic profile and enhance physiological stability or/and biocompatibility for overcoming in vivo delivery barriers. Various studies have shown that nanotechnology-based targeted drug delivery strategies can overcome the blood-brain barrier hampering treatment for different CNS disorders (Kamaly et al. 2016; Zielińska et al. 2020) (Fig. 2.1). However, a series of nano-DDSs have shown tremendous potentials for drug delivery, but their comparative efficacy has not been successfully applied in patients. It may mainly be attributed to the choice of

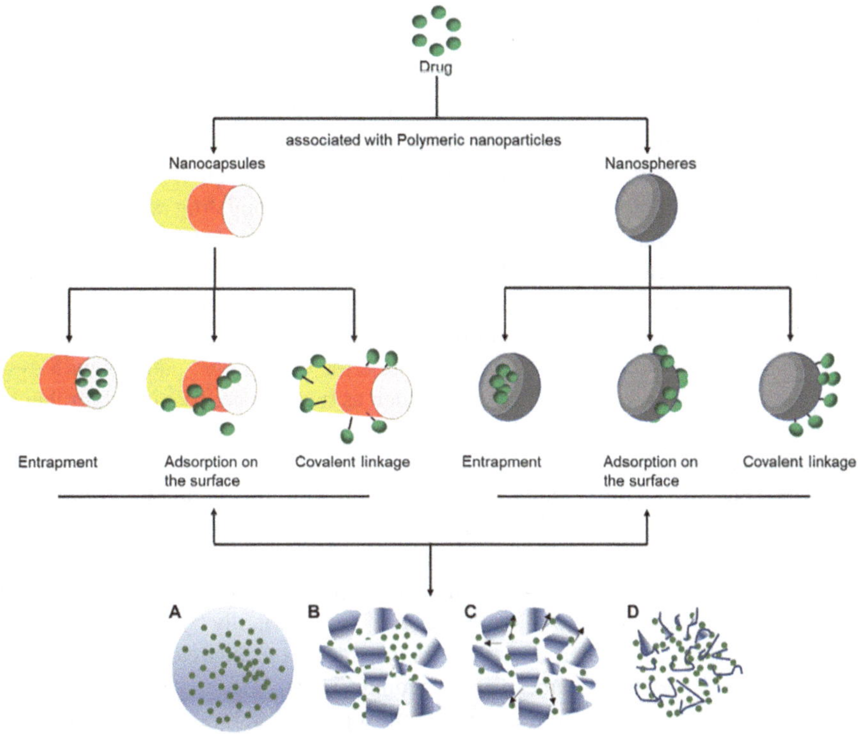

Fig. 2.1 Polymeric NPs as drug carriers, including nanocapsules and nanospheres, which differ with their morphology, can associate with drug in three different possibilities. Four mechanisms by which drugs are released from polymeric NPs: (**a**) diffusion through the polymer matrix, (**b**) diffusion through water-filled pores, (**c**) osmotic pumping, and (**d**) erosion (Kamaly et al. 2016)

preclinical model lacking clinically relevant systems. Nevertheless, nano-DDSs can provide superior pharmacokinetic profiles and high loading efficiency in the coming years.

2.3.2 *Sustained-Release DDS*

As the principal obstacle in drug delivery pertains to managing the burst release of drugs, which occurs rapidly and instantaneously post-administration, it is essential to explore strategies for circumventing this issue. Sustained-release delivery systems, a kind of pharmaceutical formulation, can prolong the release rate of therapeutic drugs by controlling the released amount of a drug during the designated time intervals. It can reduce the number of injections and eliminate the mental and physical pain brought by frequent injections, which can improve patient compliance markedly. The main suitable drugs are antipsychotics, antibiotics, polypeptides, proteins, vaccines, and anticancer drugs. Since the 1970s, three classical intrapacket drug delivery approaches, including film/slab/chip/membrane-type, fiber type, and gel type formulations, have been injected directly into the periodontal pocket to obtain sustained-release DDSs. As a result of the progress in pharmaceutical technologies, the controlled release platforms have been developed rapidly, one of which is associated with oral DDSs and subcutaneous drug delivery system.

Due to patient compliance and ease of administration, oral drug delivery systems are the most commonly used for many drugs (Haznar-Garbacz et al. 2011). The oral liquid-controlled release system (OLCRS) can be taken directly in liquid form or temporarily prepared into liquid form containing dispersed particles, such as microcapsules, microspheres, and milk drops. As a sustained-release DDS, OLCRS is a new oral controlled-release system for children, the elderly, and patients with dysphagia. At present, several OLCRS drugs, such as methadone, codeine chlorphenamine, amphetamine theophylline, and pseudoephedrine, have been developed (Maudens et al. 2018).

At present, subcutaneous injection is the most common method for administering peptides in clinics. However, frequent injections may result in discomfort and other risks. Sustained-release and controlled-release injections may provide the required therapeutic drug concentration over extended periods and improve patient compliance. For instance, small molecules are cleared rapidly by lymphatic drainage within a short time when injected intra-articularly for the treatment of osteoarthritis (Cao et al. 2019; Steinberg and Friedman 2020). Therefore, an appropriate formulation is required to solve this problem. At present, several materials supporting this DDS, such as cellulose acetate, ethylcellulose, methacrylic acid copolymer, silicone elastomer, ethylene-vinyl acetate, polyvinyl alcohol, and cross-linked alginate, have been studied. These kinds of DDSs have been studied in intraocular and periodontal diseases (Braghiroli et al. 2020).

2.3.3 Pulsatile Release DDS

Due to the circadian rhythm (24-h cycle), which regulated by the suprachiasmatic nucleus, symptoms of numerous diseases vary from day and night, such as hypertension, asthma, cardiovascular ailments, and chronic obstructive pulmonary disease (Kriplani et al. 2020). More frequent exacerbations and even higher mortality happen during night. Thus, time scheduled drug release systems are required to achieve floating drug release to reduce the death rate (Ciancia et al. 2020). Pulsatile release drug delivery systems are designed to control the release patterns of drugs in terms of transience, stepwise, or sustained release in a certain amount during suitable time periods according to the circadian rhythm of disease states (Lin 2020). These smart DDSs with intelligent materials are particularly sensitive to the disease state and therefore efficiently delivery of the drug at the right dose, place, and time.

The first pulsed DDSs that achieved pulsatile release were developed in the early 1990s and have attracted increasing interest recently to exploit novel chronotherapeutic approaches. Currently, pulsatile drug-release delivery systems have been divided into single-unit and multiple-unit systems. Multiple-unit pulsatile release systems have greater advantages of than single-unit systems; thus, they have garnered more attention in recent years. Single-unit systems usually use osmosis control and polymer erosion control. However, multiple-unit pulsatile release systems utilize various stimuli, including either exogenous or endogenous variables, to trigger drug release by endowing their surface coatings with functional polymeric materials. And these materials are pH sensitive, temperature sensitive, or can be provoked by enzymatic reaction, magnetic field excitation, ultrasonic excitation, and electric field excitation (Jain et al. 2011).

In detail, pulsatile release drug delivery systems are categorized into three types, including time-controlled pulsatile DDSs, stimuli-induced pulsatile release system, and externally regulated pulsatile release system (Jain et al. 2011). Time-controlled pulsatile DDSs employ various formulations, such as an erodible coating layer, a rupturable coating layer, and a capsule-shaped system that endows it with a release-controlling plug, to enclose the drug reservoir and release the drug at a specified time when in contact with the dissolution fluid. Stimuli-induced pulsatile release systems release the drugs depending on the type of stimuli of the biological environment which could include thermo-responsive pulsatile release and chemical stimuli-induced pulsatile release (pH, enzyme, or other stimuli). For example, pH-dependent DDSs for delivering glucose can release drugs in a glucose-rich environment (pH 5.8) after a meal. However, externally regulated pulsatile release systems utilize external equipment to accurately control the internal environment to achieve precision therapy. Currently, the following systems, electro-responsive pulsatile release, ultrasonically stimulated pulsatile release, and magnetically induced pulsatile release, have been used to regulate drugs pulsatile release.

With various advantages, including reducing the dose frequency and adapting to circadian rhythms of the body functions, pulsatile DDSs have a fast advancement in recent years. Currently, numerous marketed technologies, including PulsincapTM, Diffucap®, three-dimensional printing®, CODAS®, OROS®, IPDAS®, GEOCLOCK®, Ritalina®, Uniphyl®, and Opana®ER, have been developed for pulsatile drug delivery (Yu et al. 2018; Majeed and Thabit 2018).

2.3.4 Feedback-Regulated DDS

Modern chronopharmacological research has pointed out that diseases with rhythmic changes, such as blood pressure, hormone secretion, and gastric acid, can adjust the drug delivery behaviors according to changes in biological rhythms. Currently, most drugs lack the sophistication to regulate their release accurately according to the local environment. For example, the drugs treating hypertension should choose intermediate end-points for dosage adjustment to prevent stroke or hypotension. Likely, inhalation anesthetic should be estimated accurately according to patients' characteristics and clinical conditions to reach optimum conditions during surgery.

Thus, a feedback-controlled drug delivery system is required to deliver active drug to specific disease environment of the patients by adaption. In fact, feedback control systems have been utilized in clinical pharmacokinetics for at least 40 years, which can release therapeutic drugs in an adjustable amount at an appropriate time, thereby controlling the biological signals related to pathological processes.

In the past few decades, closed-loop DDSs have been explored for homeostasis regulation as a promising alternative to traditional DDSs (Veiseh et al. 2015; Breton et al. 2020; Xu et al. 2020) (Fig. 2.2). A closed-loop DDS usually consists of a monitoring component and an executive component. It can be activated when the biosignal level overpass a certain biosignal concentration or threshold to release drugs, while it is inhibited when the biosignal level reaches the normal range. Accordingly, a closed-loop controlled-release strategy has been developed to deliver recombinant hirudin, which is an anticoagulant, using self-regulating nano-scale polymer gels. The safe and effective anticoagulant DDSs, which simulate the

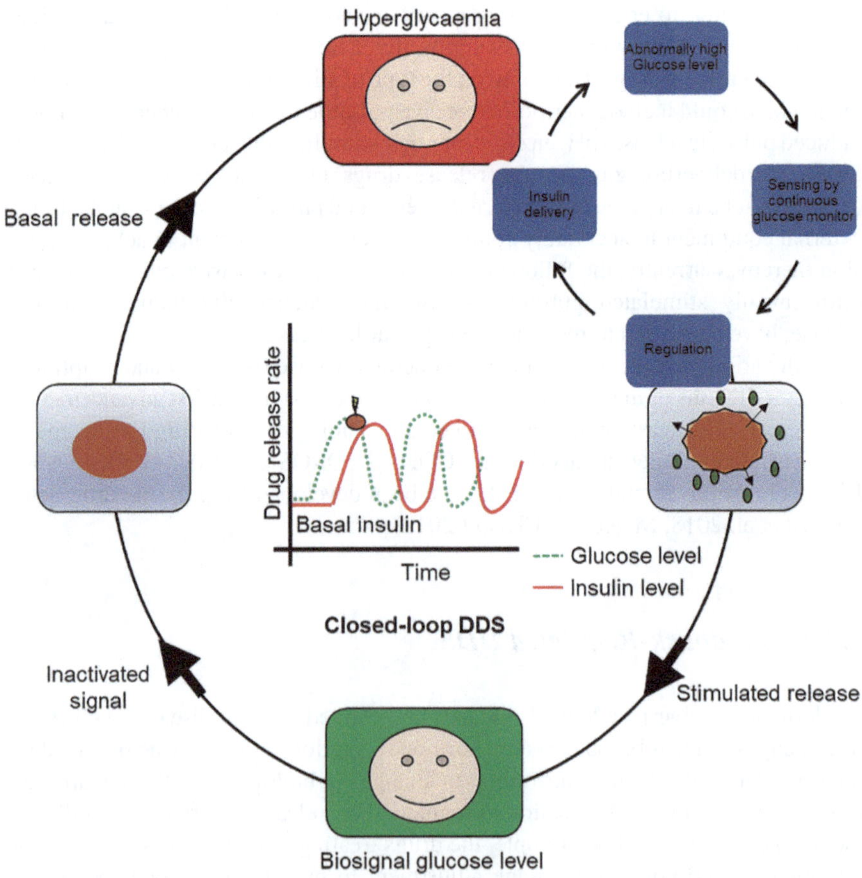

Fig. 2.2 Schematic of the mechanism of closed-loop DDS of insulin therapy for diabetes which can adjust the release rate to the biosignal level. (Reused with permission from Yu et al. 2018)

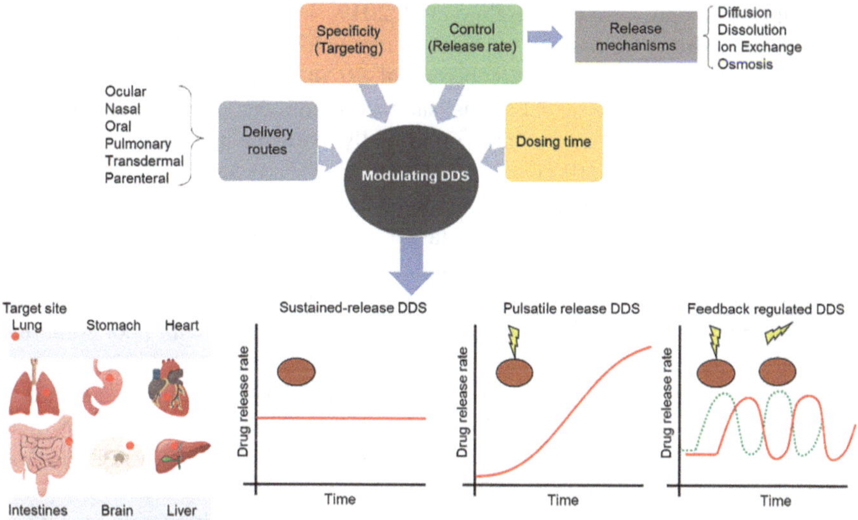

Fig. 2.3 The main components of modulating DDSs and four drug release patterns. The target DDS can deliver drugs to the target sites according to the organs' architectural feature or microenvironment. In a sustained-release DDS, the release occurs according to a preset pattern, regardless of the physical condition. The release of the pulsatile release DDSs is triggered by some internal or external stimuli. The feedback-regulated release system adjusts the release rate to the level of the biomarker, which triggers the release of the drug when the level of the biomarker rises again. (Reused with permission from Alvarez-Lorenzo and Concheiro 2014)

physiological negative feedback mechanism by responding to changes in coagulation signals, provide hope and potential for anticoagulation therapy (Satav et al. 2010).

Accordingly, feedback drug delivery vehicles use physiological responses as signals to regulate the release of drugs from the carrier (i.e., trigger, slow down, or stop drug release). Therefore, a feedback-regulated DDS based on a toxicity biomarker strategy has been successfully modeled, which may reduce the risk associated with drug overdose (Alvarez-Lorenzo and Concheiro 2014; Park 2014) (Fig. 2.3).

2.4 Delivery Strategies for Biologics

According to the reports in 2006 by the Center for Biologics Evaluation and Research (CBER) website and US FDA's Orange Book, there are 1204 chemical small-molecule drugs and 166 biological drugs after removing adjuvants such as duplicate active ingredients, salt forms, supplements, and imaging agents (Li et al. 2016d; Zhuang et al. 2016). Nowadays, biological agents are important parts of existing medical treatments (He et al. 2016; Li et al. 2016a, b, c; Pu et al. 2017). Due to their clinical success, biologics are rapidly growing classes and have become the dominant therapeutic agents (Anselmo et al. 2019; Wu et al. 2019b). For instance, peptide-based drugs from small peptides to monoclonal antibodies have been the most successful biologics to date (Roberts et al. 2020; Zhou et al. 2018; Huang et al. 2018). However, new biological forms, such as oligonucleotides (DNA- or RNA-based therapy and viral gene therapy), are rapidly maturing and being widely used in the clinic (Wang et al. 2018).

In fact, both chemical drugs and biomolecular drugs are facing the same challenges: how can we deliver them to the target sites and maximize their functions in the human body. Compared with traditional small-molecule drugs, the reason for the increasing success rate of biologics lies in their safety, target specificity, and favorable pharmacokinetics (Hu et al. 2019). Chemical drugs with poor solubility have various adverse side and toxic effects as mentioned before, while biomolecular drugs are easily degraded by enzymes in vivo, and biomolecular drugs are hard to penetrate cell membranes (António et al. 2019). Therefore, ideal DDSs play extremely important roles to ensure the function of the drug via protecting drug molecules from degradation, improving their solubility, reducing their toxicity and side effects, controlling the drug release to maintain the body circulation time when the drugs can easily penetrate in vivo barriers such as cell membranes cells and blood-brain barrier. In the following sections, we will elaborate on the DDSs of biological drugs chiefly.

2.4.1 DDSs for Proteins and Peptides

Proteins are the basic building blocks of all living cells, thought to be a complex set of organic macromolecules that perform enzyme catalysis, signal transduction and regulation, and other basic functions. Since the first recombinant protein insulin was produced and successfully entered the market, some protein-based pharmaceutical products have been developed and approved for use in humans (Anselmo et al. 2019). According to these, the intracellular delivery of functional proteins has important therapeutic significances in biological applications. At present, proteins and peptides which are made up of one or more amino acid chains, account for the majority of the treatments for various diseases. More than 60 peptides have been approved by the FDA for various indications, predominantly in oncology and immunology conditions (Sarmento 2010). Protein drugs are important parts of the drugs that are currently widely approved to treat or alleviate many metabolic and tumor-related symptoms (Agyei et al. 2017). In particular, protein and peptide drugs accounted for one-third of the top 15 drugs in 2015, which earned $174.7 billion on the global market (Lei et al. 2019; Li et al. 2018). In addition, many antibody-based therapeutics have been undergoing clinical trials and awaiting approvals since the first muromonab-CD3 was approved in 1986. Above all, our understanding of the biological fate and biodistribution mechanisms of protein and peptide drugs has advanced to the stage where the mechanisms could be affected by the selection of methods (Jain et al. 2013).

The scale-up of the production of therapeutic proteins has made rapid progress. However, the development of effective methods to conveniently and effectively deliver these drugs into the body remains a great challenge. Specifically, proteins and peptide drugs have particular physiochemical properties, such as their fragile and complex structures, potential immunogenicity responses, high molecular weight, and the spatial characteristics, which have limited their delivery systems. In addition, they are not suitable for oral administration, and they also have limitations when targeting intracellular targets. Thus, most of them target in the nervous system due to their characteristics. Moreover, repeated administration can cause allergic reactions and anaphylactic shock, which are potentially dangerous. Thus, appropriate delivery devices or protein carriers are needed for the implementation of proteins as therapeutic agents in vivo.

Since the DDS of peptide and protein drugs has been studied for several decades, different routes of administrations have been used to effectively deliver protein/peptide drugs to obtain more successful therapeutic effects (Drucker 2020). Given the patient compliance, oral administration is the most preferred route generally. Nevertheless, parenteral routes may be more suitable for peptide/protein drugs because of the absorption barriers of the gastrointestinal tract, including enzymatic degradation by proteases and low permeability of biofilms (Mullard 2021). Administration by injection is a good method for the absorption of protein/peptide drugs, but daily injections using an external pump and an intravenous line in long-term therapeutics is another challenge for medication adherence. Various peptide and protein drugs act outside the cell through receptor interactions (Torchilin and Lukyanov 2003). Therefore, extensive efforts have been dedicated to explore the feasibility of delivering protein/peptide drugs through other methods.

In addition to the traditional way, several novel DDSs have been explored to deliver protein/peptide drugs, such as conjugating polypeptides and proteins with various soluble polymers to enhance the permeability and retention effect in target sites. Various advantages of peptide and protein-polymer conjugates have been shown. These include decreased renal filtration and lower immunogenicity, protection from enzymatic degradation, and selective accumulation in target sites via the enhanced permeability and the retention effect (Veronese and Harris 2002). For instance, protein/peptide drugs can be conjugated with poly(ethylene glycol) (PEG) with low toxicity, as it is inexpensive, commercially available, and has been approved for internal applications by drug regulatory authorities/agencies (Cunningham et al. 2020). At present, PEG is utilized in PEG-modified L-asparaginase (Oncospar®, an anticancer agent), PEGylated interferon α-2b for treating chronic hepatitis C. In addition, protein/peptide drugs can also be conjugated with poly (styrene-co-maleic acid anhydride) (SMA), which is a high molecular weight polymer (1.5 kDa) and can bind to plasma albumin (Pick et al. 2018; Banaszynski et al. 2008). Besides, protein/peptide drugs can be packed by liposomes, which are formed by a bilayer of concentric spherical phospholipids in aqueous medium. Therefore, liposomes are appropriate vesicles for encapsulating peptide and protein drugs that are completely biocompatible, biologically inert, and barely cause toxicity or antigenic reactions (Banik et al. 2016).

Moreover, the utilization of NPs with biological activity and biocompatibility is more conducive to the development of more effective, safer biomedical applications. Peptides or proteins conjugated with NPs provide a train of thought to solve many difficulties arising from these considerations, which benefit from the unique performance and size structure of NPs (Spicer et al. 2018; Drucker 2020). At present, the most commonly used NPs in biomedical applications are gold NPs (AuNPs), magnetic NPs (MNPs), mesoporous silica NPs (MSNPs), and polymeric NPs. All these NPs have different advantages (Spicer et al. 2018). Most of them are very attractive drug delivery vehicles, capable of encapsulating large cargoes ranging from small molecules to large proteins, and all because they have adjustable pore size, large surface area, and large pore volume (Cho et al. 2013). The structural and functional versatilities of peptides and proteins make them play key roles in regulating, stimulating, and defining the activity of NP structures. But the delivery of therapeutic proteins/peptides is limited by their fragile structures and frequent administrations (Elliott et al. 2003). Therefore, some methods are envisaged to stabilize proteins in the delivery device and biological fluid, and to design a suitable protein carrier to effectively deliver them to the target organ. When the biomedical community is trying to solve these problems, the clinical transformation of many peptide/protein-NPs binding technologies will be a huge leap (Drucker 2020; Brayden et al. 2020) (Fig. 2.4).

2.4.2 DDSs for Nucleotide Drugs

As we all know, nucleic acids carrying hereditary information are regarded as the core genetic biological molecules in living systems. The highly specific base-pairing properties of DNA play important roles in many fundamental biological processes, especially in replication and transcription. However, the characteristics of nucleic acids, such as inherent biological properties, sequence programmability, and biocompatibility as well as its physical and chemical properties, make it more attractive for use as functional construction materials in the so-called DNA origami. DNA has been demonstrated as a multifunctional molecular building block (Angelbello et al. 2018). Consisting of four nucleotide bases, DNA is different from proteins. Oligonucleotides are nucleic acid polymers of about 20 nucleotides (including the nucleotides in DNA or RNA) with the potential to treat or manage a wide range of diseases. Oligonucleotides can easily bind to their complementary strands, so they are often used as probes to determine the sequence of DNA or RNA traditionally (Dowdy 2017; Bhatia et al. 2012; Pickar-Oliver and Gersbach 2019). The development of oligonucleotides as therapeutic drug candidates began about 30 years ago and include antisense oligonucleotides (ASOS), aptamers, and siRNAs of various types in the last 15 years and CRISPR-Cas9 in 2012 (Stein and Castanotto 2017). By the end of 2017, a total of 6 kinds of oligonucleotide drugs had been approved by the US FDA for marketing application, among which five are antisense oligonucleotide drugs, including Macugen (Pegaptanib), Kynamro (Mipomersen),

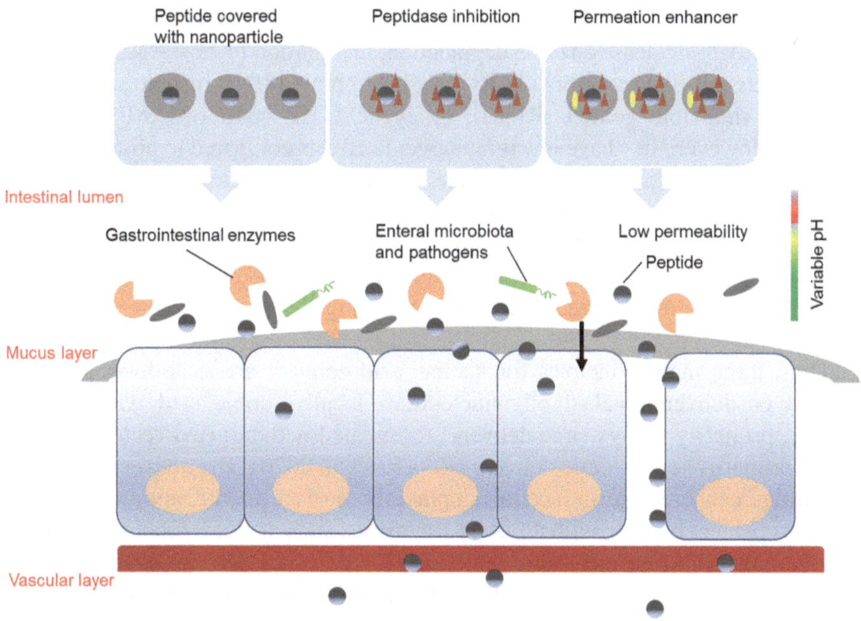

Fig. 2.4 Schematic representation of intestinal structure and the associated challenges to oral absorption of peptides. The ingested protein is effectively broken down into amino acids by a series of pH-sensitive proteases, and these amino acids are then actively transported across the mucosa through a complex family of transporters. A similar digestion process ensures that enzymes break down complex carbohydrates and lipids into smaller, more easily absorbed molecules. Besides, peptide drug can be encapsulated in nanoparticles, which releases peptides near the epithelium or targets the receptor and the uptake of epithelial particles will be quantified. In addition, peptidase inhibitors can be contained in DDS, but their long-term safety is questionable. Permeation enhancers can weakly bind to peptides through non-covalent bonds and produce a detergent-like effect on epithelial cells. (Reused with permission from Brayden et al. 2020; Drucker 2020)

Defitelio (Defibrotide), Spinraza (Nusinersen), and Exondys 51 (Eteplirsen) (Dowdy 2017). Currently, there are a large number of drugs that are in clinical studies at different stages. Compared to traditional drugs, oligonucleotide drugs have various advantages, such as selective targeting toward single genes, easily staying away from miss genes, and combining with unknown off-target which can pharma coevolve their sequence according to disease (either cancer or pandemic influenza) (Neveu et al. 2013; Yu et al. 2014; Xie et al. 2013).

Although nucleic acids have always been considered as potential drugs with unparalleled selectivity, an effective method for delivering these nucleic acids to the target sites, especially crossing the lipid bilayer, has remained a great obstacle to be solved (Dowdy 2017; Blain and Szostak 2014; Wilbie et al. 2019). Nucleic acids utilized as drugs have the same problem as protein/peptide drugs. Nucleic acids are much larger than conventional low-molecular weight drugs, which can affect their tissue distribution. The therapeutic nucleic acids are macromolecules (range from 4 to 10 kDa) with high charge, so they cannot be able to easily cross lipids and be uptaken by host cells. In addition, they suffer from in vivo stability issues, including susceptibility to nuclease enzyme degradation, recognition by host RNA/DNA sensors such as toll-like receptors (TLRs), double-stranded RNA receptors PKR, retinoic acid inducible gene I (RIG-I), or cyclic GMP-AMP (cGAMP) synthase (cGAS). However, they have low immunogenic effects compared to proteins (Revia et al. 2019; Wang et al. 2014b; Yang et al. 2016). Notably, nucleases presented in the bloodstream and intracellular environment can cause RNA to degrade within seconds in blood plasma, while DNA can degrade after 1 h. Therefore, nucleic acid-based therapeutics require chemical modification or the use of appropriate vectors to protect them from degradation before reaching the site of action (Smith and Zain 2019).

Thus, three main principles for nucleic acid delivery are as follows: injected/inhaled or delivered naked oligonucleotides locally; conjugated with targeted ligands, primarily for systemic delivery; formulated with targeted vectors, such as lipids, primarily for systemic delivery (Roberts et al. 2020). The delivery of nucleic acid drugs can promote intracellular uptake by surface modification in terms of directly covalent binding with different fragments, such as lipids (cholesterol in the circulation), peptides, aptamers, antibodies, and sugars (N-acetylgalactosamine (GalNAc)), to target specific cells/tissues or reduce circulation clearance (van den Boorn et al. 2011; Fu et al. 2015). Ideally, the pharmacokinetic properties of bioconjugates are clear because they are unique and single-component molecular entities composed of precise stoichiometry. At present, nucleic acids have been packaged into carriers such as viruses, polycation polyethyleneimine (PEI) NPs, immunoconjugates, and lipid formulations such as LNPs, which are affected by the incidental defects of each packaging agent (Shaffer 2020). For instance, Alnylam Onpattro was used to treat transthyretin amyloidosis as the first siRNA therapeutic drug approved by the FDA, which is packaged in the cationic amino MC3 lipid NPs to target liver cells (Shaffer 2020). Moreover, Acorda Therapeutics focuses on the research of inhaled RNA in response to respiratory pathogens including COVID-19 (Chen et al. 2014b). So far, only a very limited number of possible modification methods have been studied, and even fewer have been used in clinical practice.

Over the last 30 years, the rapid development of nanoscience and nanotechnology has raised new questions about how DNA and DNA structures interact with inorganic nanomaterials or cells and animals, which should be important for their biological and biomedical applications (Seeman 1982). In 1982, Seeman proposed that DNA can be locked into a knot in a stable state via breaking the sequence symmetry of the four strands that make up the Holliday knot (Hu and Niemeyer 2019; Chao et al. 2019). This theory supports the development of DNA nanotechnology, which is widely accepted in materials science community and used to construct more complex micro- and nanostructures (Yatsunyk et al. 2014). DNA has the ability to self-recognize and self-assemble depending on the formation of base-specific interactions allowing the chains to adopt the structure in a controlled manner (Bradley 2018; Hu and Niemeyer 2019).

Due to the potential of using DNA molecules as substrates for the production of materials with biological functions, the use of DNA-based nanorobotic systems for

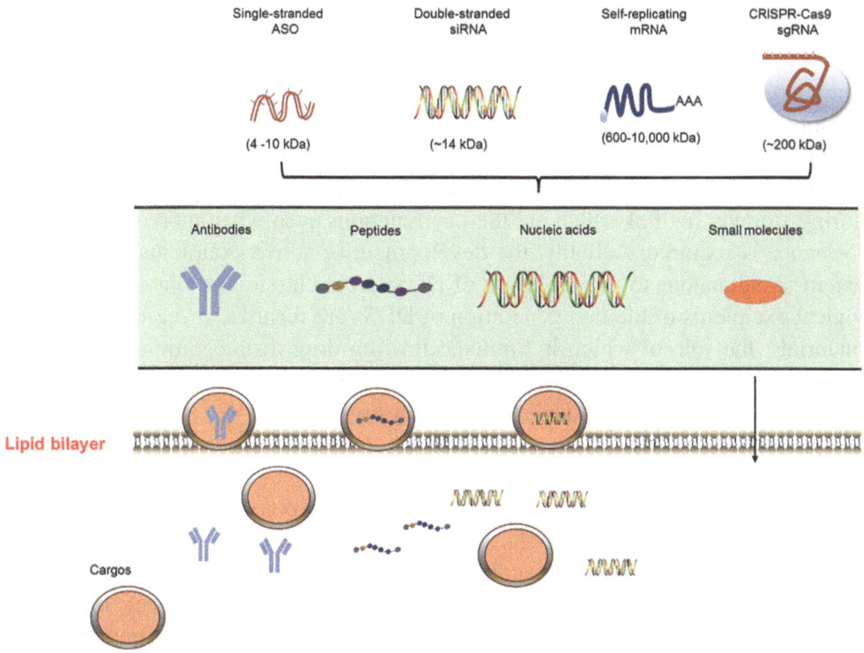

Fig. 2.5 Many typical biomolecules, such as protein peptides, DNA, and RNA, have been developed as drugs. However, all of them as drugs have the same problem that they have limited bioavailability. For example, the entry across the lipid bilayer of these drugs is very difficult, and they suffer from in vivo stability issues, including susceptibility to enzyme degradation and recognition by host RNA/DNA sensors. Therefore, these drugs require proprietary cargos or DDS to the target sites and achieve the therapeutic effect. (Reused with permission from Dowdy 2017)

the targeted delivery of therapeutic payloads to tumors became a hot topic of attention. For example, DNA origami was used to produce thrombin-bound DNA nanotubes to insulate the protease from the circulating substrate and also hybridize with predesigned DNA aptamers that recognize nucleoli, which are selectively expressed on tumor-associated vascular endothelial cells (Wagenbauer et al. 2017). The DNA sequence information is used to encode the shape of individual DNA origami components, and then the geometry and details of the interactions between these components will control their copy number, position, and orientation in higher-order components (Niemeyer 2017). Therefore, DNA nanocapsules provide another tool to basic biomedical research, which are developed for in vivo analysis of intracellular pH values, chloride or important messenger molecules, as well as being used for the targeted transport of molecular payloads and cell surface evaluation (Dowdy 2017; Stewart et al. 2018) (Fig. 2.5). Generally, DNA nanocapsules may be a compatible carrier for use as drug delivery systems.

2.5 Different Components of DDS

The therapeutic efficiency in a great extent depends on the release rate of the active substances in the target site. Therefore, the active excipients composed of the DDSs play key roles to achieve higher therapeutic efficiency. Traditionally, the excipients are utilized to facilitate the preparation of drugs and to ensure the stability of drugs during storage, both of which are the key functions even when more complex performance is required. Actually, the development of active excipients is to a great extent simultaneous to the evolution of DDSs from classic formulations. The biological excipients of the first generation of DDSs are regarded as biologically inert materials, the role of which is limited, allowing drug disintegration once it contacted with physiological fluids. These kinds of biomaterials are still utilized in a variety of DDSs such as immediate oral release systems.

In order to overcome these shortcomings of the first generation of excipients, materials science is integrated with engineering principles under the biological and biomedical standards to produce good biological responsive materials, so-called second generation of biomaterials, which can better interact with surrounding tissues during temporary (biodegradable) or permanent contact (non-biodegradable). Furthermore, the third generation of biomaterials can assist to overcome the limitations and allow the drug to reach the optimal therapeutic plasma/tissue concentration by regulating drug release behaviors and pharmacokinetics, such as using lipids and polymers as the components of smart DDSs, which combine diagnostic and drug delivery functions (Linko et al. 2015). DNA molecules, for instance, a truly programmable material, can be formed into customer-predesigned shapes, so-called DNA origami nanostructures, serving as smart drug delivery vehicles (Wu et al. 2016). Importantly, the therapeutic and diagnostic applications of nanomedicine are playing increasingly important roles in human health, including various types of synthetic NPs such as liposomes, micelles, and other nanotherapeutic platforms and conjugates. Blood cells, such as red blood cells, have also been developed as permeable carriers to release drugs to diseased tissues or act as biological factors, allowing enzymes to degrade pathological substrates (Alvarez-Lorenzo and Concheiro 2014).

Generally, advanced knowledge about human body composition facilitates the development of various novel biomaterials with good biocompatibility in the human body, which can actively interact with biological components and detect certain signals from the body to change the behaviors and functions of DDSs. However, the safety and excretion/elimination assessments must be guaranteed when the design of the biomaterials are intended to reach profound tissues and cell structures. Accordingly, drugs can be used in various diseases with different DDSs.

2.6 Concluding Remarks

In the past few decades, pharmaceutical and bio-related industries have undergone significant transformations, and have developed various small synthetic drugs successfully, which have great impact on our society and economy. More importantly, the effective therapeutic effects of drugs mainly rely on their DDSs, which are designed meticulously according to the internal variables and external stimuli. The DDSs adopt multidisciplinary approaches to effectively deliver the drug to the target site, thereby adjusting the drug's metabolic kinetics, pharmacodynamics, toxicity, and immunogenicity. With more requirements of the quality and safety for patients, specialized treatments for specific patient groups with unmet needs attract more attentions. In this new context, novel and smart DDSs, such as target nanocarriers and theranostic systems, have emerged in the past decades to play key roles in the clinic.

Compared with traditional formulations, the novel DDSs have the following advantages: improving the stability of the drug, reducing the degradation of the drug, reducing the toxic and side effects of the drug, improving the bioavailability of the drug, maintaining a stable and effective drug concentration, avoiding drug concentration fluctuations in the blood, and increasing the concentration of the drug at the target site. At present, researchers have developed and established various types of new DDSs, and with significant research investment, the drug and market share continue to grow rapidly.

Meanwhile, the quality control of new drug delivery excipients should be guaranteed along with their bioavailability, while their cytotoxicity, genotoxicity, antigenicity, and clearance should be reduced as observed and measured in the target tissues or cells. Although some novel DDSs have been successfully applied in the pharmaceutical industry, there are still many problems, such as the safety regarding the interaction between the delivery vectors and the cells or biological barriers, the stability of the delivery vectors in the biological environment, and targeted technology of drug delivery and release profile. Thus, great efforts should be dedicated to explore the underground mechanisms. And there are still many uncovered findings, which can promote the development and gradual maturity of the new DDSs and technologies.

Despite the concerns mentioned above, the development of the novel DDSs is stepping toward optimized design, easy production, and more accurate evaluation of effective formulations in vivo. In the future, DDS is bound to demonstrate broader development prospects, showing a wide range of applications and benefiting a vast number of disease patients.

Acknowledgments Dr. Xuefeng Li is sponsored by National Natural Science Foundation of China (81972204), Natural Science Foundation of Guangdong Province (2019A1515011097), Innovation Program of Shenzhen (Grant No. JCYJ20180508165208399), Science and Technology Planning Project of Guangzhou (201904010089), the grant from the State Key Lab of Respiratory Disease, Guangzhou Medical University (SKLRD-Z-202002), and the 111 Project (D18010) from the Ministry of Education of China. The authors thank Huan Zhang from School of Life Sciences, Jilin University, for editorial assistance.

References

Agyei D, Ahmed I, Akram Z et al (2017) Protein and peptide biopharmaceuticals: an overview. Protein Pept Lett 24:94–101. https://doi.org/10.2174/0929866523666161222150444

Alvarez-Lorenzo C, Concheiro A (2014) Smart drug delivery systems: from fundamentals to the clinic. Chem Commun (Camb) 50:7743–7765. https://doi.org/10.1039/c4cc01429d

Angelbello AJ, Chen JL, Childs-Disney JL et al (2018) Using genome sequence to enable the design of medicines and chemical probes. Chem Rev 118:1599–1663. https://doi.org/10.1021/acs.chemrev.7b00504

Anselmo AC, Gokarn Y, Mitragotri S (2019) Non-invasive delivery strategies for biologics. Nat Rev Drug Discov 18:19–40. https://doi.org/10.1038/nrd.2018.183

António JPM, Russo R, Carvalho CP et al (2019) Boronic acids as building blocks for the construction of therapeutically useful bioconjugates. Chem Soc Rev 48:3513–3536. https://doi.org/10.1039/c9cs00184k

Banaszynski LA, Sellmyer MA, Contag CH et al (2008) Chemical control of protein stability and function in living mice. Nat Med 14:1123–1127. https://doi.org/10.1038/nm.1754

Banik BL, Fattahi P, Brown JL (2016) Polymeric nanoparticles: the future of nanomedicine. Wiley Interdiscip Rev Nanomed Nanobiotechnol 8:271–299. https://doi.org/10.1002/wnan.1364

Bhatia D, Chakraborty S, Krishnan Y (2012) Gene delivery: designer DNA give RNAi more spine. Nat Nanotechnol 7:344–346. https://doi.org/10.1038/nnano.2012.87

Blain JC, Szostak JW (2014) Progress toward synthetic cells. Annu Rev Biochem 83:615–640. https://doi.org/10.1146/annurev-biochem-080411-124036

Bradley CA (2018) Drug delivery: DNA nanorobots - seek and destroy. Nat Rev Drug Discov 17:242. https://doi.org/10.1038/nrd.2018.40

Braghiroli A, Braido F, Piraino A et al (2020) Day and night control of COPD and role of pharmacotherapy: a review. Int J Chron Obstruct Pulmon Dis 15:1269–1285. https://doi.org/10.2147/copd.s240033

Brayden DJ, Hill TA, Fairlie DP et al (2020) Systemic delivery of peptides by the oral route: formulation and medicinal chemistry approaches. Adv Drug Deliv Rev 157:2–36. https://doi.org/10.1016/j.addr.2020.05.007

Breton MD, Kanapka LG, Beck RW et al (2020) A randomized trial of closed-loop control in children with type 1 diabetes. N Engl J Med 383:836–845. https://doi.org/10.1056/NEJMoa2004736

Cao Y, Samy KE, Bernards DA et al (2019) Recent advances in intraocular sustained-release drug delivery devices. Drug Discov Today 24:1694–1700. https://doi.org/10.1016/j.drudis.2019.05.031

Chao J, Wang J, Wang F et al (2019) Solving mazes with single-molecule DNA navigators. Nat Mater 18:273–279. https://doi.org/10.1038/s41563-018-0205-3

Chen J, Li X, Wu X et al (2014a) Au-Silica nanowire nanohybrid as a hyperthermia agent for photothermal therapy in the near-infrared region. Langmuir 30:9514–9523. https://doi.org/10.1021/la5020434

Chen N, Li J, Song H et al (2014b) Physical and biochemical insights on DNA structures in artificial and living systems. Acc Chem Res 47:1720–1730. https://doi.org/10.1021/ar400324n

Chen Z, Chen Z, Zhang A et al (2016) Electrospun nanofibers for cancer diagnosis and therapy. Biomater Sci 4:922–932. https://doi.org/10.1039/c6bm00070c

Chen W, He J, Li H et al (2020a) A quinolone derivative-based organoplatinum (II) metallacycle supramolecular self-delivery nanocarrier for combined cancer therapy. Supramol Chem 32:1–8

Chen W, Li X, Liu C et al (2020b) beta-Cyclodextrin modified Pt(II) metallacycle-based supramolecular hyperbranched polymer assemblies for DOX delivery to liver cancer cells. Proc Natl Acad Sci USA 117:30942–30948. https://doi.org/10.1073/pnas.2007798117

Cho EJ, Holback H, Liu KC et al (2013) Nanoparticle characterization: state of the art, challenges, and emerging technologies. Mol Pharm 10:2093–2110. https://doi.org/10.1021/mp300697h

Ciancia S, Cafarelli A, Zahoranova A et al (2020) Pulsatile drug delivery system triggered by acoustic radiation force. Front Bioeng Biotechnol 8:317. https://doi.org/10.3389/fbioe.2020.00317

Cochrane SA, Li X, He S et al (2015) Synthesis of Tridecaptin-antibiotic conjugates with *in vivo* activity against gram-negative bacteria. J Med Chem 58:9779–9785. https://doi.org/10.1021/acs.jmedchem.5b01578

Cohen EA, Ball P (1998) Made to measure: new materials for the 21st century. Foreign Aff (Council on Foreign Relations) 77:136

Cunningham RD, Kopf AH, Elenbaas BOW et al (2020) Iterative RAFT-mediated copolymerization of styrene and maleic anhydride toward sequence- and length-controlled copolymers and their applications for solubilizing lipid membranes. Biomacromolecules 21:3287–3300. https://doi.org/10.1021/acs.biomac.0c00736

DeWeerdt S (2016) Drug development: through the barrier. Nature 537:S154–S157. https://doi.org/10.1038/537S154a

Dowdy SF (2017) Overcoming cellular barriers for RNA therapeutics. Nat Biotechnol 35:222–229. https://doi.org/10.1038/nbt.3802

Drucker DJ (2020) Advances in oral peptide therapeutics. Nat Rev Drug Discov 19:277–289. https://doi.org/10.1038/s41573-019-0053-0

Elliott S, Lorenzini T, Asher S et al (2003) Enhancement of therapeutic protein *in vivo* activities through glycoengineering. Nat Biotechnol 21:414–421. https://doi.org/10.1038/nbt799

Fan M, Liu J, Zhao B et al (2020) Indirect comparison of NSAIDs for ankylosing spondylitis: network meta-analysis of randomized, double-blinded, controlled trials. Exp Ther Med 19:3031–3041. https://doi.org/10.3892/etm.2020.8564

Fan M, Liu J, Zhao B et al (2021) Comparison of efficacy and safety of urate-lowering therapies for hyperuricemic patients with gout: a meta-analysis of randomized, controlled trials. Clin Rheumatol 40:683–692. https://doi.org/10.1007/s10067-020-05272-4

Fu Q, Wang Y, Ma Y et al (2015) Programmed hydrolysis in designing paclitaxel prodrug for nanocarrier assembly. Sci Rep 5. https://doi.org/10.1038/srep12023

Gao HL, Jiang XG (2017) The progress of novel drug delivery systems. Yao Xue Xue Bao 52:181–188

Gao H, Pang Z, Jiang X (2013) Targeted delivery of nano-therapeutics for major disorders of the central nervous system. Pharm Res 30:2485–2498. https://doi.org/10.1007/s11095-013-1122-4

Gong P, Wang Y, Zhang P et al (2020) Immunocyte membrane-coated nanoparticles for cancer immunotherapy. Cancers (Basel) 13. https://doi.org/10.3390/cancers13010077

Hashida M (2020) Role of pharmacokinetic consideration for the development of drug delivery systems: a historical overview. Adv Drug Deliv Rev 157:71. https://doi.org/10.1016/j.addr.2020.06.015

Haznar-Garbacz D, Garbacz G, Eisenächer F et al (2011) An oral-controlled release drug delivery system for liquid and semisolid drug formulations. AAPS PharmSciTech 12:1183–1185. https://doi.org/10.1208/s12249-011-9689-8

He S, Li X, Li R et al (2016) Annexin A2 modulates ROS and impacts inflammatory response via IL-17 signaling in polymicrobial sepsis mice. PLoS Pathog 12:e1005743. https://doi.org/10.1371/journal.ppat.1005743

Heinemann L (2008) The failure of exubera: are we beating a dead horse? J Diabetes Sci Technol 2:518–529. https://doi.org/10.1177/193229680800200325

Helfand WH, Cowen DL (1983) Evolution of pharmaceutical oral dosage forms. Pharm Hist 25:3–18

Hu Y, Niemeyer CM (2019) From DNA nanotechnology to material systems engineering. Adv Mater 31:e1806294. https://doi.org/10.1002/adma.201806294

Hu Q, Li H, Wang L et al (2019) DNA nanotechnology-enabled drug delivery systems. Chem Rev 119:6459–6506. https://doi.org/10.1021/acs.chemrev.7b00663

Huang F, Zhang J, Yang D et al (2018) MicroRNA expression profile of whole blood is altered in adenovirus-infected pneumonia children. Mediat Inflamm 2018:2320640. https://doi.org/10.1155/2018/2320640

Jain KK (2014) Current status and future prospects of drug delivery systems. Methods Mol Biol 1141:1–56. https://doi.org/10.1007/978-1-4939-0363-4_1

Jain D, Raturi R, Jain V et al (2011) Recent technologies in pulsatile drug delivery systems. Biomatter 1:57–65. https://doi.org/10.4161/biom.1.1.17717

Jain A, Gulbake A, Shilpi S et al (2013) Peptide and protein delivery using new drug delivery systems. Crit Rev Ther Drug Carrier Syst 30:293–329. https://doi.org/10.1615/critrevtherdrugcarriersyst.2013006955

Kamaly N, Yameen B, Wu J et al (2016) Degradable controlled-release polymers and polymeric nanoparticles: mechanisms of controlling drug release. Chem Rev 116:2602–2663. https://doi.org/10.1021/acs.chemrev.5b00346

Kang C, Sun Y, Zhu J et al (2016) Delivery of nanoparticles for treatment of brain tumor. Curr Drug Metab 17:745–754. https://doi.org/10.2174/1389200217666160728152939

Khafagy el S, Morishita M, Onuki Y et al (2007) Current challenges in non-invasive insulin delivery systems: a comparative review. Adv Drug Deliv Rev 59:1521–1546. https://doi.org/10.1016/j.addr.2007.08.019

Kriplani P, Pathak K, Philip A (2020) Preparation and evaluation of '3 Cap' pulsatile drug delivery system of Ramipril. Cardiovasc Hematol Agents Med Chem 19:50. https://doi.org/10.2174/1871525718666200528140527

Kumar R, Mondal K, Panda PK et al (2020) Core-shell nanostructures: perspectives towards drug delivery applications. J Mater Chem B 8:8992. https://doi.org/10.1039/d0tb01559h

Lan H, Zhang W, Jin K et al (2020) Modulating barriers of tumor microenvironment through nanocarrier systems for improved cancer immunotherapy: a review of current status and future perspective. Drug Deliv 27:1248–1262. https://doi.org/10.1080/10717544.2020.1809559

Langer R (2001) Drug delivery. Drugs on target. Science 293:58–59. https://doi.org/10.1126/science.1063273

Lei Y, Wang K, Li X et al (2019) Cell-surface translocation of annexin A2 contributes to bleomycin-induced pulmonary fibrosis by mediating inflammatory response in mice. Clin Sci (Lond) 133:789–804. https://doi.org/10.1042/CS20180687

Li X, Zhou X, Ye Y et al (2014a) Lyn regulates inflammatory responses in Klebsiella pneumoniae infection via the p38/NF-kappaB pathway. Eur J Immunol 44:763–773. https://doi.org/10.1002/eji.201343972

Li Y, Gan CP, Zhang S et al (2014b) FIP200 is involved in murine pseudomonas infection by regulating HMGB1 intracellular translocation. Cell Physiol Biochem 33:1733–1744. https://doi.org/10.1159/000362954

Li X, Ye Y, Zhou X et al (2015) Atg7 enhances host defense against infection via downregulation of superoxide but upregulation of nitric oxide. J Immunol 194:1112–1121. https://doi.org/10.4049/jimmunol.1401958

Li R, Fang L, Pu Q et al (2016a) Lyn prevents aberrant inflammatory responses to Pseudomonas infection in mammalian systems by repressing a SHIP-1-associated signaling cluster. Signal Transduct Target Ther 1:16032. https://doi.org/10.1038/sigtrans.2016.32

Li R, Fang L, Tan S et al (2016b) Type I CRISPR-Cas targets endogenous genes and regulates virulence to evade mammalian host immunity. Cell Res 26:1273–1287. https://doi.org/10.1038/cr.2016.135

Li X, He S, Li R et al (2016c) Pseudomonas aeruginosa infection augments inflammation through miR-301b repression of c-Myb-mediated immune activation and infiltration. Nat Microbiol 1:16132. https://doi.org/10.1038/nmicrobiol.2016.132

Li X, He S, Zhou X et al (2016d) Lyn delivers bacteria to lysosomes for eradication through TLR2-initiated autophagy related phagocytosis. PLoS Pathog 12:e1005363. https://doi.org/10.1371/journal.ppat.1005363

Li R, Fang L, Pu Q et al (2018) MEG3-4 is a miRNA decoy that regulates IL-1beta abundance to initiate and then limit inflammation to prevent sepsis during lung infection. Sci Signal 11. https://doi.org/10.1126/scisignal.aao2387

Liang H, Deng X, Li X et al (2014) Molecular mechanisms of master regulator VqsM mediating quorum-sensing and antibiotic resistance in Pseudomonas aeruginosa. Nucleic Acids Res 42:10307–10320. https://doi.org/10.1093/nar/gku586

Lin SY (2020) Thermoresponsive gating membranes embedded with liquid crystal(s) for pulsatile transdermal drug delivery: an overview and perspectives. J Control Release 319:450–474. https://doi.org/10.1016/j.jconrel.2019.12.046

Linko V, Ora A, Kostiainen MA (2015) DNA nanostructures as smart drug-delivery vehicles and molecular devices. Trends Biotechnol 33:586–594. https://doi.org/10.1016/j.tibtech.2015.08.001

Lu X, Miao L, Gao W et al (2020) Engineered PLGA microparticles for long-term, pulsatile release of STING agonist for cancer immunotherapy. Sci Transl Med 12:eaaz6606. https://doi.org/10.1126/scitranslmed.aaz6606

Ma Y, Dong S, Li X et al (2020) Extracellular vesicles: an emerging nanoplatform for cancer therapy. Front Oncol 10:606906. https://doi.org/10.3389/fonc.2020.606906

Majeed W, Thabit H (2018) Closed-loop insulin delivery: current status of diabetes technologies and future prospects. Expert Rev Med Devices 15:579–590. https://doi.org/10.1080/17434440.2018.1503530

Maudens P, Jordan O, Allémann E (2018) Recent advances in intra-articular drug delivery systems for osteoarthritis therapy. Drug Discov Today 23:1761–1775. https://doi.org/10.1016/j.drudis.2018.05.023

Mullard A (2021) FDA approves first anti-ANGPTL3 antibody, for rare cardiovascular indication. Nat Rev Drug Discov 20:251. https://doi.org/10.1038/d41573-021-00047-1

Neveu M, Kim HJ, Benner SA (2013) The "strong" RNA world hypothesis: fifty years old. Astrobiology 13:391–403. https://doi.org/10.1089/ast.2012.0868

Niemeyer CM (2017) DNA nanotechnology: on-command molecular Trojans. Nat Nanotechnol 12:1117–1119. https://doi.org/10.1038/nnano.2017.222

Pan S, Xing H, Fu X et al (2018) The effect of photothermal therapy on osteosarcoma with polyacrylic acid-coated gold nanorods. Dose Response 16:1559325818789841. https://doi.org/10.1177/1559325818789841

Park K (2014) Controlled drug delivery systems: past forward and future back. J Control Release 190:3–8. https://doi.org/10.1016/j.jconrel.2014.03.054

Pick H, Alves AC, Vogel H (2018) Single-vesicle assays using liposomes and cell-derived vesicles: from modeling complex membrane processes to synthetic biology and biomedical applications. Chem Rev 118:8598–8654. https://doi.org/10.1021/acs.chemrev.7b00777

Pickar-Oliver A, Gersbach CA (2019) The next generation of CRISPR-Cas technologies and applications. Nat Rev Mol Cell Biol 20:490–507. https://doi.org/10.1038/s41580-019-0131-5

Prausnitz MR, Mitragotri S, Langer R (2004) Current status and future potential of transdermal drug delivery. Nat Rev Drug Discov 3:115–124. https://doi.org/10.1038/nrd1304

Pu Q, Gan C, Li R et al (2017) Atg7 deficiency intensifies inflammasome activation and pyroptosis in pseudomonas sepsis. J Immunol 198:3205–3213. https://doi.org/10.4049/jimmunol.1601196

Revia RA, Stephen ZR, Zhang M (2019) Theranostic nanoparticles for RNA-based cancer treatment. Acc Chem Res 52:1496–1506. https://doi.org/10.1021/acs.accounts.9b00101

Roberts TC, Langer R, Wood MJA (2020) Advances in oligonucleotide drug delivery. Nat Rev Drug Discov 19:673–694. https://doi.org/10.1038/s41573-020-0075-7

Sarmento B (2010) Nanomedicines for delivery of therapeutic proteins and biopharmaceuticals. Ther Deliv 1:231–235. https://doi.org/10.4155/tde.10.39

Satav SS, Bhat S, Thayumanavan S (2010) Feedback regulated drug delivery vehicles: carbon dioxide responsive cationic hydrogels for antidote release. Biomacromolecules 11:1735–1740. https://doi.org/10.1021/bm1005454

Seeman NC (1982) Nucleic acid junctions and lattices. J Theor Biol 99:237–247. https://doi.org/10.1016/0022-5193(82)90002-9

Shaffer C (2020) Mist begins to clear for lung delivery of RNA. Nat Biotechnol 38:1110–1112. https://doi.org/10.1038/s41587-020-0692-z

Smith CIE, Zain R (2019) Therapeutic oligonucleotides: state of the art. Annu Rev Pharmacol Toxicol 59:605–630. https://doi.org/10.1146/annurev-pharmtox-010818-021050

Soppimath KS, Aminabhavi TM, Dave AM et al (2002) Stimulus-responsive "smart" hydrogels as novel drug delivery systems. Drug Dev Ind Pharm 28:957–974. https://doi.org/10.1081/ddc-120006428

Spicer CD, Jumeaux C, Gupta B et al (2018) Peptide and protein nanoparticle conjugates: versatile platforms for biomedical applications. Chem Soc Rev 47:3574–3620. https://doi.org/10.1039/c7cs00877e

Stayton PS, El-Sayed ME, Murthy N et al (2005) 'Smart' delivery systems for biomolecular therapeutics. Orthod Craniofac Res 8:219–225. https://doi.org/10.1111/j.1601-6343.2005.00336.x

Stein CA, Castanotto D (2017) FDA-approved oligonucleotide therapies in 2017. Mol Ther 25:1069–1075. https://doi.org/10.1016/j.ymthe.2017.03.023

Steinberg D, Friedman M (2020) Sustained-release delivery of antimicrobial drugs for the treatment of periodontal diseases: fantasy or already reality? Periodontol 84:176–187. https://doi.org/10.1111/prd.12341

Stewart MP, Langer R, Jensen KF (2018) Intracellular delivery by membrane disruption: mechanisms, strategies, and concepts. Chem Rev 118:7409–7531. https://doi.org/10.1021/acs.chemrev.7b00678

Sun J, Kormakov S, Liu Y et al (2018) Recent progress in metal-based nanoparticles mediated photodynamic therapy. Molecules 23. https://doi.org/10.3390/molecules23071704

Sun M, Qian Q, Shi L et al (2019) Amphiphilic drug-drug conjugate for cancer therapy with combination of chemotherapeutic and antiangiogenesis drugs. Science China Chem 63:35–41. https://doi.org/10.1007/s11426-019-9602-4

Sun X, Yu K, Zhou Y et al (2021) Self-assembled pH-sensitive polymeric nanoparticles for the inflammation-targeted delivery of Cu/Zn-superoxide dismutase. ACS Appl Mater Interfaces 13:18152. https://doi.org/10.1021/acsami.1c03589

Torchilin VP, Lukyanov AN (2003) Peptide and protein drug delivery to and into tumors: challenges and solutions. Drug Discov Today 8:259–266. https://doi.org/10.1016/s1359-6446(03)02623-0

van den Boorn JG, Schlee M, Coch C et al (2011) SiRNA delivery with exosome nanoparticles. Nat Biotechnol 29:325–326. https://doi.org/10.1038/nbt.1830

Veiseh O, Tang BC, Whitehead KA et al (2015) Managing diabetes with nanomedicine: challenges and opportunities. Nat Rev Drug Discov 14:45–57. https://doi.org/10.1038/nrd4477

Veronese FM, Harris JM (2002) Introduction and overview of peptide and protein pegylation. Adv Drug Deliv Rev 54:453–456. https://doi.org/10.1016/s0169-409x(02)00020-0

Wagenbauer KF, Sigl C, Dietz H (2017) Gigadalton-scale shape-programmable DNA assemblies. Nature 552:78–83. https://doi.org/10.1038/nature24651

Wang W, Ye Y, Li J et al (2014a) Lyn regulates cytotoxicity in respiratory epithelial cells challenged by cigarette smoke extracts. Curr Mol Med 14:663–672. https://doi.org/10.2174/1566524014666140603095027

Wang X, Huang X, Yang Z et al (2014b) Targeted delivery of tumor suppressor microRNA-1 by transferrin-conjugated lipopolyplex nanoparticles to patient-derived glioblastoma stem cells. Curr Pharm Biotechnol 15:839–846. https://doi.org/10.2174/1389201015666141031105234

Wang K, Zhang T, Lei Y et al (2018) Identification of ANXA2 (annexin A2) as a specific bleomycin target to induce pulmonary fibrosis by impeding TFEB-mediated autophagic flux. Autophagy 14:269–282. https://doi.org/10.1080/15548627.2017.1409405

Wang L, Dong S, Liu Y et al (2020) Fabrication of injectable, porous hyaluronic acid hydrogel based on an in-situ bubble-forming hydrogel entrapment process. Polymers (Basel) 12. https://doi.org/10.3390/polym12051138

Wilbie D, Walther J, Mastrobattista E (2019) Delivery aspects of CRISPR/Cas for *in Vivo* genome editing. Acc Chem Res 52:1555–1564. https://doi.org/10.1021/acs.accounts.9b00106

Wu M, Li X (2015) Klebsiella pneumoniae and Pseudomonas aeruginosa. In: Molecular medical microbiology. Elsevier, pp 1547–1564

Wu YW, Goubran H, Seghatchian J et al (2016) Smart blood cell and microvesicle-based Trojan horse drug delivery: merging expertise in blood transfusion and biomedical engineering in the field of nanomedicine. Transfus Apher Sci 54:309–318. https://doi.org/10.1016/j.transci.2016.04.013

Wu K, Li Y, Zhang S et al (2019a) Preferential homing of tumor-specific and functional CD8+ stem cell-like memory T cells to the bone marrow. J Immunother 42:197–207. https://doi.org/10.1097/CJI.0000000000000273

Wu K, Wang F, Guo G et al (2019b) CD4(+) TSCMs in the bone marrow assist in maturation of antibodies against influenza in mice. Mediat Inflamm 2019:3231696. https://doi.org/10.1155/2019/3231696

Wu K, Zhang S, Zhang X et al (2019c) IL-21 expands HIV-1-specific CD8(+) T memory stem cells to suppress HIV-1 replication *in vitro*. J Immunol Res 2019:1801560. https://doi.org/10.1155/2019/1801560

Xie J, Teng L, Yang Z et al (2013) A polyethylenimine-linoleic acid conjugate for antisense oligonucleotide delivery. Biomed Res Int 2013:710502. https://doi.org/10.1155/2013/710502

Xie J, Yang Z, Zhou C et al (2016) Nanotechnology for the delivery of phytochemicals in cancer therapy. Biotechnol Adv 34:343–353. https://doi.org/10.1016/j.biotechadv.2016.04.002

Xie J, Bi Y, Zhang H et al (2020) Cell-penetrating peptides in diagnosis and treatment of human diseases: from preclinical research to clinical application. Front Pharmacol 11:697. https://doi.org/10.3389/fphar.2020.00697

Xu X, Huang X, Zhang Y et al (2020) Self-regulated hirudin delivery for anticoagulant therapy. Sci Adv 6. https://doi.org/10.1126/sciadv.abc0382

Xue B, Zhang Y, Xu M et al (2019) Curcumin-silk fibroin nanoparticles for enhanced anti-candida albicans activity *in vitro* and *in vivo*. J Biomed Nanotechnol 15:769–778. https://doi.org/10.1166/jbn.2019.2722

Yang Z, Xie J, Zhu J et al (2016) Functional exosome-mimic for delivery of siRNA to cancer: *in vitro* and *in vivo* evaluation. J Control Release 243:160–171. https://doi.org/10.1016/j.jconrel.2016.10.008

Yang C, Xue B, Song W et al (2018) Reducing the toxicity of amphotericin B by encapsulation using methoxy poly(ethylene glycol)-b-poly(l-glutamic acid-co-l-phenylalanine). Biomater Sci 6:2189–2196. https://doi.org/10.1039/c8bm00506k

Yang Y, Guo J, Huang L (2020a) Tackling TAMs for cancer immunotherapy: it's nano time. Trends Pharmacol Sci 41:701–714. https://doi.org/10.1016/j.tips.2020.08.003

Yang Z, Shi J, Xie J et al (2020b) Large-scale generation of functional mRNA-encapsulating exosomes via cellular nanoporation. Nat Biomed Eng 4:69–83. https://doi.org/10.1038/s41551-019-0485-1

Yang X, Ma Y, Xie H et al (2021) Extracellular vesicles in the treatment of Parkinson's disease: a review. Curr Med Chem 28:6375. https://doi.org/10.2174/0929867328666210113170941

Yatsunyk LA, Mendoza O, Mergny JL (2014) "Nano-oddities": unusual nucleic acid assemblies for DNA-based nanostructures and nanodevices. Acc Chem Res 47:1836–1844. https://doi.org/10.1021/ar500063x

Ye Y, Li X, Wang W et al (2014) Atg7 deficiency impairs host defense against Klebsiella pneumoniae by impacting bacterial clearance, survival and inflammatory responses in mice. Am J Physiol Lung Cell Mol Physiol 307:L355–L363. https://doi.org/10.1152/ajplung.00046.2014

Youan BB (2010) Chronopharmaceutical drug delivery systems: hurdles, hype or hope? Adv Drug Deliv Rev 62:898–903. https://doi.org/10.1016/j.addr.2010.04.010

Yu B, Wang X, Zhou C et al (2014) Insight into mechanisms of cellular uptake of lipid nanoparticles and intracellular release of small RNAs. Pharm Res 31:2685–2695. https://doi.org/10.1007/s11095-014-1366-7

Yu J, Zhang Y, Yan J et al (2018) Advances in bioresponsive closed-loop drug delivery systems. Int J Pharm 544:350–357. https://doi.org/10.1016/j.ijpharm.2017.11.064

Yuan K, Xie K, Lan T et al (2020) TXNDC12 promotes EMT and metastasis of hepatocellular carcinoma cells via activation of beta-catenin. Cell Death Differ 27:1355–1368. https://doi.org/10.1038/s41418-019-0421-7

Yun YH, Lee BK, Park K (2015) Controlled drug delivery: historical perspective for the next generation. J Control Release 219:2–7. https://doi.org/10.1016/j.jconrel.2015.10.005

Zhang S, Yu M, Guo Q et al (2015) Annexin A2 binds to endosomes and negatively regulates TLR4-triggered inflammatory responses via the TRAM-TRIF pathway. Sci Rep 5:15859. https://doi.org/10.1038/srep15859

Zhao Z, Ukidve A, Kim J et al (2020) Targeting strategies for tissue-specific drug delivery. Cell 181:151–167. https://doi.org/10.1016/j.cell.2020.02.001

Zhou C, Yang Z, Teng L (2014a) Nanomedicine based on nucleic acids: pharmacokinetic and pharmacodynamic perspectives. Curr Pharm Biotechnol 15:829–838. https://doi.org/10.2174/1389201015666141020155620

Zhou P, Sun X, Zhang Z (2014b) Kidney-targeted drug delivery systems. Acta Pharm Sin B 4:37–42. https://doi.org/10.1016/j.apsb.2013.12.005

Zhou X, Li X, Ye Y et al (2014c) MicroRNA-302b augments host defense to bacteria by regulating inflammatory responses via feedback to TLR/IRAK4 circuits. Nat Commun 5:3619. https://doi.org/10.1038/ncomms4619

Zhou X, Ye Y, Sun Y et al (2015) Transient receptor potential channel 1 deficiency impairs host defense and proinflammatory responses to bacterial infection by regulating protein kinase Calpha signaling. Mol Cell Biol 35:2729–2739. https://doi.org/10.1128/MCB.00256-15

Zhou X, Li X, Wu M (2018) miRNAs reshape immunity and inflammatory responses in bacterial infection. Signal Transduct Target Ther 3:14. https://doi.org/10.1038/s41392-018-0006-9

Zhuang Y, Li Y, Li X et al (2016) Atg7 knockdown augments concanavalin A-induced acute hepatitis through an ROS-mediated p38/MAPK pathway. PLoS One 11:e0149754. https://doi.org/10.1371/journal.pone.0149754

Zielińska A, Carreiró F, Oliveira AM et al (2020) Polymeric nanoparticles: production, characterization, toxicology and ecotoxicology. Molecules 25. https://doi.org/10.3390/molecules25163731

Chapter 3
Drug Delivery Strategies That Target Tumors

Shiyan Dong, Jon Ashley, and Lesheng Teng

Abstract Chemotherapy is an important cancer treatment strategy and one of the few available options for treating metastatic tumors or lesions that cannot be removed by surgery. Unfortunately, classic chemotherapy drugs have limited therapeutic potential and often cause serious side effects (Yan et al., Dose Response 18:1559325820936161, 2020). Because of the limitations of conventional chemotherapies, the cancer chemotherapy paradigm has now shifted from conventional methods to tumor-specific targeting methods, which may maximize treatment efficiency and minimize related toxic effects of therapeutic molecules.

Using nanocarriers to deliver chemotherapeutic drugs is a promising chemotherapy strategy. Nanoformulations used to deliver chemotherapeutic agents include liposomes, polymeric nanoparticles, and polymeric micelles (Fig. 3.1). The use of nanocarriers as a delivery vehicle for chemotherapeutic drugs has the following advantages: (1) Improved solubility of poorly soluble drugs, which avoids using toxic solubilizers to dissolve poorly soluble chemotherapeutic drugs (Bernabeu et al., Int J Pharm 526:474-495, 2017). (2) Increased circulation time and improved stability of chemotherapy drugs. (3) Preferentially aggregated nanocarriers at the tumor site due to the enhanced permeability and retention effect, thereby improving the efficacy and reducing the side effects of chemotherapy.

Keywords Tumor targeting · Drug delivery strategies · Passive targeting · Active targeting

---

S. Dong
School of Life Sciences, Jilin University, Changchun, Jilin, China

Department of Radiation Oncology, The University of Texas Southwestern Medical Center, Dallas, TX, USA

J. Ashley
Department of Health Technology, Technical University of Denmark, Lyngby, Denmark

L. Teng (✉)
School of Life Sciences, Jilin University, Changchun, Jilin, China
e-mail: tenglesheng@jlu.edu.cn

L. Teng et al. (eds.), *Drug Delivery to Tumors*,
https://doi.org/10.1007/978-981-19-8930-8_3

3.1 Passive Targeting and Irrational Application of EPR Effect

The delivery of nanocarriers mediated by the enhanced permeability and retention (EPR) effect is the biggest breakthrough in targeted anticancer therapies (Torchilin 2011). The EPR effect was first reported by Maeda and Matsumura in 1986 (Matsumura and Maeda 1986). Rapid growth is one of cancer's key characteristics. Tumor cells need to obtain large amounts of nutrients and oxygen from the blood to maintain rapid growth. Studies have shown that most solid tumors usually produce extensive amounts of different vascular permeability factors, resulting in vascular structural defects, and thus enhancing vascular permeability (Maeda 2001a; Iyer et al. 2006; Maeda et al. 2006). The unique anatomical-pathophysiological properties of tumor blood vessels facilitate the transportation of macromolecules to tumor tissues. This vascular phenomenon is called the EPR effect, which allows macromolecules larger than 40 kDa to selectively leak out of tumor blood vessels and accumulate in tumor tissues. In contrast, tight junctions in normal endothelial cells prevent EPR effect-driven drug delivery from occurring in normal tissues (Maeda 2001b). Therefore, the EPR effect provides tumor-targeted drug delivery, which is a landmark principle in tumor-targeted chemotherapy and a promising paradigm for developing anticancer drugs (Figs. 3.1 and 3.2).

Nanocarriers used as drug delivery vehicles (such as liposomes, polymeric nanoparticles, micelles) can leak out of tumor blood vessels through the EPR effect and deliver drugs to the tumor through passive accumulation (He et al. 2021; Ma et al. 2021; Yang et al. 2016, 2019, 2021). So far, different nanocarriers have been developed based on the EPR effect, which have either been approved for clinical

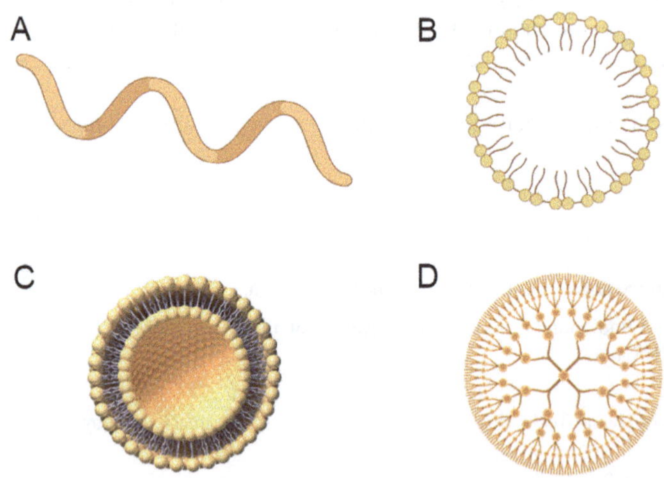

Fig. 3.1 Nanomaterials are commonly used in biomedical applications. (**a**) Polymer conjugate, (**b**) micelle, (**c**) liposome, (**d**) dendrimer. (This image was made by figdraw.com and exported with ID: WOTSR0009e)

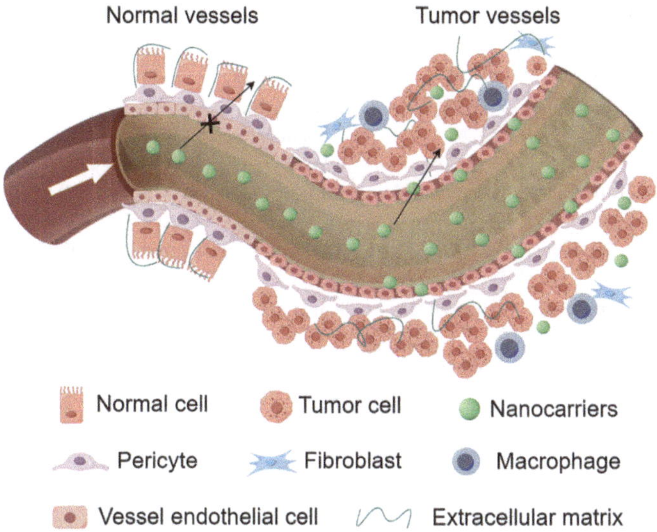

Fig. 3.2 Accumulation of nanocarriers in tumor tissues via the EPR effect. (This image was made by figdraw.com and exported with ID: UAAWU4d00c)

treatment of cancer or are being evaluated in different clinical trials (Tables 3.1 and 3.2). Doxil®, a liposomal preparation of doxorubicin (DOX) hydrochloride, is the first FDA-approved nano-drug. Encapsulating the antitumor drug DOX in polyethylene glycol (PEG)-modified liposomes has shown high efficacy in EPR-based tumor treatment, and significantly reduced the side effects of free DOX (Barenholz 2012). In a clinical trial of 15 cancer patients, Gabizon et al. reported that the clearance of Doxil® (0.1 L/h) is much lower than that of free DOX (45 L/h), allowing tumor cells to be exposed to antitumor drugs for longer periods. Therefore, at the same dose, Doxil® can improve DOX's chemotherapeutic efficacy. Importantly, due to the EPR effect, Doxil® changes the distribution of DOX and significantly reduces the irreversible cardiotoxicity of free DOX. So far, Doxil® has been widely used in clinical chemotherapy (Gabizon et al. 1994). Clinical data has shown that PEGylated DOX liposomes effectively treat unresectable hepatocellular carcinoma (Schmidinger et al. 2001), metastatic breast cancer (Perez et al. 2002), lymphoma (Wollina et al. 2003), and sarcoma (Skubitz 2003). So far, in addition to Doxil®, liposomal preparations of daunorubicin (Allen and Martin 2004), cytarabine (Glantz et al. 1999), vincristine (Krishna et al. 2001), and irinotecan (Thomas and Pommier 2019) have also been approved by the FDA and used in clinical chemotherapy. Anticancer drug micellar formulations developed by exploiting the passive EPR effect can further increase the tumor targeting and efficacy of anticancer drugs (Torchilin 2001). Polymer-drug conjugates have multiple advantages, such as improved treatment efficiency, reduced side effects, flexible administration, and improved patient compliance. Polymer-drug conjugates combine poorly water-soluble drugs with water-soluble natural or synthetic polymeric carriers through

Table 3.1 List of nanoformulations based on the EPR effect approved for clinical use

Product name	Formulation type	Drug	Indication(s)	Approval year	References
Doxil	Liposomes	DOX	Kaposi's Sarcoma; ovarian cancer; multiple myeloma	1995	Barenholz (2012)
DaunoXome	Liposomes	Daunorubicin	Karposi's sarcoma	1996	Krauss et al. (2019)
Myocet	Liposomes	DOX	Metastatic breast cancer	2005	Mross et al. (2004)
Onivyde	Liposomes	Irinotecan	Pancreatic cancer	2015	Drummond et al. (2010)
Marqibo	Liposomes	Vincristine	Non-Hodgkin's lymphoma	2012	Silverman and Deitcher (2013)
DepoCyt	Liposomes	Cytarabine	Lymphomatous meningitis	1996	Koller-Lucae et al. (1997)
Abraxane	Albumin nanoparticles	Paclitaxel	Breast cancer; non-small cell lung cancer; pancreatic cancer	2005	Nyman et al. (2005)
Vyxeos	Liposomes	Daunorubicin and Cytarabine	Newly diagnosed therapy-related acute myeloid leukemia (t-AML) or AML with myelodysplasia-related changes	2017	Blair (2018)
Genexol-PM	Polymeric micelles	Paclitaxel	Recurrent or metastatic breast cancer; locally advanced or metastatic non-small cell lung cancer (combination with cisplatin); combination therapy in the treatment of ovarian cancer with other chemotherapeutic agents	2007	Oerlemans et al. (2010)

biodegradable covalent bonds, passively accumulating into the tumor microenvironment through the EPR effect (Matsumura and Maeda 1986). Genexol-PM is a paclitaxel formulation based on polymeric micelles, which is now used in the clinic (Kim et al. 2004). Genexol-PM's efficacy and safety in patients with histologically confirmed metastatic breast cancer (MBC) were evaluated in a phase II clinical trial (Lee et al. 2008). Traditional polyoxyethylene castor oil (CrEL)-based paclitaxel as a first-line treatment has a remission rate of 21–54% for MBC patients (Paridaens et al. 2000), while the overall remission rate for the Genexol-PM group is 58.5%. Genexol-PM does not contain CrEL, a toxic solubilizer, which is associated with allergic reactions. In fact, the applicable dose of paclitaxel based on CrEL is higher

Table 3.2 A list of nanocarriers based on the EPR effect that are in clinical trials

Product name	Formulation type	Drug	Indication(s)	Approval status	References
NK-105	Polymeric micelles	Paclitaxel	Advanced stomach cancer and breast cancer	Phase-III clinical trial	Oerlemans et al. (2010)
NC-6004	Polymeric micelles	Cisplatin	Head cancer and neck cancer in the 5-fluorouracil and cetuximab combination	Phase-III clinical trial	Oerlemans et al. (2010) and Subbiah et al. (2018)
Thermodox	Liposomes	DOX hydrochloride	Recurrent chest wall breast cancer and liver cancer used in combination of hyperthermia approach (thermosensitive nanoformulation)	Phase-III clinical trial	Lyon et al. (2018)
Lipoplatin	Liposomes	Cisplatin	Non-small cell lung cancer	Phase-III clinical trial	Boulikas (2009)
Aroplatin	Liposomes	Structural analog of oxaliplatin	Metastatic colorectal cancer	Phase II clinical trial	Dragovich et al. (2006)

(Smith et al. 1999). Everything considered, nanocarriers, which are based on the EPR effect, can significantly reduce chemotherapy's side effects and toxicity to healthy tissues, and improve their therapeutic effect.

Since its discovery, great progress has been made to use the EPR effect to specifically deliver antitumor drugs to tumor tissues. To obtain a more effective delivery system, the mechanisms and factors affecting the EPR effect should be studied further. The EPR effect is regulated by factors such as bradykinin (Wang et al. 2014; Yu et al. 2014), matrix metalloproteinase (MMP, collagenase) (Szabo et al. 2007), vascular epidermal growth factor (VEGF) (Terman and Stoletov 2001; Goel and Mercurio 2013), tumor necrosis factor-α (Wang and Lin 2008), and nitric oxide (Burke et al. 2013; Vahora et al. 2016).

3.1.1 Bradykinin

Bradykinin is an active peptide generated by the kallikrein-kinin system and has been found in stomach cancer, liver cancer, glioma, prostate cancer, and lung cancer. Clinical studies have shown that high levels of bradykinin derivatives are found in plasma, peritoneum, and pleural effusions of patients with advanced cancer (Matsumura et al. 1991). Bradykinin can interact with bradykinin 2 receptor (B2R) to promote the expression of VEGF and enhance angiogenesis, thereby augmenting the vascular permeability of tissues (Matsumura et al. 1988). Previous studies reported that B2R is highly expressed in tumor tissues, so bradykinin can mediate and enhance the EPR effect in solid tumors (Bhoola et al. 2001; Kalyane et al. 2019). In addition, bradykinin can also activate nitric oxide synthase, leading to increased nitric oxide (NO), which is an important mediator for regulating vascular permeability (Kou et al. 2002). Therefore, bradykinin can directly or indirectly regulate the EPR effect.

3.1.2 Nitric Oxide

NO is a type of free radical gas and an important signal transmission molecule in the living body. NO is involved in various cancer development processes. For example, NO is produced by leukocytes that infiltrate in tumor tissues, and NO synthase, where it plays a major role. Studies have shown that tumor cell growth becomes slower in NO synthase gene knockout mice. Studies have confirmed that NO has a vasodilatory effect, which is very important for maintaining oxygen and nutrient supplies for tumor tissues (Kisley et al. 2002). NO plays an important role in enhancing the EPR effect (Doi et al. 1996). Its effect in enhancing vascular permeability can augment the delivery of therapeutic and diagnostic agents to tumor tissues. Nitroglycerin (NG), an NO producer, enhances the EPR effect. NO_2^- is firstly released from NG and then converted to NO in the hypoxic and low pH environments of tumor tissues, thereby causing vasodilation and increasing blood flow (Fig. 3.3). This mechanism has been confirmed in mouse tumor models. NG can increase tumor delivery of nanocarriers (the EPR effect) and enhance the therapeutic effect of antitumor drugs (Seki et al. 2009; Fang et al. 2011).

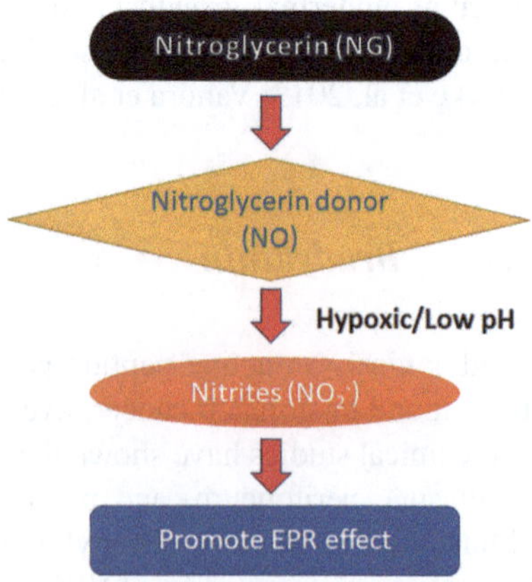

Fig. 3.3 The mechanism of nitroglycerin producing NO. Nitroglycerin (NG), an NO donor, increases permeability by generating nitrite (NO_2^-). Under hypoxic and acidic conditions, the conversion of NG to NO increases blood flow of solid tumors through vasodilation, which, in turn, leads to enhanced delivery of therapeutic molecules at the tumor site

3.1.3 *Vascular Endothelial Growth Factor*

VEGF is a signal protein produced by different cell types that stimulates the formation of blood vessels. VEGF has previously been identified as a vascular permeability factor by Dvorak et al. (1999, Senger et al. 1983). Tumor cells grow rapidly and need to consume large amounts of oxygen and nutrients. VEGF maintains angiogenesis and helps tumor growth. Studies have confirmed that VEGF induces increased vascular permeability (EPR effect), and its effect is designed to upregulate NO production (Papapetropoulos et al. 1997; Murohara et al. 1998). Overexpression of VEGF in tumor tissues can also lead to gap formation between vascular endothelial cells, an important cause of tumor metastasis. VEGF inhibitors have been regarded as effective treatments against solid tumors; for example, bevacizumab has been used in clinical practice.

3.1.4 *Tumor Necrosis Factor-α*

Tumor necrosis factor (TNF) is a multifunctional cytokine that plays an important role in cell survival, proliferation, differentiation, and death. Previous studies have found that TNF-α has both tumor-promoting activities and anticancer effects (Wang and Lin 2008; Waters et al. 2013). On the one hand, TNF-α acts on TNF receptors (TNFRs) to activate mitogen-activated protein kinase (MAPK) and nuclear factor κB (NF-κB). NF-κB is the main cell survival signal and helps tumor cells grow, proliferate, and resist apoptosis. MAPK is closely related to tumor progression and can stimulate cancer cell invasion and metastasis and tumor angiogenesis (Wang and Lin 2008). On the other hand, TNF-α may kill cancer, thus making it a potential cancer therapeutic agent (Wang and Lin 2008). TNF-α is toxic when administered systemically by itself, but when combined with nanocarriers (e.g., liposomes, gold nanoparticles), side effects are reduced, and administration safety is improved. At the same time, TNF-α acts as a vascular destructive agent to promote the EPR effect and increase the permeability of nanocarriers at the tumor sites, thereby augmenting the accumulation of drugs in tumor tissues (De Wilt et al. 2000; Kalyane et al. 2019). Currently, tumor treatment with TNF-conjugated liposomes (Li et al. 2001) and gold nanoparticles (Shenoi et al. 2013) is being investigated at the preclinical level.

3.2 Active Targeting

Nanocarriers with active targeting properties have been developed by exploiting passive targeting. Active targeting is based on the molecular recognition processes, which actively guide antitumor drugs to the tumor site by coupling ligands or

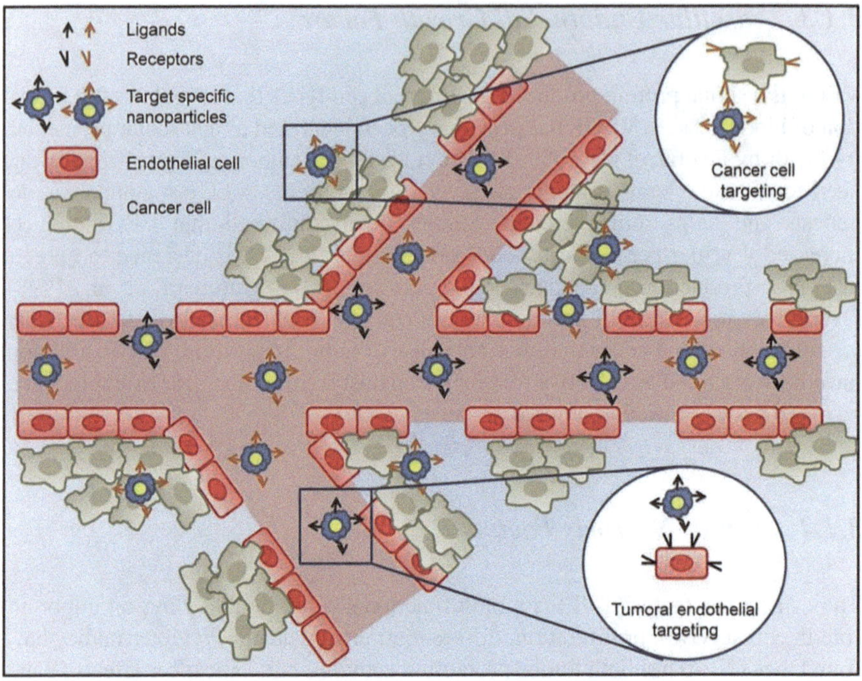

Fig. 3.4 In active targeting, two cellular targets can be distinguished: targeting tumor endothelium and targeting cancer cells (Yingchoncharoen et al. 2016)

monoclonal antibodies recognized by tumor cells on the surface of the nano-drug delivery vehicles (Xie et al. 2013, 2016, 2020; Zhou et al. 2014). Commonly used active targeting strategies include targeting tumor endothelium and targeting tumor cells (Fig. 3.4).

In 1971, Judah Folkman proposed that the endothelium's destruction in solid tumors can lead to tumor cell death caused by lacking oxygen and nutrients. Therefore, tumor growth can be inhibited by preventing the formation of new tumor blood vessels (Folkman 1971). The main targets of tumor endothelium include VEGF (Shadidi and Sioud 2003), vascular cell adhesion molecule-1 (Dienst et al. 2005), MMPs (Genís et al. 2006), and αvβ3 integrin receptor (Desgrosellier and Cheresh 2010). The advantages of tumor endothelial targeting are the following: (1) Different types of tumors express similar endothelial cell markers, so this strategy has a wide range of applications (Gosk et al. 2008); (2) genetic stability of endothelial cells leads to low drug resistance; and (3) the process of targeting tumor endothelium is simple, so that nanocarriers can directly bind to the receptor after entering the circulatory system.

The strategy for targeting tumor cells is to directly kill them through nanocarriers that target ligands. A tumor cell target should have the following characteristics: (1) Sufficient amount of targets (overexpression) on the tumor cells' surface and low expression in normal tissues; (2) the cell surface receptor should be easily recognized by nanocarriers to improve active targeting efficiency (Cho et al. 2008). The next section will describe the classification of different targets, introducing principles, advantages, and disadvantages of active targeting of nanocarriers developed for molecules that actively target tumor cells.

Active targeting can provide better selectivity and thus show better efficacy than passive targeting. However, no nanoformulations with active targeting are currently used in clinics for antitumor applications. Although many authors have provided evidence of efficacy for this strategy in preclinical models, only a handful of clinical trials have been conducted (details will be shared later). On the other hand, a large number of preclinical studies have been published using various nanocarriers and targeting ligands (Table 3.3).

3.2.1 Folate Receptor

FR, also known as high-affinity membrane folate binding protein, with a molecular weight of 38–40 kDa, is a glycosylphosphatidylinositol-linked membrane glycoprotein (Antony 1996). FA is an essential substance for life support and is essential for DNA replication and cell division. Most cells generally express reduced folate carriers. Under physiological conditions, exogenous reduced folate is mainly transported to the cytoplasm through folate carriers with low affinity (K_d ~ 1–5 μM). Oxidized folate (folate acid) can also be preferentially transported to the cytoplasm with high affinity (K_d ~ 100pM) through FR-mediated endocytosis (Maziarz et al. 1999; Sudimack and Lee 2000). FR's normal tissue distribution is relatively limited to a low number of polarized epithelia (only low-level expression in the choroid plexus, placenta, lung, thyroid, and kidney), and its expression is only localized on the top surface of polarized cells (Low and Antony 2004), thereby preventing its exposure to the loop. Under malignant conditions, intercellular junctions are lost, and FR loses its polarized cell position, thus localizing on the entire tumor cell surface (Fig. 3.5) (Ledermann et al. 2015; Biffi et al. 2019). Increased FR expression is often observed in various human cancers. Studies have shown that FR is overexpressed in more than 90% of ovarian cancers (Ross et al. 1994). High-frequency FR overexpression has also been found in endometrial cancer, colorectal cancer, breast cancer, lung cancer, renal cell carcinoma, and neuroendocrine cancer (Garin-Chesa et al. 1993; Cheung et al. 2016).

Therefore, FR is considered a highly selective tumor marker. Two specific drug targeting strategies have been developed for FR: (1) Coupling with folic acid or its analogs; (2) coupling with a monoclonal antibody against FR (e.g. MOv18) (Sudimack and Lee 2000). The active targeted nanocarriers based on FR have been evaluated in preclinical studies (Kularatne and Low 2010; Teng et al. 2012; Kato et al. 2018). FA ligands are widely used for targeting because they are non-toxic, non-immunogenic, inexpensive, stable in storage and in circulation, and easy to bind to carriers with high binding affinity. Yating et al. prepared an FR-targeted human serum albumin nanoparticle formulation of cabazitaxel

Table 3.3 Representative selection of nanocarriers active targeting to cancer cells

Specific cell surface moieties	Targeting ligands	Cancer types	Comments	References
Folate receptor (FR)	Folate (FA)	Human cervical cancer (HeLa); Human lung cancer (A549)	In vitro, FA-albumin nanoparticles have higher cytotoxicity and cell uptake activity than free paclitaxel (PTX) and nontargeted PTX albumin nanoparticles; in vivo, FA-albumin nanoparticles can accumulate at the tumor site and provide the best treatment	Sun et al. (2019)
FR	FA	Human squamous cell oral carcinoma (KB)	In vivo, FA-PEG-liposomal ursolic acid (UA) exhibited greater area under curve (AUC) and half-life than free UA by sixfold and 9.8-fold, respectively. FA-PEG-liposomal UA reduced tumor volume by 55% compared with the control. Animal lifespan was 56, 47, and 42 days for FA-PEG-liposomal UA, PEG-liposomal UA, or free UA, respectively	Yang et al. (2014a)
FR	FA	Mouse lymphoma expressing FR (J6456-FR)	In vivo, levels in J6456-FR tumors were 17-fold higher for FA-liposomes than for PEG-liposomes. DOX levels in ascitic fluid (2.25-fold) and plasma fluid (14-fold) were also lower when DOX was delivered by F-liposomes than by PEG-liposomes	Shmeeda et al. (2006)
FR	FA	Mouse breast cancer cell line (4T1.2), human breast cancer cell line (MCF-7), and drug-resistant cancer cell line (NCI/ADR-RES)	Compared with free DOX, the maximum tolerated dose was increased by about 1.5 times; compared with free DOX and Doxil, FA-coupled DOX-loaded nanomicelle carriers showed significantly enhanced tumor suppression with minimal toxicity	Lu et al. (2014)
Transferrin receptor (TfR)	Holo-transferrin	Hepatocellular carcinoma (HepG2)	In vitro, Tf-liposomes have higher cytotoxicity than PEG-liposomes and free DOX. In vivo, Tf-liposomes had the best therapeutic effect. After 96 hours, the tumor AUC was Tf-liposome> PEG-liposome> free DOX	Li et al. (2009)

(continued)

Table 3.3 (continued)

Specific cell surface moieties	Targeting ligands	Cancer types	Comments	References
TfR	Holo-transferrin	Human gastric cancer (MKN45P)	In vivo, cisplatin levels in tumor cells increased significantly when treated with Tf-PEG-liposomes, not with nontargeted liposomes and free cisplatin. Treating tumor-bearing mice xenografts with TfR-targeted formulation, but not nontargeted liposomes and free cisplatin, significantly prolonged animal survival	Iinuma et al. (2002)
Epidermal growth factor receptor (EGFR)	Anti-EGFR antibody	Human breast cancer (MDA-MB-468), human glioblastoma (U87)	In vivo, anti-EGFR-liposomes were internalized more efficiently than nontargeted liposomes (92% vs. 5%). Anti-EGFR-liposomes, but not nontargeted liposomes and free drugs, improved the anticancer efficacy of various drugs in tumor-bearing mice	Mamot et al. (2005)
Human epidermal growth factor receptor 2 (HER2)	Anti-HER2	Human breast cancer (MCF-7)	In animal models, the cure rate of anti-HER2 immunoliposome-DOX reached 50%, and anti-HER2 immunoliposome-DOX was also better than combinations consisting of free MAb plus free DOX or free MAb plus liposomal DOX	Park et al. (2002)
HER2	Anti-HER2	Human breast cancer (SKBR3)	In vitro, PTX/rapamycin (RAP) cytotoxicity to immunoliposomes was found to be higher which may be due to the increased uptake mediated by HER2 binding; immunoliposomes could control tumor growth in vivo, with corresponding tumor volume of 25.27%, 44.38%, and 47.78% for untreated control, PTX/RAP solution, and control liposomes, respectively	Eloy et al. (2017)

(FA-nanoparticles-CTX). Preclinical studies showed that FA-nanoparticles-CTX significantly increases accumulation at tumor sites and improves CTX's antitumor effects in vivo and in vitro. In addition, FA-nanoparticles can improve CTX's safety and tolerability and have great potential for targeted drug delivery (Sun et al. 2019;

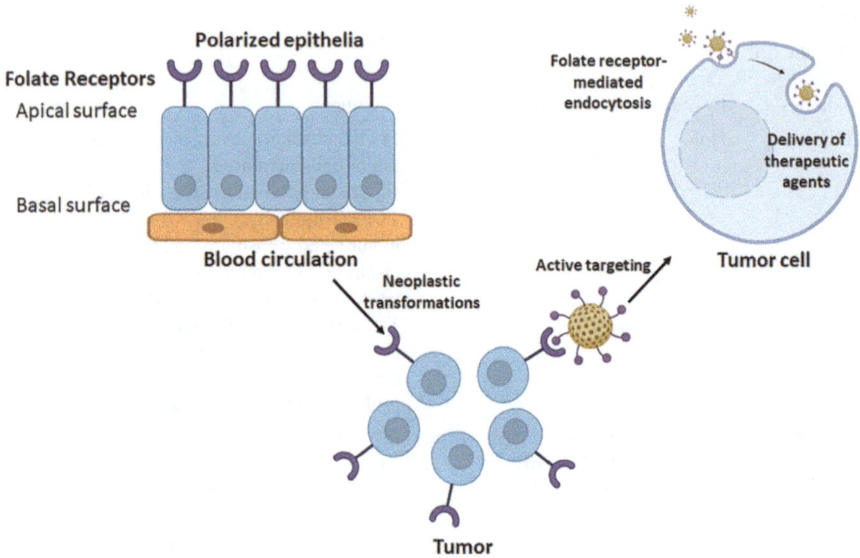

Fig. 3.5 The role of FR in active cancer targeting nanocarriers. FR tissue distribution is mainly limited to a limited number of polarized epithelia, and its expression is only localized on the top surface of polarized cells, thereby preventing exposure to circulation. When a tumor develops, the cell-to-cell connection disappears, and FR loses its polarized cell position, thereby positioning on the entire tumor cell surface. This process allows FR to be utilized by actively targeted nanoparticles in the blood circulation. Once they bind to FR, folic acid–targeted nanoparticles will be internalized by the cell through the FR-mediated endocytic pathway. The targeting effect of nanoparticles on FR provides the potential for selective delivery of therapeutic agents to tumor tissues

Sun et al. 2020). Fangyuan et al. reported the preparation of fluorescent gold nanoclusters (GNCs) conjugated with cisplatin prodrug and FA (FA-GNC-Pt). In preclinical studies, FA-GNC-Pt was used for breast cancer fluorescence imaging and targeted chemotherapy. FA modification significantly accelerated the cellular uptake of mouse 4T1 breast cancer cells and increased the cytotoxicity of GNC-Pt nanoparticles. FA-GNC-Pt nanoparticles selectively aggregate in 4T1 tumors because of the FA's tumor targeting effect and generate strong fluorescence signals. In addition, FA-GNC-Pt nanoparticles significantly inhibited growth and lung metastasis in an orthotopically implanted 4T1 breast tumor model. GNC based on FR targeting had great potential in fluorescence tumor imaging and cancer treatment (Zhou et al. 2016). Recent studies have shown that nanoparticles coupled with folate analogs such as methotrexate (MTX) also have active targeting effects. Fei et al. prepared MTX-conjugated mixed micelles (M-MTXs) to co-deliver MTX and siRNA to tumor cells. Studies have confirmed that MTX can be used both as a targeting ligand and as a therapeutic agent. When conjugated with a dendrimer, MTX can directly target FR, be internalized into the cell, and then come into effect (Wong and Choi 2015; Hao et al. 2019).

Despite the encouraging preclinical results, so far, no nanocarriers based on FR targeting have been used to treat cancer in the clinic. The study by Yu et al. showed that patients with high FR expression in serous ovarian cancer tissues had lower chemotherapy response and poorer overall survival, which may be related to FR-induced anti-apoptosis (Chen et al. 2012).

3.2.2 *Transferrin Receptor*

Transferrin receptor protein 1 (TfR1), also known as CD71, is a transmembrane glycoprotein involved in iron homeostasis and cell growth regulation. It consists of two monomers connected by disulfide bonds. Each monomer binds a holo-transferrin (Tf) molecule to create an iron-Tf-ferritin complex, which enters the cell through clathrin-mediated endocytosis (Speeckaert et al. 2010; Shen et al. 2018). A second transferrin receptor (TfR2) in humans is expressed mainly in erythroid precursor cells and hepatocytes (Kawabata 2019). Most importantly, TfR1 was found to be expressed at abnormally high levels in different cancers including breast cancer, liver cancer, lung cancer, brain cancer, prostate cancer, colon cancer, and ovarian cancer. TfR1's expression in cancer cells may be 100 times higher than its average expression in normal cells (Daniels et al. 2006). Therefore, TfR1 is regarded as a molecular target for active tumor targeting (Yang et al. 2014b, 2015). Because TfR1 is expressed in brain capillary endothelial cells of the blood-brain barrier, nanocarriers that target TfR1 have become a promising strategy for treating brain cancer (Ying et al. 2010; Kang et al. 2016; Biffi et al. 2019; Wang et al. 2019). Nanomedicine strategies based on active targeting of TfR1 are being evaluated in clinical trials (Van Der Meel et al. 2013).

SGT-53 is a nanocomposite in which the payload (normal human wt p53 cDNA) is encapsulated in cationic liposomes, and its surface is modified with an antibody targeting TfR1 (TfRscFv). In a phase 1a clinical trial of patients with advanced solid tumors, SGT-53 showed tumor-specific targeting and was also proven to be well tolerated. At the same time, the combination of SGT-53 and docetaxel showed enhanced antitumor activity (Pirollo et al. 2016). In addition, currently, more further studies aim to conjugate holo-Tf to nano-drug delivery vehicles as a targeting ligand for TfR1. Xue et al. prepared Tf-modified daunorubicin liposomes for treating glioma. In vitro experiments showed that TF-modified liposomes can increase drug uptake by tumor cells and increase the cytotoxicity of antitumor drugs. In vitro experiments showed that Tf modification significantly increased the transportation of liposomes across the blood-brain barrier and at the same time increased the delivery of antitumor drugs to glioma (Ying et al. 2010). In short, TfR-mediated active targeting is a potential tumor targeting strategy, especially in the field of brain tumor–targeted delivery. The nanocarriers developed based on this strategy are currently being evaluated in clinical trials and have achieved encouraging experimental results (Zuckerman et al. 2014; Jhaveri et al. 2018).

3.2.3 Epidermal Growth Factor Receptor Family

EGFR family includes a group of transmembrane proteins, such as the extracellular protein ligand of epidermal growth factor (EGF). EGFR belongs to four different receptor families, including EGFR (ErbB-1; HER1]), HER2 (c-ErbB2), HER-3 (c-ErbB3), and HER4 (c-ErbB4) (Scaltriti and Baselga 2006). EGFR and its signaling pathways play an important role in the pathogenesis and progression of various cancers. It was observed clinically that HER2 is expressed in 14–91% of breast cancer patients (Scaltriti and Baselga 2006; Acharya et al. 2009). EGFR is usually overexpressed in epithelial tumors, including lung adenocarcinoma (40% of cases), glioblastoma (50%), and head and neck cancer (80–100%) (Nicholas et al. 2006; Lurje and Lenz 2009). Therefore, the EGFR family is considered an attractive receptor for cancer treatment targets. Known endogenous ligands of EGFR include EGF, transforming growth factor-α (TGF-α), β-cellulose, amphiregulin, heparin-binding EGF (HB-EGF), and upregulated protein (Laskin and Sandler 2004). These ligands all have the potential to target EGFR as targeting moieties. Among them, TGF-α and EGF are most commonly observed proteins in humans; hence, they hold the most promise. The current cancer-targeted treatment method for EGFR is the monoclonal antibody cetuximab, which targets EGFR's extracellular domain (Grunwald and Hidalgo 2003; Mendelsohn and Baselga 2003). Cetuximab is usually administered as a combination therapy to improve antitumor efficacy. An effective treatment strategy is to use cetuximab as the targeting part of nano-drug delivery carriers to deliver antitumor drugs. Francesco et al. developed a cetuximab-guided Avidin-Nucleic-Acid-Nano-Assemblies (ANANAs) loaded with DOX to treat breast cancer. Experimental results showed that cetuximab conjugated with ANANAS carrying DOX can be more effectively internalized into tumor cells than antibody-conjugated drugs. Results from in vivo experiments showed that when the DOX dose was 1/10 of the dose traditionally used in the in vivo test of the DOX nanocarriers (0.56 mg/kg), cetuximab-guided ANANAS resulted in 43% tumor inhibition (Roncato et al. 2018). HER2 is an attractive target for mAb-based targeted therapy strategies. Herceptin monoclonal antibody is a monoclonal antibody approved by the FDA designed to antagonize the function of HER2 (Harries and Smith 2002). It can be used as a targeting ligand to be coupled with nanocarriers to deliver chemotherapy drugs to tumor cells with high HER2 expression. Clinical research results showed that MM-302 (an antibody-liposome adriamycin conjugate targeting HER2) monotherapy combined with trastuzumab and cyclophosphamide is safe and effective for patients with advanced HER2-positive breast cancer. HER2-mediated delivery of MM-302 to HER2-overexpressing tumor cells was confirmed in multiple evaluable biopsies after MM-302 treatment, by the co-localization of anti-PEG, anti-HER2, and anti-cytokeratin antibodies (Munster et al. 2018). In a recent study, Hengfeng et al. designed a multivalent bi-specific nanobioconjugate engager (mBiNE) to promote selective, immune-mediated eradication of cancer cells. By simultaneously targeting HER2 expressed by cancer cells and the pro-phagocytosis signal mediated by calreticulin, mBiNE stimulates phagocytosis against HER2 and produces a durable antitumor immune response against tumors that express HER2. mBiNE represents a new tool that stimulates innate and adaptive immunity and promotes innovative broad-based active immunotherapy (Yuan et al. 2017).

3.3 Emerging Strategy for Tumor Targeting

With the advancement of nanomaterial sciences, a new targeting strategy has been proposed including the development of "activatable" or "activated" stimulation-sensitive nanocarriers. The nanocarrier maintains its stealth function during the circulation process. After reaching the tumor site, nanocarrier transformation is triggered by the unique tumor cell environment (such as hypoxia or acidic environments), thereby allowing the drug to be released or interact with specific targets. Drug retention in nanocarriers can also be solved by applying external stimuli (e.g., magnetic fields, light, ultrasound), thereby allowing controlled and selective cell targeting.

3.3.1 Hypoxia Response Nanocarriers

The hypoxic microenvironment is common in most solid tumors because of the imbalance between the increased oxygen consumption of fast-growing tumor cells and the lack of oxygen supply to abnormal vasculature (Wilson and Hay 2011; Petrova et al. 2018). Nevertheless, the incidence and severity of hypoxia vary in each patient population. Because of multiple contributions to chemoresistance, radio resistance, angiogenesis, metastasis, invasiveness, resistance to cell death, changes in metabolism, and genomic instability, hypoxia is a negative prognosis and prediction factor (Rankin and Giaccia 2008; Semenza 2010). Hypoxia represents a convincing treatment goal because it plays a major role in tumor development and resistance to treatment. Also, hypoxia is more severe in most tumors than in normal tissues (Wilson and Hay 2011).

Recently, several hypoxia-responsive nanoparticles (HRNPs) were developed to promote controlled drug release and enhance cellular uptake for cancer treatment. In the study by Yujing et al., dual hypoxia-targeting RNAi nanomedicine was prepared for precise cancer treatment. Under tumors' hypoxic environment, liposomes prepared from peptides modified with the hypoxia-sensitive group nitroimidazole (NI) will be reduced by a series of nitro reductases catalyzed by a series of single-electron reduction. This results in converting the hydrophobic NI group to the hydrophilic 2-aminoimidazole (AI), which leads to the decomposition of liposomes and subsequent rapid reaction to release drugs at the tumor site. In this study, hypoxia-responsive liposomes showed prolonged blood circulation, high tumor aggregation, and powerful antitumor efficacy, providing a promising advanced strategy for precise cancer treatment (Li et al. 2020). In the study by Soyoung et al., black hole quencher 3 (BHQ3) and carboxymethyl dextran (CMD) were coupled through an azo bond to prepare a hypoxia-sensitive polymer conjugate. This polymer conjugate can self-assemble into nanoparticles (CMD-BHQ3 nanoparticles) and the antitumor drug DOX can be loaded onto the NP hydrophobic core. Under tumors' hypoxic environment, azobenzene derivatives are reduced to aniline derivatives by a large number of reductases in the presence of electron donors, leading to the breaking of azo bonds through electron transfer and ultimately to drug release. Cell experiments showed that the toxicity of hypoxia-sensitive NP increases in hypoxic environments; in vivo biodistribution studies have shown that CMD-BHQ3 nanoparticles preferentially accumulate in tumors after systemic administration. Overall, CMD-BHQ3 nanoparticles may be a promising carrier for selective drug release in hypoxic tumors (Son et al. 2018).

3.3.2 pH-Sensitive Nanoparticles

Due to poor tissue perfusion and high cell metabolism, interstitial acidification is a sign of solid tumor tissue. Studies have shown that tumor acidosis helps tumor cells survive, resist apoptosis, and promote tumor migration (Fang et al. 2008; Estrella et al. 2013; Sun et al. 2021). In addition, a chemical barrier is formed in the acidic environment of tumor tissues, which limits the effects and activities of many anticancer drugs (Gerweck et al. 2006; Trédan et al. 2007). Although the intracellular pH of healthy tissue and tumor cells is similar, the extracellular pH of tumors is lower than that of normal tissues. The average extracellular tumor pH is between 6.0 and 7.0, while in normal tissues and blood, the extracellular pH is about 7.4 (Danhier et al. 2010). pH-sensitive nano-drug delivery systems have now been developed by using nanoparticles that contain acid-labile chemical bonds (commonly used acid-labile chemical bonds are shown in Table 3.4). Under an acidic tumor environment, pH-sensitive nanoparticles can specifically trigger the release of drugs, thereby improving the efficiency of cancer treatment.

Chunlai et al. reported that the hydrophobic aliphatic dendritic polyester BOLTORN® H40 (H40) and hydrophilic polyethylene glycol (MPEG) were connected through an acetal bond to obtain a pH-responsive copolymer H40-star-MPEG that can self-assemble into micelles in water and encapsulate DOX in a hydrophobic core. The acetal bond will break in low pH solutions, leading to rapid hydrolysis of H40-star-MPEG. Experimental results showed that in an acidic environment, the release of the drug in DOX-loaded micelles is greatly accelerated. DOX-loaded micelles can be effectively internalized by cancer cells, thereby effectively inhibiting the proliferation of cancer cells. H40-star-MPEG with acetal bonds provides a promising pH-responsive drug carrier for cancer treatment (Tu et al. 2013). Recently, Xin et al. reported that a pH-sensitive bromelain nanoparticle enhances the penetration of DOX in solid tumors through ortho ester cross-linking. Experimental results showed that the nanoparticles exhibited DOX release triggered

Table 3.4 Examples of acid-labile chemical bonds and their degradation products

Name	Acid-labile chemical bonds	Degradation product	References
Ortho ester	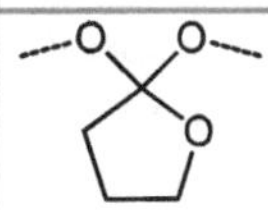	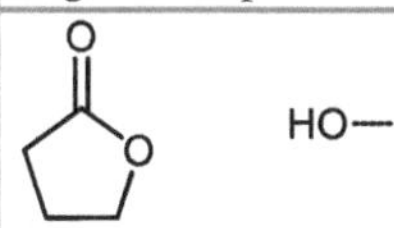	Wang et al. (2020)
Hydrazone	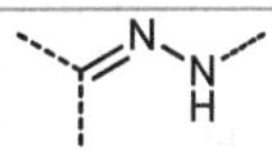	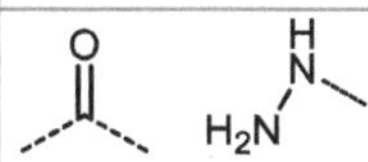	Du et al. (2011)
Cis-aconityl	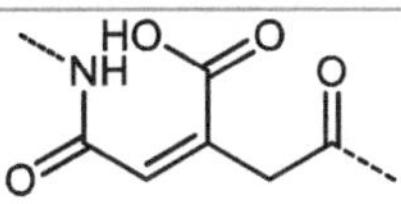	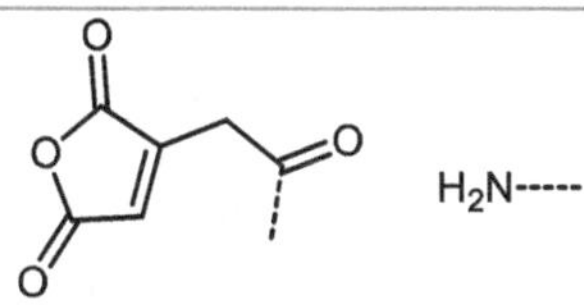	Tahvilian et al. (2016)
Imine	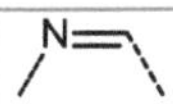	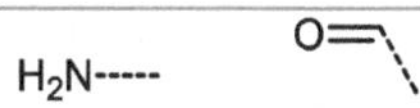	Xie et al. (2018)
Ester	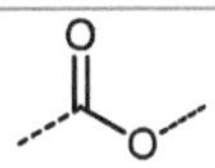	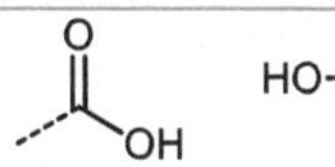	Shenoy et al. (2005)
Acetal/ketal	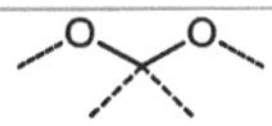		Gu et al. (2013)

by pH, based on the cleavage of the ortho ester bond. The DOX release rate of pH-responsive nanoparticles is higher than that of non-sensitive nanoparticles, and the DOX drug released reaches 86% within 120 h at a pH of 5.5. In vivo experiments showed that these pH-sensitive nanoparticles transport more drugs to the tumor site with excellent antitumor effects (Wang et al. 2020). It was proved that using the acidic pH of the cancer environment is an effective method in nano-systems for drug delivery. In the near future, we expect that multifunctional drug delivery nano-systems for cancer treatment will be developed for practical clinical applications.

3.3.3 External Stimulus-Responsive Nanoparticles

External stimulus-responsive nanoparticles usually contain heat-sensitive, light-sensitive, or magnetically sensitive materials. These nanoparticles can remain stable in circulation and achieve controlled drug release while showing selective targeting effects by applying external stimuli.

Ana M. et al. found that local tumor hyperthermia can increase the permeability of tumor blood vessels and significantly enhance the aggregation of nanocarriers in solid tumors (Ponce et al. 2006; Torchilin 2009). Temperature-sensitive nanocarriers contain lower critical solution temperature (LCST) polymerization, which is slightly higher than the physiological temperature. Poly(N-isopropylacrylamide)

(NIPAM) is the most widely used polymer for thermosensitive nanocarriers (Schild 1992; Kono et al. 2002). When the tumor's internal temperature exceeds the LCST, these polymers will dissolve, leading to the destruction of the nanocarrier structure and ultimately the release of the drug (Kono 2001). By using specific lipids, polymers, or other modifiers, local hyperthermia has become the most widely used stimulus to trigger liposomal drug release (Lindner et al. 2004). At present, ultrasound is most commonly used to locally increase tumor temperature (Dromi et al. 2007; Grüll and Langereis 2012). In addition, the thermomagnetic effect of magnetic materials in a magnetic field (Hilger et al. 2002) and the thermal effect caused by laser (Zhou et al. 2017) can also be used to achieve local heating of the tumor.

Of all stimuli, light has recently received much attention because light-induced stimulation is not only positioned in time and space, but most of the photochemical processes involved do not require additional reagents (Yavlovich et al. 2010; Chatani et al. 2014). Photo-responsive polymers (PRPs) are molecules that change their properties when irradiated with light of the appropriate wavelength. Drug-PRP conjugates are often used for antitumor drug delivery, and the active drug in the conjugate will be released when the tumor is irradiated (Sun et al. 2018). Qiao et al. reported a light-triggered biocompatible drug delivery system based on coumarin-functionalized block copolymers. The antitumor drug 5-Fluorouracil molecule is covalently linked to the coumarin-functionalized block copolymers by light irradiation, and self-assembled to form micelles under aqueous conditions. Under UV conditions (254 nm), the controlled release of 5-FU from micelle-drug conjugates was light-triggered under physiological conditions (Jin et al. 2011). Ultraviolet and visible wavelengths of light poorly penetrate human tissues such as skin, leading to limitations in the ultraviolet-excited light-sensitive nano-drug delivery system. For this reason, other polymers that are sensitive to infrared or near infrared light have been studied (Jiang et al. 2006; Gong et al. 2013; Feng et al. 2017).

Magnetic nanoparticles (MNPs) are also commonly used nanoscale materials that were first designed for passive targeting of magnetic resonance imaging. Superparamagnetic iron oxide nanoparticles (SPIONs), such as Combidex®, Feridex®, and Ferumoxytol®, were approved for clinical medical imaging (Veiseh et al. 2010; Mahmoudi et al. 2011). Freeman et al. first proposed the theory of magnetically guided drug targeting. Magnetically guided drug targeting (MGDT) mainly uses a high-gradient external magnetic field generated by a rare earth permanent magnet (usually an NdFeB magnet) to guide MNPs loaded with antitumor drugs to the tumor site (Freeman et al. 1960). Based on this theory, MNPs were developed for antitumor drug delivery with encouraging results in animal experiments (Kubo et al. 2000; Goodwin et al. 2001; Alexiou et al. 2003). However, this MNP-based drug delivery method has obvious limitations. Because of the magnetic field's characteristics, magnetic particle capture at the target is effective for targets close to the organism's surface. However, the strength of the magnetic field decreases rapidly as the distance increases, so deeper positions in the body become more difficult to aim. Current research on MNP mainly focuses on modifying the targeting ligand and loading antitumor drugs based on SPION. The local temperature increase achieved by using SPION in an alternating magnetic field can eliminate tumors.

This principle is called "magnetic thermal ablation" (Hilger et al. 2002). In addition, increasing the tumor's local temperature can accelerate the release of the drug in MNPs and improve the delivery efficiency of antitumor drugs (Wust et al. 2006; Chen et al. 2010; Thomas et al. 2010).

3.4 Conclusions

Nanocarriers with tumor targeting effects have received unprecedented attention in cancer chemotherapy. The discovery of the EPR effect represents an important milestone in the history of cancer-targeted chemotherapy. The EPR effect allows the nanocarrier to passively target solid tumors, thus improving the chemotherapy effect and reducing adverse reactions. Based on passive targeting, nanocarriers with active targeting were developed. Conjugating the targeting moiety to the nanocarrier gives it specific targeting ability. The advantage of this active targeting strategy is to enable fine positioning of drug delivery and reduce side effects on healthy tissues, further improving the efficiency of drug delivery to tumors. The use of stimuli-sensitive nanocarriers as delivery vehicles for chemotherapeutic drugs is a promising potential tumor targeting strategy. The advantage of this strategy is the controlled release and selective targeting of antitumor drugs. There is no doubt that nanocarriers have unparalleled advantages as delivery vehicles for antitumor drugs. With the development of nano-oncology and materials science, we expect that using tumor-targeted nanoparticles as drug carriers can be more widely applied in tumor chemotherapy.

Acknowledgments The authors thank Damiana Chiavolini (Department of Radiation Oncology) of UT Southwestern Medical Center for scientific editing.

References

Acharya S, Dilnawaz F, Sahoo SK (2009) Targeted epidermal growth factor receptor nanoparticle bioconjugates for breast cancer therapy. Biomaterials 30(29):5737–5750. https://doi.org/10.1016/j.biomaterials.2009.07.008

Alexiou C, Jurgons R, Schmid RJ et al (2003) Magnetic drug targeting—biodistribution of the magnetic carrier and the chemotherapeutic agent mitoxantrone after locoregional cancer treatment. J Drug Target 11(3):139–149. https://doi.org/10.3109/1061186031000150791

Allen TM, Martin FJ (2004) Advantages of liposomal delivery systems for anthracyclines. Semin Oncol 31(6):5–15. https://doi.org/10.1053/j.seminoncol.2004.08.001

Antony AC (1996) Folate receptors. Annu Rev Nutr 16(1):501–521. https://doi.org/10.1146/annurev.nu.16.070196.002441

Barenholz Y (2012) Doxil(R)–the first FDA-approved nano-drug: lessons learned. J Control Release 160(2):117–134. https://doi.org/10.1016/j.jconrel.2012.03.020

Bernabeu E, Cagel M, Lagomarsino E et al (2017) Paclitaxel: what has been done and the challenges remain ahead. Int J Pharm 526(1–2):474–495. https://doi.org/10.1016/j.ijpharm.2017.05.016

Bhoola K, Ramsaroop R, Plendl J et al (2001) Kallikrein and kinin receptor expression in inflammation and cancer. Biol Chem 382(1):77–89. https://doi.org/10.1515/BC.2001.013

Biffi S, Voltan R, Bortot B et al (2019) Actively targeted nanocarriers for drug delivery to cancer cells. Expert Opin Drug Deliv 16(5):481–496. https://doi.org/10.1080/17425247.2019.1604679

Blair HA (2018) Daunorubicin/cytarabine liposome: a review in acute myeloid leukaemia. Drugs 78(18):1903–1910. https://doi.org/10.1007/s40265-018-1022-3

Boulikas T (2009) Clinical overview on lipoplatin: a successful liposomal formulation of cisplatin. Expert Opin Investig Drugs 18(8):1197–1218. https://doi.org/10.1517/13543780903114168

Burke AJ, Sullivan FJ, Giles FJ et al (2013) The yin and yang of nitric oxide in cancer progression. Carcinogenesis 34(3):503–512. https://doi.org/10.1093/carcin/bgt034

Chatani S, Kloxin CJ, Bowman CN (2014) The power of light in polymer science: photochemical processes to manipulate polymer formation, structure, and properties. Polym Chem 5(7):2187–2201. https://doi.org/10.1039/c3py01334k

Chen Y, Bose A, Bothun GD (2010) Controlled release from bilayer-decorated magnetoliposomes via electromagnetic heating. ACS Nano 4(6):3215–3221. https://doi.org/10.1021/nn100274v

Chen YL, Chang MC, Huang CY et al (2012) Serous ovarian carcinoma patients with high alpha-folate receptor had reducing survival and cytotoxic chemo-response. Mol Oncol 6(3):360–369. https://doi.org/10.1016/j.molonc.2011.11.010

Cheung A, Bax HJ, Josephs DH et al (2016) Targeting folate receptor alpha for cancer treatment. Oncotarget 7(32):52553. https://doi.org/10.18632/oncotarget.9651

Cho K, Wang X, Nie S et al (2008) Therapeutic nanoparticles for drug delivery in cancer. Clin Cancer Res 14(5):1310–1316. https://doi.org/10.1158/1078-0432.CCR-07-1441

Danhier F, Feron O, Préat V (2010) To exploit the tumor microenvironment: passive and active tumor targeting of nanocarriers for anti-cancer drug delivery. J Control Release 148(2):135–146. https://doi.org/10.1016/j.jconrel.2010.08.027

Daniels TR, Delgad T, Helguera G et al (2006) The transferrin receptor part II: targeted delivery of therapeutic agents into cancer cells. Clin Immunol 121(2):159–176. https://doi.org/10.1016/j.clim.2006.06.006

De Wilt JHW, Ten Hagen TLM, De Boeck G et al (2000) Tumour necrosis factor alpha increases melphalan concentration in tumour tissue after isolated limb perfusion. Br J Cancer 82(5):1000–1003. https://doi.org/10.1054/bjoc.1999.1032

Desgrosellier JS, Cheresh DA (2010) Integrins in cancer: biological implications and therapeutic opportunities. Nat Rev Cancer 10(1):9–22. https://doi.org/10.1038/nrc2748

Dienst A, Grunow A, Unruh M et al (2005) Specific occlusion of murine and human tumor vasculature by VCAM-1–targeted recombinant fusion proteins. J Natl Cancer Inst 97(10):733–747. https://doi.org/10.1093/jnci/dji130

Doi K, Akaike T, Horie H et al (1996) Excessive production of nitric oxide in rat solid tumor and its implication in rapid tumor growth. Cancer 77(8):1598–1604. https://doi.org/10.1002/(SICI)1097-0142(19960415)77:8+<1598::AID-CNCR3>3.0.CO;2-A

Dragovich T, Mendelson D, Kurtin S et al (2006) A Phase 2 trial of the liposomal DACH platinum L-NDDP in patients with therapy-refractory advanced colorectal cancer. Cancer Chemother Pharmacol 58(6):759–764. https://doi.org/10.1007/s00280-006-0235-4

Dromi S, Frenkel V, Luk A et al (2007) Pulsed-high intensity focused ultrasound and low temperature–sensitive liposomes for enhanced targeted drug delivery and antitumor effect. Clin Cancer Res 13(9):2722–2727. https://doi.org/10.1158/1078-0432.CCR-06-2443

Drummond DC, Noble CO, Guo Z et al (2010) Development of a highly stable and targetable nanoliposomal formulation of topotecan. J Control Release 141(1):13–21. https://doi.org/10.1016/j.jconrel.2009.08.006

Du JZ, Du XJ, Mao CQ et al (2011) Tailor-made dual pH-sensitive polymer–doxorubicin nanoparticles for efficient anticancer drug delivery. J Am Chem Soc 133(44):17560–17563. https://doi.org/10.1021/ja207150n

Dvorak H, Nagy J, Feng D et al (1999) Vascular permeability factor/vascular endothelial growth factor and the significance of microvascular hyperpermeability in angiogenesis. In: Vascular growth factors and angiogenesis, pp 97–132
Eloy JO, Petrilli R, Chesca DL et al (2017) Anti-HER2 immunoliposomes for co-delivery of paclitaxel and rapamycin for breast cancer therapy. Eur J Pharm Biopharm 115:159–167. https://doi.org/10.1016/j.ejpb.2017.02.020
Estrella V, Chen T, Lloyd M et al (2013) Acidity generated by the tumor microenvironment drives local invasion. Cancer Res 73(5):1524–1535. https://doi.org/10.1158/0008-5472.CAN-12-2796
Fang JS, Gillies RD, Gatenby RA (2008) Adaptation to hypoxia and acidosis in carcinogenesis and tumor progression. Semin Cancer Biol 18(5):330–337. https://doi.org/10.1016/j.semcancer.2008.03.011
Fang J, Nakamura H, Maeda H (2011) The EPR effect: unique features of tumor blood vessels for drug delivery, factors involved, and limitations and augmentation of the effect. Adv Drug Deliv Rev 63(3):136–151. https://doi.org/10.1016/j.addr.2010.04.009
Feng Q, Zhang Y, Zhang W et al (2017) Programmed near-infrared light-responsive drug delivery system for combined magnetic tumor-targeting magnetic resonance imaging and chemo-phototherapy. Acta Biomater 49:402–413. https://doi.org/10.1016/j.actbio.2016.11.035
Folkman J (1971) Transplacental carcinogenesis by stilbestrol. N Engl J Med 285(7):404–405. https://doi.org/10.1056/NEJM197108122850711
Freeman M, Arrott A, Watson J (1960) Magnetism in medicine. J Appl Phys 31(5):404–405
Gabizon A, Catane R, Uziely B et al (1994) Prolonged circulation time and enhanced accumulation in malignant exudates of doxorubicin encapsulated in polyethylene-glycol coated liposomes. Cancer Res 54(4):987–992
Garin-Chesa P, Campbell I, Saigo P et al (1993) Trophoblast and ovarian cancer antigen LK26. Sensitivity and specificity in immunopathology and molecular identification as a folate-binding protein. Am J Pathol 142(2):557
Genís L, Gálvez BG, Gonzalo P et al (2006) MT1-MMP: universal or particular player in angiogenesis? Cancer Metastasis Rev 25(1):77–86. https://doi.org/10.1007/s10555-006-7891-z
Gerweck LE, Vijayappa S, Kozin S (2006) Tumor pH controls the in vivo efficacy of weak acid and base chemotherapeutics. Mol Cancer Ther 5(5):1275–1279. https://doi.org/10.1158/1535-7163. MCT-06-0024
Glantz MJ, Jaeckle KA, Chamberlain MC et al (1999) A randomized controlled trial comparing intrathecal sustained-release cytarabine (DepoCyt) to intrathecal methotrexate in patients with neoplastic meningitis from solid tumors. Clin Cancer Res 5(11):3394–3402
Goel HL, Mercurio AM (2013) VEGF targets the tumour cell. Nat Rev Cancer 13(12):871–882. https://doi.org/10.1038/nrc3627
Gong H, Cheng L, Xiang J et al (2013) Near-infrared absorbing polymeric nanoparticles as a versatile drug carrier for cancer combination therapy. Adv Funct Mater 23(48):6059–6067. https://doi.org/10.1002/adfm.201301555
Goodwin SC, Bittner CA, Peterson CL et al (2001) Single-dose toxicity study of hepatic intra-arterial infusion of doxorubicin coupled to a novel magnetically targeted drug carrier. Toxicol Sci 60(1):177–183. https://doi.org/10.1093/toxsci/60.1.177
Gosk S, Moos T, Gottstein C et al (2008) VCAM-1 directed immunoliposomes selectively target tumor vasculature in vivo. Biochim Biophys Acta Biomembr 1778(4):854–863. https://doi.org/10.1016/j.bbamem.2007.12.021
Grüll H, Langereis S (2012) Hyperthermia-triggered drug delivery from temperature-sensitive liposomes using MRI-guided high intensity focused ultrasound. J Control Release 61(2):317–327. https://doi.org/10.1016/j.jconrel.2012.04.041
Grunwald V, Hidalgo M (2003) Developing inhibitors of the epidermal growth factor receptor for cancer treatment. J Natl Cancer Inst 95(12):851–867. https://doi.org/10.1093/jnci/95.12.851
Gu Y, Zhong Y, Meng F et al (2013) Acetal-linked paclitaxel prodrug micellar nanoparticles as a versatile and potent platform for cancer therapy. Biomacromolecules 14(8):2772–2780. https://doi.org/10.1021/bm400615n

Hao F, Lee RJ, Yang C et al (2019) Targeted co-delivery of siRNA and methotrexate for tumor therapy via mixed micelles. Pharmaceutics 11(2):92. https://doi.org/10.3390/pharmaceutics11020092

Harries M, Smith I (2002) The development and clinical use of trastuzumab (Herceptin). Endocr Relat Cancer 9(2):75–85. https://doi.org/10.1677/erc.0.0090075

He X, Ma Y, Xie H et al (2021) Biomimetic nanostructure platform for cancer diagnosis based on tumor biomarkers. Front Bioeng Biotechnol 9:687664. https://doi.org/10.3389/fbioe.2021.687664

Hilger I, Hiergeist R, Hergt R et al (2002) Thermal ablation of tumors using magnetic nanoparticles: an in vivo feasibility study. Investig Radiol 37(10):580–586. https://doi.org/10.1097/00004424-200210000-00008

Iinuma H, Maruyama K, Okinaga K (2002) Intracellular targeting therapy of cisplatin-encapsulated transferrin-polyethylene glycol liposome on peritoneal dissemination of gastric cancer. Int J Cancer 99(1):130–137. https://doi.org/10.1002/ijc.10242

Iyer AK, Khaled G, Fang J et al (2006) Exploiting the enhanced permeability and retention effect for tumor targeting. Drug Discov Today 11(17–18):812–818. https://doi.org/10.1016/j.drudis.2006.07.005

Jhaveri A, Deshpande P, Pattni B et al (2018) Transferrin-targeted, resveratrol-loaded liposomes for the treatment of glioblastoma. J Control Release 277:89–101

Jiang J, Tong X, Morris D et al (2006) Toward photocontrolled release using light-dissociable block copolymer micelles. Macromolecules 39(13):4633–4640. https://doi.org/10.1021/ma060142z

Jin Q, Mitschang F, Agarwal S (2011) Biocompatible drug delivery system for photo-triggered controlled release of 5-fluorouracil. Biomacromolecules 12(10):3684–3691. https://doi.org/10.1021/bm2009125

Kalyane D, Raval N, Maheshwari R et al (2019) Employment of enhanced permeability and retention effect (EPR): nanoparticle-based precision tools for targeting of therapeutic and diagnostic agent in cancer. Mater Sci Eng C Mater Biol Appl 98:1252–1276. https://doi.org/10.1016/j.msec.2019.01.066

Kang C, Sun Y, Zhu J et al (2016) Delivery of nanoparticles for treatment of brain tumor. Curr Drug Metab 17(8):745–754. https://doi.org/10.2174/1389200217666160728152939

Kato T, Jin CS, Lee D et al (2018) Preclinical investigation of folate receptor-targeted nanoparticles for photodynamic therapy of malignant pleural mesothelioma. Int J Oncol 53(5):2034–2046. https://doi.org/10.3892/ijo.2018.4555

Kawabata H (2019) Transferrin and transferrin receptors update. Free Radic Biol Med 133:46–54. https://doi.org/10.1016/j.freeradbiomed.2018.06.037

Kim TY, Kim DW, Chung JY et al (2004) Phase I and pharmacokinetic study of Genexol-PM, a cremophor-free, polymeric micelle-formulated paclitaxel, in patients with advanced malignancies. Clin Cancer Res 10(11):3708–3716. https://doi.org/10.1158/1078-0432.ccr-03-0655

Kisley LR, Barrett BS, Bauer AK et al (2002) Genetic ablation of inducible nitric oxide synthase decreases mouse lung tumorigenesis. Cancer Res 62(23):6850–6856

Koller-Lucae SK, Schott H, Schwendener RA (1997) Interactions with human blood in vitro and pharmacokinetic properties in mice of liposomal N4-octadecyl-1-Beta-D--arabinofuranosylcytosine, a new anticancer drug. J Pharmacol Exp Ther 282(3):1572–1580

Kono K (2001) Thermosensitive polymer-modified liposomes. Adv Drug Deliv Rev 53(3):307–319. https://doi.org/10.1016/S0169-409X(01)00204-6

Kono K, Yoshino K, Takagishi T (2002) Effect of poly (ethylene glycol) grafts on temperature-sensitivity of thermosensitive polymer-modified liposomes. J Control Release 80(1–3):321–332. https://doi.org/10.1016/S0168-3659(02)00018-4

Kou R, Greif D, Michel T (2002) Dephosphorylation of endothelial nitric-oxide synthase by vascular endothelial growth factor: implications for the vascular responses to cyclosporina. J Biol Chem 277(33):29669–29673. https://doi.org/10.1074/jbc.M204519200

Krauss AC, Gao X, Li L et al (2019) FDA approval summary: (daunorubicin and cytarabine) liposome for injection for the treatment of adults with high-risk acute myeloid leukemia. Clin Cancer Res 25(9):2685–2690. https://doi.org/10.1158/1078-0432.ccr-18-2990

Krishna R, Webb MS, Onge GS et al (2001) Liposomal and nonliposomal drug pharmacokinetics after administration of liposome-encapsulated vincristine and their contribution to drug tissue distribution properties. J Pharmacol Exp Ther 298(3):1206–1212

Kubo T, Sugita T, Shimose S et al (2000) Targeted delivery of anticancer drugs with intravenously administered magnetic liposomes in osteosarcoma-bearing hamsters. Int J Oncol 17(2):309–324

Kularatne SA, Low PS (2010) Targeting of nanoparticles: folate receptor. Cancer Nanotechnol 624:249–265. https://doi.org/10.1007/978-1-60761-609-2_17

Laskin JJ, Sandler AB (2004) Epidermal growth factor receptor: a promising target in solid tumours. Cancer Treat Rev 30(1):1–17. https://doi.org/10.1016/j.ctrv.2003.10.002

Ledermann JA, Canevari S, Thigpen T (2015) Targeting the folate receptor: diagnostic and therapeutic approaches to personalize cancer treatments. Ann Oncol 26(10):2034–2043. https://doi.org/10.1093/annonc/mdv250

Lee KS, Chung HC, Im SA et al (2008) Multicenter phase II trial of Genexol-PM, a Cremophor-free, polymeric micelle formulation of paclitaxel, in patients with metastatic breast cancer. Breast Cancer Res Treat 108(2):241–250. https://doi.org/10.1007/s10549-007-9591-y

Li YP, Pei YY, Zhou ZH et al (2001) Stealth polycyanoacrylate nanoparticles as tumor necrosis factor-α carriers: pharmacokinetics and anti-tumor effects. Biol Pharm Bull 24(6):662–665. https://doi.org/10.1248/bpb.24.662

Li X, Ding L, Xu Y et al (2009) Targeted delivery of doxorubicin using stealth liposomes modified with transferrin. Int J Pharm 373(1–2):116–123. https://doi.org/10.1016/j.ijpharm.2009.01.023

Li Y, Ding J, Xu X et al (2020) Dual hypoxia-targeting RNAi nanomedicine for precision cancer therapy. Nano Lett 20(7):4857–4863. https://doi.org/10.1021/acs.nanolett.0c00757

Lindner LH, Eichhorn ME, Eibl H et al (2004) Novel temperature-sensitive liposomes with prolonged circulation time. Clin Cancer Res 10(6):2168–2178. https://doi.org/10.1158/1078-0432.CCR-03-0035

Low PS, Antony AC (2004) Folate receptor-targeted drugs for cancer and inflammatory diseases-preface. Adv Drug Deliv Rev 56(8):1055–1058. https://doi.org/10.1016/j.addr.2004.02.003

Lu J, Zhao W, Huang Y et al (2014) Targeted delivery of doxorubicin by folic acid-decorated dual functional nanocarrier. Mol Pharm 11(11):4164–4178. https://doi.org/10.1021/mp500389v

Lurje G, Lenz HJ (2009) EGFR signaling and drug discovery. Oncology (Basel) 77(6):400–410. https://doi.org/10.1159/000279388

Lyon PC, Gray MD, Mannaris C et al (2018) Safety and feasibility of ultrasound-triggered targeted drug delivery of doxorubicin from thermosensitive liposomes in liver tumours (TARDOX): a single-centre, open-label, phase 1 trial. Lancet Oncol 19(8):1027–1039. https://doi.org/10.1016/s1470-2045(18)30332-2

Ma Y, Yang Z, Huntoon K et al (2021) Advanced immunotherapy approaches for glioblastoma. Adv Ther 4(8):2100046. https://doi.org/10.1002/adtp.202100046

Maeda H (2001a) The enhanced permeability and retention (EPR) effect in tumor vasculature: the key role of tumor-selective macromolecular drug targeting. Adv Enzym Regul 41:189–207. https://doi.org/10.1016/S0065-2571(00)00013-3

Maeda H (2001b) SMANCS and polymer-conjugated macromolecular drugs: advantages in cancer chemotherapy. Adv Drug Deliv Rev 46(1–3):169–185. https://doi.org/10.1016/S0169-409X(00)00134-4

Maeda H, Greish K, Fang J (2006) The EPR effect and polymeric drugs: a paradigm shift for cancer chemotherapy in the 21st century. Polym Ther II Adv Polym Sci 193:103–121. https://doi.org/10.1007/12_026

Mahmoudi M, Sant S, Wang B et al (2011) Superparamagnetic iron oxide nanoparticles (SPIONs): development, surface modification and applications in chemotherapy. Adv Drug Deliv Rev 63(1–2):24–46. https://doi.org/10.1016/j.addr.2010.05.006

Mamot C, Drummond DC, Noble CO et al (2005) Epidermal growth factor receptor-targeted immunoliposomes significantly enhance the efficacy of multiple anticancer drugs in vivo. Cancer Res 65(24):11631–11638. https://doi.org/10.1158/0008-5472.can-05-1093

Matsumura Y, Maeda H (1986) A new concept for macromolecular therapeutics in cancer chemotherapy: mechanism of tumoritropic accumulation of proteins and the antitumor agent smancs. Cancer Res 46(12 Part 1):6387–6392

Matsumura Y, Kimura M, Yamamoto T et al (1988) Involvement of the kinin-generating cascade in enhanced vascular permeability in tumor tissue. Jpn J Cancer Res 79(12):1327–1334. https://doi.org/10.1111/j.1349-7006.1988.tb01563.x

Matsumura Y, Maruo K, Kimura M et al (1991) Kinin-generating cascade in advanced cancer patients and in vitro study. Jpn J Cancer Res 82(6):732–741. https://doi.org/10.1111/j.1349-7006.1991.tb01910.x

Maziarz KM, Monaco HL, Shen F et al (1999) Complete mapping of divergent amino acids responsible for differential ligand binding of folate receptors α and β. J Biol Chem 274(16):11086–11091. https://doi.org/10.1074/jbc.274.16.11086

Mendelsohn J, Baselga J (2003) Status of epidermal growth factor receptor antagonists in the biology and treatment of cancer. J Clin Oncol 21(14):2787–2799. https://doi.org/10.1200/JCO.2003.01.504

Mross K, Niemann B, Massing U et al (2004) Pharmacokinetics of liposomal doxorubicin (TLC-D99; Myocet) in patients with solid tumors: an open-label, single-dose study. Cancer Chemother Pharm 54(6):514–524. https://doi.org/10.1007/s00280-004-0825-y

Munster P, Krop IE, LoRusso P et al (2018) Safety and pharmacokinetics of MM-302, a HER2-targeted antibody–liposomal doxorubicin conjugate, in patients with advanced HER2-positive breast cancer: a phase 1 dose-escalation study. Br J Cancer 119(9):1086–1093. https://doi.org/10.1038/s41416-018-0235-2

Murohara T, Horowitz JR, Silver M et al (1998) Vascular endothelial growth factor/vascular permeability factor enhances vascular permeability via nitric oxide and prostacyclin. Circulation 97(1):99–107. https://doi.org/10.1161/01.CIR.97.1.99

Nicholas MK, Lukas RV, Jafri NF et al (2006) Epidermal growth factor receptor–mediated signal transduction in the development and therapy of gliomas. Clin Cancer Res 12(24):7261–7270. https://doi.org/10.1158/1078-0432.CCR-06-0874

Nyman DW, Campbell KJ, Hersh E et al (2005) Phase I and pharmacokinetics trial of ABI-007, a novel nanoparticle formulation of paclitaxel in patients with advanced nonhematologic malignancies. J Clin Oncol 23(31):7785–7793. https://doi.org/10.1200/jco.2004.00.6148

Oerlemans C, Bult W, Bos M et al (2010) Polymeric micelles in anticancer therapy: targeting, imaging and triggered release. Pharm Res 27(12):2569–2589. https://doi.org/10.1007/s11095-010-0233-4

Papapetropoulos A, García-Cardeña G, Madri JA et al (1997) Nitric oxide production contributes to the angiogenic properties of vascular endothelial growth factor in human endothelial cells. J Clin Invest 100(12):3131–3139. https://doi.org/10.1172/JCI119868

Paridaens R, Biganzoli L, Bruning P et al (2000) Paclitaxel versus doxorubicin as first-line single-agent chemotherapy for metastatic breast cancer: a european organization for research and treatment of cancer randomized study with cross-over. J Clin Oncol 18(4):724–724. https://doi.org/10.1200/jco.2000.18.4.724

Park JW, Hong KL, Kirpotin DB et al (2002) Anti-HER2 immunoliposomes: enhanced efficacy attributable to targeted delivery. Clin Cancer Res 8(4):1172–1181

Perez AT, Domenech GH, Frankel C et al (2002) Pegylated liposomal doxorubicin (Doxil®) for metastatic breast cancer: the Cancer Research Network, Inc., experience. Cancer Investig 20(sup2):22–29. https://doi.org/10.1081/CNV-120014883

Petrova V, Annicchiarico-Petruzzelli M, Melino G et al (2018) The hypoxic tumour microenvironment. Oncogenesis 7(1):1–13. https://doi.org/10.1038/s41389-017-0011-9

Pirollo KF, Nemunaitis J, Leung PK et al (2016) Safety and efficacy in advanced solid tumors of a targeted nanocomplex carrying the p53 gene used in combination with docetaxel: a phase 1b study. Mol Ther 24(9):1697–1706. https://doi.org/10.1038/mt.2016.135

Ponce AM, Vujaskovic Z, Yuan F et al (2006) Hyperthermia mediated liposomal drug delivery. Int J Hyperth 22(3):205–213. https://doi.org/10.1080/02656730600582956

Rankin E, Giaccia A (2008) The role of hypoxia-inducible factors in tumorigenesis. Cell Death Differ 15(4):678–685. https://doi.org/10.1038/cdd.2008.21

Roncato F, Rruga F, Porcu E et al (2018) Improvement and extension of anti-EGFR targeting in breast cancer therapy by integration with the avidin-nucleic-acid-nano-assemblies. Nat Commun 9(1):4070. https://doi.org/10.1038/s41467-018-06602-6

Ross JF, Chaudhuri PK, Ratnam M (1994) Differential regulation of folate receptor isoforms in normal and malignant tissues in vivo and in established cell lines. Physiologic and clinical implications. Cancer 73(9):2432–2443. https://doi.org/10.1002/1097-0142(19940501)73:9<2432::AID-CNCR2820730929>3.0.CO;2-S

Scaltriti M, Baselga J (2006) The epidermal growth factor receptor pathway: a model for targeted therapy. Clin Cancer Res 12(18):5268–5272. https://doi.org/10.1158/1078-0432.ccr-05-1554

Schild HG (1992) Poly (N-isopropylacrylamide): experiment, theory and application. Prog Polym Sci 17(2):163–249. https://doi.org/10.1016/0079-6700(92)90023-R

Schmidinger M, Wenzel C, Locker G et al (2001) Pilot study with pegylated liposomal doxorubicin for advanced or unresectable hepatocellular carcinoma. Br J Cancer 85(12):1850–1852. https://doi.org/10.1054/bjoc.2001.2149

Seki T, Fang J, Maeda H (2009) Enhanced delivery of macromolecular antitumor drugs to tumors by nitroglycerin application. Cancer Sci 100(12):2426–2430. https://doi.org/10.1111/j.1349-7006.2009.01323.x

Semenza GL (2010) Defining the role of hypoxia-inducible factor 1 in cancer biology and therapeutics. Oncogene 29(5):625–634. https://doi.org/10.1038/onc.2009.441

Senger DR, Galli SJ, Dvorak AM et al (1983) Tumor cells secrete a vascular permeability factor that promotes accumulation of ascites fluid. Science 219(4587):983–985. https://doi.org/10.1126/science.6823562

Shadidi M, Sioud M (2003) Selective targeting of cancer cells using synthetic peptides. Drug Resist Update 6(6):363–371. https://doi.org/10.1016/j.drup.2003.11.002

Shen Y, Li X, Dong D et al (2018) Transferrin receptor 1 in cancer: a new sight for cancer therapy. Am J Cancer Res 8(6):916–931

Shenoi MM, Iltis I, Choi J et al (2013) Nanoparticle delivered vascular disrupting agents (VDAs): use of TNF-alpha conjugated gold nanoparticles for multimodal cancer therapy. Mol Pharm 10(5):1683–1694. https://doi.org/10.1021/mp300505w

Shenoy D, Little S, Langer R et al (2005) Poly (ethylene oxide)-modified poly (β-amino ester) nanoparticles as a pH-sensitive system for tumor-targeted delivery of hydrophobic drugs: part 2. In vivo distribution and tumor localization studies. Pharm Res 22(12):2107–2114

Shmeeda H, Mak L, Tzemach D et al (2006) Intracellular uptake and intracavitary targeting of folate-conjugated liposomes in a mouse lymphoma model with up-regulated folate receptors. Mol Cancer Ther 5(4):818–824. https://doi.org/10.1158/1535-7163.MCT-05-0543

Silverman JA, Deitcher SR (2013) Marqibo® (vincristine sulfate liposome injection) improves the pharmacokinetics and pharmacodynamics of vincristine. Cancer Chemother Pharm 71(3):555–564. https://doi.org/10.1007/s00280-012-2042-4

Skubitz KM (2003) Phase II trial of Pegylated-liposomal doxorubicin (Doxil™) in sarcoma. Cancer Investig 21(2):167–176. https://doi.org/10.1081/CNV-120016412

Smith RE, Brown AM, Mamounas EP et al (1999) Randomized trial of 3-hour versus 24-hour infusion of high-dose paclitaxel in patients with metastatic or locally advanced breast cancer: national surgical adjuvant breast and bowel project protocol B-26. J Clin Oncol 17(11):3403–3411. https://doi.org/10.1200/JCO.1999.17.11.3403

Son S, Rao NV, Ko H et al (2018) Carboxymethyl dextran-based hypoxia-responsive nanoparticles for doxorubicin delivery. Int J Biol Macromol 110:399–405. https://doi.org/10.1016/j.ijbiomac.2017.11.048

Speeckaert MM, Speeckaert R, Delanghe JR (2010) Biological and clinical aspects of soluble transferrin receptor. Crit Rev Clin Lab Sci 47(5–6):213–228. https://doi.org/10.3109/10408363.2010.550461

Subbiah V, Grilley-Olson JE, Combest AJ et al (2018) Phase Ib/II Trial of NC-6004 (nanoparticle cisplatin) plus gemcitabine in patients with advanced solid tumors. Clin Cancer Res 24(1):43–51. https://doi.org/10.1158/1078-0432.CCR-17-1114

Sudimack J, Lee RJ (2000) Targeted drug delivery via the folate receptor. Adv Drug Deliv Rev 41(2):147–162. https://doi.org/10.1016/S0169-409X(99)00062-9

Sun J, Kormakov S, Liu Y et al (2018) Recent progress in metal-based nanoparticles mediated photodynamic therapy. Molecules 23(7):1704. https://doi.org/10.3390/molecules23071704

Sun Y, Zhao Y, Teng S et al (2019) Folic acid receptor-targeted human serum albumin nanoparticle formulation of cabazitaxel for tumor therapy. Int J Nanomedicine 14:135–148. https://doi.org/10.2147/IJN.S181296

Sun Y, Lee RJ, Meng F et al (2020) Microfluidic self-assembly of high cabazitaxel loading albumin nanoparticles. Nanoscale 12(32):16928–16933. https://doi.org/10.1039/c9nr10941b

Sun X, Yu K, Zhou Y et al (2021) Self-assembled pH-sensitive polymeric nanoparticles for the inflammation-targeted delivery of Cu/Zn-superoxide dismutase. ACS Appl Mater Interfaces 13(15):18152–18164. https://doi.org/10.1021/acsami.1c03589

Szabo C, Ischiropoulos H, Radi R (2007) Peroxynitrite: biochemistry, pathophysiology and development of therapeutics. Nat Rev Drug Discov 6(8):662–680. https://doi.org/10.1038/nrd2222

Tahvilian R, Tajani B, Sadrjavadi K et al (2016) Preparation and characterization of pH-sensitive camptothecin-cis-aconityl grafted chitosan oligosaccharide nanomicelles. Int J Biol Macromol 92:795–802. https://doi.org/10.1016/j.ijbiomac.2016.07.100

Teng L, Xie J, Teng L et al (2012) Clinical translation of folate receptor-targeted therapeutics. Expert Opin Drug Deliv 9(8):901–908. https://doi.org/10.1517/17425247.2012.694863

Terman BI, Stoletov KV (2001) VEGF and tumor angiogenesis. Einstein Quart J Biol Med 18:59–66

Thomas A, Pommier Y (2019) Targeting topoisomerase I in the era of precision medicine. Clin Cancer Res 25(22):6581–6589

Thomas CR, Ferris DP, Lee JH et al (2010) Noninvasive remote-controlled release of drug molecules in vitro using magnetic actuation of mechanized nanoparticles. J Am Chem Soc 132(31):10623–10625. https://doi.org/10.1021/ja1022267

Torchilin VP (2001) Structure and design of polymeric surfactant-based drug delivery systems. J Control Release 73(2–3):137–172. https://doi.org/10.1016/S0168-3659(01)00299-1

Torchilin V (2009) Multifunctional and stimuli-sensitive pharmaceutical nanocarriers. Eur J Pharm Biopharm 71(3):431–444. https://doi.org/10.1016/j.ejpb.2008.09.026

Torchilin V (2011) Tumor delivery of macromolecular drugs based on the EPR effect. Adv Drug Deliv Rev 63(3):131–135. https://doi.org/10.1016/j.addr.2010.03.011

Trédan O, Galmarini CM, Patel K et al (2007) Drug resistance and the solid tumor microenvironment. J Natl Cancer Inst 99(19):1441–1454. https://doi.org/10.1093/jnci/djm135

Tu C, Zhu L, Qiu F et al (2013) Facile PEGylation of Boltorn® H40 for pH-responsive drug carriers. Polymer 54(8):2020–2027. https://doi.org/10.1016/j.polymer.2012.12.029

Vahora H, Khan MA, Alalami U et al (2016) The potential role of nitric oxide in halting cancer progression through chemoprevention. J Cancer Prev 21(1):1–12. https://doi.org/10.15430/JCP.2016.21.1.1

Van Der Meel R, Vehmeijer LJ, Kok RJ et al (2013) Ligand-targeted particulate nanomedicines undergoing clinical evaluation: current status. Adv Drug Deliv Rev 65(10):1284–1298. https://doi.org/10.1016/j.addr.2013.08.012

Veiseh O, Gunn JW, Zhang M (2010) Design and fabrication of magnetic nanoparticles for targeted drug delivery and imaging. Adv Drug Deliv Rev 62(3):284–304. https://doi.org/10.1016/j.addr.2009.11.002

Wang X, Lin Y (2008) Tumor necrosis factor and cancer, buddies or foes? Acta Pharmacol Sin 29(11):1275–1288. https://doi.org/10.1111/j.1745-7254.2008.00889.x

Wang G, Ye Y, Zhang X et al (2014) Bradykinin stimulates IL-6 production and cell invasion in colorectal cancer cells. Oncol Rep 32(4):1709–1714. https://doi.org/10.3892/or.2014.3366

Wang X, Zhao Y, Dong S et al (2019) Cell-penetrating peptide and transferrin co-modified liposomes for targeted therapy of glioma. Molecules 24(19):3540. https://doi.org/10.3390/molecules24193540

Wang X, Xu J, Xu X et al (2020) pH-sensitive bromelain nanoparticles by ortho ester crosslinkage for enhanced doxorubicin penetration in solid tumor. Mater Sci Eng C Mater Biol Appl 113:111004. https://doi.org/10.1016/j.msec.2020.111004

Waters JP, Pober JS, Bradley JR (2013) Tumour necrosis factor and cancer. J Pathol 230(3):241–248. https://doi.org/10.1002/path.4188

Wilson WR, Hay MP (2011) Targeting hypoxia in cancer therapy. Nat Rev Cancer 11(6):393–410. https://doi.org/10.1038/nrc3064

Wollina U, Dummer R, Brockmeyer NH et al (2003) Multicenter study of pegylated liposomal doxorubicin in patients with cutaneous T-cell lymphoma. Cancer 98(5):993–1001. https://doi.org/10.1002/cncr.11593

Wong PT, Choi SK (2015) Mechanisms and implications of dual-acting methotrexate in folate-targeted nanotherapeutic delivery. Int J Mol Sci 16(1):1772–1790. https://doi.org/10.3390/ijms16011772

Wust P, Gneveckow U, Wust P et al (2006) Magnetic nanoparticles for interstitial thermotherapy–feasibility, tolerance and achieved temperatures. Int J Hyperth 22(8):673–685. https://doi.org/10.1080/02656730601106037

Xie J, Teng L, Yang Z et al (2013) A polyethylenimine-linoleic acid conjugate for antisense oligonucleotide delivery. Biomed Res Int 2013:1–7. https://doi.org/10.1155/2013/710502

Xie J, Yang Z, Zhou C et al (2016) Nanotechnology for the delivery of phytochemicals in cancer therapy. Biotechnol Adv 34(4):343–353. https://doi.org/10.1016/j.biotechadv.2016.04.002

Xie J, Fan Z, Li Y et al (2018) Design of pH-sensitive methotrexate prodrug-targeted curcumin nanoparticles for efficient dual-drug delivery and combination cancer therapy. Int J Nanomedicine 13:1381–1398. https://doi.org/10.2147/IJN.S152312

Xie J, Bi Y, Zhang H et al (2020) Cell-penetrating peptides in diagnosis and treatment of human diseases: from preclinical research to clinical application. Front Pharmacol 11. https://doi.org/10.3389/fphar.2020.00697

Yan L, Shen J, Wang J et al (2020) Nanoparticle-based drug delivery system: a patient-friendly chemotherapy for oncology. Dose Response 18(3):1559325820936161. https://doi.org/10.1177/1559325820936161

Yang G, Yang T, Zhang W et al (2014a) In vitro and in vivo antitumor effects of folate-targeted ursolic acid stealth liposome. J Agric Food Chem 62(10):2207–2215. https://doi.org/10.1021/jf405675g

Yang Z, Yu B, Zhu J et al (2014b) A microfluidic method to synthesize transferrin-lipid nanoparticles loaded with siRNA LOR-1284 for therapy of acute myeloid leukemia. Nanoscale 6(16):9742–9751. https://doi.org/10.1039/c4nr01510j

Yang X, Yang S, Chai H et al (2015) A novel isoquinoline derivative anticancer agent and its targeted delivery to tumor cells using transferrin-conjugated liposomes. PLoS One 10(8). https://doi.org/10.1371/journal.pone.0136649

Yang Z, Xie J, Zhu J et al (2016) Functional exosome-mimic for delivery of siRNA to cancer: in vitro and in vivo evaluation. J Control Release 243:160–171. https://doi.org/10.1016/j.jconrel.2016.10.008

Yang Z, Ma Y, Zhao H et al (2019) Nanotechnology platforms for cancer immunotherapy. Nanomed Nanobiotechnol 12(2). https://doi.org/10.1002/wnan.1590

Yang X, Ma Y, Xie H et al (2021) Extracellular vesicles in the treatment of Parkinson's disease: a review. Curr Med Chem 28(31):6375–6394. https://doi.org/10.2174/0929867328666210113170941

Yavlovich A, Smith B, Gupta K et al (2010) Light-sensitive lipid-based nanoparticles for drug delivery: design principles and future considerations for biological applications. Mol Membr Biol 27(7):364–381. https://doi.org/10.3109/09687688.2010.507788

Ying X, Wen H, Lu WL et al (2010) Dual-targeting daunorubicin liposomes improve the therapeutic efficacy of brain glioma in animals. J Control Release 141(2):183–192. https://doi.org/10.1016/j.jconrel.2009.09.020

Yingchoncharoen P, Kalinowski DS, Richardson DR (2016) Lipid-based drug delivery systems in cancer therapy: what is available and what is yet to come. Pharmacol Rev 68(3):701–787. https://doi.org/10.1124/pr.115.012070

Yu HS, Wang SW, Chang AC et al (2014) Bradykinin promotes vascular endothelial growth factor expression and increases angiogenesis in human prostate cancer cells. Biochem Pharmacol 87(2):243–253. https://doi.org/10.1016/j.bcp.2013.10.016

Yuan H, Jiang W, Von Roemeling CA et al (2017) Multivalent bi-specific nanobioconjugate engager for targeted cancer immunotherapy. Nat Nanotechnol 12(8):763–769. https://doi.org/10.1038/NNANO.2017.69

Zhou C, Yang Z, Teng L (2014) Nanomedicine based on nucleic acids: pharmacokinetic and pharmacodynamic perspectives. Curr Pharm Biotechnol 15(9):829–838. https://doi.org/10.2174/1389201015666141020155620

Zhou F, Feng B, Yu H et al (2016) Cisplatin prodrug-conjugated gold nanocluster for fluorescence imaging and targeted therapy of the breast cancer. Theranostics 6(5):679–687. https://doi.org/10.7150/thno.14556

Zhou F, Feng B, Wang T et al (2017) Programmed multiresponsive vesicles for enhanced tumor penetration and combination therapy of triple-negative breast cancer. Adv Funct Mater 27(20):1606530. https://doi.org/10.1002/adfm.201606530

Zuckerman JE, Gritli I, Tolcher A et al (2014) Correlating animal and human phase Ia/Ib clinical data with CALAA-01, a targeted, polymer-based nanoparticle containing siRNA. Proc Natl Acad Sci USA 111(31):11449–11454. https://doi.org/10.1073/pnas.1411393111

Chapter 4
Combined Therapy in Cancer Treatment

Yifan Ma, Jingjing Zhang, Yuan Yuan, Harshvardhan Modh, and Zhaogang Yang

Abstract Despite the extensive application of conventional therapies, most of these strategies, especially in the forms of monotherapy, are found to be inefficient in cancer therapy in clinic. Combined therapy, a polytherapy modality that combines multiple treatment strategies and therapeutic agents, has become a more promising approach with improved outcomes in anti-cancer treatment. This approach could solve the difficulties and challenges caused by the intrinsic heterogeneity and complexity of cancer, such as multiple drug resistance (MDR) and high metastatic potential, which is generally considered as the most crucial cause of treatment failure. So far, three types of combined therapy for cancer treatment, including the combination of chemical and nucleic acid drugs, chemo-immunotherapy, and combined therapy of immunotherapy with photothermal therapy (PTT) or photodynamic therapy (PDT), have been developed and extensively evaluated in both preclinical and clinical stages. This session summarizes recent advances in utilizing the combination of multiple therapies and therapeutics to improve the outcomes of cancer treatment.

Y. Ma
Department of Biomedical Engineering, The Ohio State University, Columbus, OH, USA

J. Zhang
Department of Chemical and Biomolecular Engineering, The Ohio State University, Columbus, OH, USA

Y. Yuan
Key Laboratory for Ultrafine Materials of Ministry of Education, and School of Materials Science and Engineering, East China University of Science and Technology, Shanghai, China

Frontiers Science Center for Materiobiology and Dynamic Chemistry, and Engineering Research Center for Biomedical Materials of Ministry of Education, East China University of Science and Technology, Shanghai, China

H. Modh
Department of Pharmacy, National University of Singapore, Singapore, Singapore

Z. Yang (✉)
School of Life Sciences, Jilin University, Changchun, China
e-mail: zhaogangyang@jlu.edu.cn

L. Teng et al. (eds.), *Drug Delivery to Tumors*,
https://doi.org/10.1007/978-981-19-8930-8_4

Keywords Combined therapy · Multiple drug resistance (MDR) · Metastatic potential · Cancer treatment

4.1 Introduction

Cancer, the second leading cause of death in the world, accounts for a total of nearly ten million deaths in 2022. Though cancer mortality rates have lowered to some extent in recent years, only a very small proportion of such improvement is contributed to the development of new drugs (Garattini 2003; Mokhtari et al. 2017). This limited contribution is due to the heavy labor and expense required, which includes countless in vitro and in vivo evaluation and several clinical trials before the final approval from Food and Drug Administration (FDA), for the development of new clinical anti-cancer drugs (Marchetti and Schellens 2007). Considering the difficulties to obtain a new drug, a combination of multiple therapeutic treatments that have been already applied in current clinical trials has become more attractive and feasible. Moreover, compared to the conventional mono-therapy approach which is currently extensive exploited in the therapy of various cancers, the combination therapy approach is much more effective.

Currently, an organization of methods between chemotherapy, nucleic acid drugs phototherapy, and immunotherapy is the most commonly used combination in the current preclinical and clinical treatment of cancers. Among them, chemotherapy usually has high toxicity to the patient with various negative responses and can also profoundly inhibit the immune system in patients by affecting bone marrow cells and increasing susceptibility to host diseases (Partridge et al. 2001; Lebaron et al. 1988). In contrast, the toxicity will be dramatically reduced if some nucleic acid drugs are used in combination therapy since different pathways could be regulated. Most importantly, the incorporation of nucleic acid drugs could exert a synergistic or additive effects, and thus, a smaller clinical dosage for each individual chemical is achieved (Albain et al. 2008; Mokhtari et al. 2013). Moreover, the incorporation of nucleic acid drugs into conventional chemical therapy may reduce the toxic effects on regular cells but produce cytotoxic effects on tumor cells simultaneously. This could be highly possible when one drug in the combination regimen is antagonistic to the other drug in regular cells, thereby protecting them from cytotoxic effects (Blagosklonny 2005). In addition to the combination of chemotherapy and nucleic acid drugs, immunotherapy has also been widely cooperated with chemotherapy and termed as chemo-immunotherapy. As is well known, single chemotherapy is insufficient to resolve the physiological complexity of cancers. Tumor recurrence is highly possible as gene mutations and clinically undetectable metastases are common in patients with cancer. Moreover, multidrug resistance (MDR), which usually happens when tumors become resistant to specific drugs with different therapeutic targets and chemical structures, is also common in multiple cancer types (Szakács et al. 2006). Accordingly, in the chemo-immunotherapy, chemotherapy mainly plays a role in inducing toxicity and death of tumor cells, while immunotherapy can prevent the relapse of cancer caused by the active invasion and tumor metastasis (Ma et al. 2021). Moreover, the chemotherapy could not solely kill the tumor cells through directly inducing apoptosis but also improve the antitumor immune responses which are primed by immunogenic cell death (ICD). Besides, phototherapy, primarily photothermal therapy (PTT) and photodynamic therapy (PDT), has also been applied to improve the therapeutic effect of cancer treatment. In contrast to the direct cytotoxic effects on tumors, phototherapies often utilize thermal energy or excitation light with molecular oxygen to kill tumor. Because of its non-invasiveness, excellent controllability, and minimum side effects, these phototherapies also attract ever-increasing attention and are considered as a promising candidate to improve the treatment outcomes of monotherapy, especially immunotherapy (Shang et al. 2020; Yang et al. 2020). According to multiple reports, phototherapy can significantly improve the outcome of immunotherapy as phototherapy generally enhances the effects of tumor-related antigens and induces immune responses by ICD, which leads to synergistic anti-cancer effect in cancer therapy. In turn, combination with immunotherapy also helps to overcome the obstacle that single phototherapy has low efficiency and efficacy in the treatment of cancer patients with metastasis and recurrence (Fan et al. 2017).

In this chapter, we summarized three types of combined therapy for cancer treatment that have been developed and extensively evaluated in both preclinical and clinical stages, including the combination of chemical and nucleic acid drugs, chemo-immunotherapy, and combined therapy of immunotherapy with PTT or PDT. After surveying the most recent literatures, we comprehensively evaluated the advances in combining multiple therapies and therapeutics to improve cancer treatment outcomes. Moreover, improvements in designing novel combined therapies and the future outlook were discussed.

4.2 Chemical and Nucleic Acid Drugs

Chemotherapy was originated from the development of a DNA alkylating chemical named nitrogen mustard during the Second World War, and it is now one type of gold-standard drug treatment against cancer (Gilman 1963; Qin et al. 2018). Currently, chemotherapeutic agents are classified into several categories with diverse mechanisms, including alkylating drugs, antimetabolites, anti-microtubule drugs, topoisomerase antagonists, and cytotoxic antibiotics. These chemical drugs, when further combined with delivery platforms, can induce potent anti-tumor effects on various cancer types and thus have been widely applied in preclinical studies and clinical trials. However, though authenticated and effective, mono-chemotherapy is found to be undesirable in clinical use, especially when facing up with the complicated physiological conditions of cancers. The gene mutations in patients that result in high immortality and activate invasion and pro-metastatic effects are considered as the key contributors to the failure of a single chemotherapeutic modality. Furthermore, MDR is another persistent difficulty in successfully treating local and metastatic cancers when chemotherapy is applied (Burrell et al. 2013). Generally, chemical drugs can induce tumor cells' apoptosis, whereas these drugs trigger both pump and non-pump cellular defense that prevents cells from death, and such a contradictory effect ultimately fails the treatment and leads to relapse or recurrence of many tumors. To circumvent these inadequacies of conventional chemotherapy, alternative therapeutic strategies that can supplement chemotherapy are of great potential. In recent decades, gene therapy, which applies functional nucleic acid drugs to knockdown the disorder genes, restore wild-type genes, or suppress undesirable genes/proteins expression during cancer therapy, has been evaluated in the combination with chemotherapy. These nucleic acid drugs, including plasmid DNA (pDNA), message RNA (mRNA), small interfering RNA (siRNA), and microRNA (miRNA), which are generally introduced with carriers for DNA/RNA protection and ensuring efficient delivery, have been demonstrated to be a potent collaborative therapy when co-administrated with tradition chemotherapeutic drugs.

4.2.1 Combination of Chemical and Nucleic Acids Drug for Improved Apoptosis

One of the hallmarks of cancer is its intrinsic resistance to apoptosis in which the tumor cells employ various means, usually self-destruction of the intrinsic apoptosis pathways, to prevent cell from death, and this process finally promotes the tumor development and progression (Hanahan and Weinberg Robert 2011). As a result of multiple gene mutations during tumor progression, the balance between pro-apoptotic and anti-apoptotic pathways becomes dysfunctional, and thus, cancer cells tend to be more anti-apoptotic. As expected, though multiple chemotherapeutic drugs used for cancer treatment are designed to directly induce the apoptosis of cancer cells, these gene defects resulted from the gene mutations in the apoptotic pathways of tumor cells could significantly undermine the efficacy of chemotherapeutic drugs and cause tumor recurrence and relapse. Therefore, the strategy using co-delivery of apoptosis-inducing chemotherapeutic drugs with nucleic acids drugs to reverse or restore the apoptotic functionality in these cells with gene mutations has been developed. Currently, these combinations often rely on the re-expression of tumor-suppressor genes, such as TP53, and apoptosis-inducing proteins including tumor necrosis factor-related apoptosis-inducing ligand (TRAIL) and tumor necrosis factor-alpha (TNF-α). Other useful strategies include the down-regulation of anti-apoptotic genes like survivin, B-cell lymphoma protein 2 (Bcl-2), or Bcl-2-associated X (BAX).

The mutated tumor-suppressor p53 gene is critical in regulating the progression and apoptosis of cancer cells. Due to the gene mutation and DNA damage which render to the failure of cancer cell apoptosis, the majority of cancer cells with loss of p53 function are disabled to activate cell-cycle arrest and DNA repair (Fisher 2001). More importantly, the p53 deficiency in the cancer cells often impairs the therapeutic effectiveness of most chemotherapeutic agents currently available. Therefore, restoration of the expression of functional p53 genes to achieve apoptotic function has been widely evaluated and identified as a promising strategy to sensitize most cancers to chemotherapeutics, and this results in an improved outcome in the inhibition of tumor growth. For example, Xu et al. introduced a co-delivery system of chitosan-p53-encoding plasmids (chi-p53) and DOX using a double-walled PLGA microsphere where chi-p53 and DOX were loaded in the shell and core phases, respectively (Xu et al. 2012). As expected, the incorporated chi-p53 were first released from nanoparticles, followed by a sustaining release of DOX from the core site. Such a sequential release strategy and P53 restoration design are supposed to improve chemotherapeutics and provide a promising combined modality for synergistic cancer treatment. In another similar study, Shi et al. used porous PLGA microparticles containing both DOX and p53 plasmid to improve tumor growth inhibition and enhance tumor cell apoptosis in the treatment of lung cancer (Shi et al. 2014). This combination delivery strategy targeting P53 restoration is designated to improve chemotherapeutics and provide a promising combined modality for synergistic cancer treatment.

In addition to the restoration of tumor suppressor genes, the introduction of pro-apoptotic genes is also an applicable approach to cooperate with chemotherapy for improving tumor apoptosis. TRAIL, also known as APO2L, is a transmembrane protein that belongs to the tumor necrosis factor family which can trigger apoptotic cell death in cancer cells (Fisher 2001). As such, the introduction of TRAIL gene can not only be a promising target for cancer treatment but also a sensitizer for tumor cells to other delivered drugs, especially chemodrugs. In the treatment of brain gliomas, Liu et al. showed that local restoration of the TRAIL gene synergized with chemotherapy could be a potent anti-glioma modality with low toxicity (Liu et al. 2012). According to their results, synergistic tumor growth inhibition could be achieved through the co-delivery of the DOX and TRAIL gene using their dendrimer delivery platform (DGDPT/pORF-hTRAIL), as confirmed by delayed cancer progression and increased life-span in the U87-MG tumor-bearing mice when compared with the single treatment of DOX or TRAIL. Other than the introduction of TRAIL gene, TRAIL agonistic antibodies and recombinant TRAIL which have shown excellent safety profiles are also promising and now evaluated in the combination therapy with chemodrugs (Snajdauf et al. 2021). Recently, dulanermin (a recombinant human TRAIL) combined with vinorelbine and cisplatin was investigated in a phase III clinical trial in those with advanced non-small-cell lung cancer (NSCLC) (Ouyang et al. 2018). In this trail, the combination of dulanermin with chemodrug could dramatically increases not only progression-free survival (PFS) but also objective response rate (ORR), which indicated favorable clinical outcomes and superior benefits for the patients receiving the combined TRAIL/chemotherapeutics therapy.

By presenting the anti-apoptotic adaptation, cancer cells can ensure their survival against anti-cancer therapies, especially chemotherapy. As such, silencing anti-apoptotic genes by RNA interference (RNAi) technology is another approach to reduce or prevent chemoresistance in multiple cancers, and such strategy could greatly improve the antitumor efficacy. Among multiple genes evaluated in this process, Bcl-2 survivin and BAX are the most common targets for gene silencing with the combination of chemotherapy for cancer treatment. For example, Nakamura et al. introduced a novel approach with both 5-fluorouracil (5-FU) and Bcl-2 siRNA assembled in PEGylated nanoparticle for treating human colorectal cancer. Since the overexpression of BCL-2 primarily accounts for the 5-FU resistance in this cancer, knockdown of its expression by siRNA would sensitize the cancer cells to chemotherapy and thus extend the patients' survival (Nakamura et al. 2011). As expected, in the xenograft animal model, the combination treatment greatly inhibited tumor growth when compared to the monotreatment. Apart from Bcl-2, survivin is another anti-apoptotic gene that is highly expressed in various carcinomas and has been extensively explored as a therapeutic target in combination therapy for improved antitumor results. In head and neck squamous cell carcinoma (HNSCC) therapy, Khan et al. combined cisplatin, a chemotherapeutics specific to HNSCC, with survivin siRNA, to achieve the enhanced cell apoptosis (Zakir et al. 2012). The results demonstrated that down-regulation of survivin by RNAi technology significantly improved the sensitivity of tumor cells to cisplatin-based chemotherapy. In another case related to the ovarian cancer therapy, Chu et al. developed a non-viral supramolecular micelles delivery platform assembled by β-cyclodextrin (β-CD) and polyethylenimine (PEI) for the carrying of Paclitaxel (PTX) and shRNAs against survivin (Chen et al. 2016a). This shRNA/PTX combination was evidenced to effectively induce significant in vitro cell apoptosis and tumor suppression in tumor-bearing mice.

4.2.2 Combination of Chemotherapy and Nucleic Acids for Overcoming Multidrug Resistance (MDR)

Drug resistance has become a main and persistent impediment to the efficient therapy of primary and metastatic cancers. During the routine chemotherapy, cancer patients administered with chemotherapeutic drugs usually develop resistance to the chemodrugs, thus rendering to the failure of cancer treatment which finally results in the recurrence of cancer. To overcome the difficulties which are caused by chemoresistance, chemotherapeutic agents combined with other types of drugs have been extensively used for chemotherapy (Li et al. 2013; Hansen et al. 2017). However, growing evidence demonstrates that tumor cells also acquire resistance to multiple drugs. In the recent decade, numerous efforts have been made to explore molecular mechanisms involved in MDR. Based on the ever-growing knowledge, multiple methods associated with the combination of chemotherapeutics and nucleic acids that are able to downregulate the efflux transporters are developed, and such combined therapy of chemotherapeutics with nucleic acids is well-recognized as a promising way to overcome MDR, thus improving the outcomes of cancer treatment.

Previously, MDR was found to be highly associated with MDR protein 1 (MDR1), which was also named as P-glycoprotein (P-gp). P-gp is a well-characterized adenosine triphosphate (ATP)-binding cassette (ABC) transporter that can actively deliver various molecules across plasma membranes (Waghray and Zhang 2018). More importantly, P-gp has been demonstrated to be overexpressed in various cancer types, where it functions as an ATP-driven drug efflux pump. This specificity makes P-gp a prime target and numerous attempts have been made to inhibit its function and restore the efficacy of anti-cancer drugs. Among them, some successful cases have verified that effective knockdown of P-gp protein through gene silencing with corresponding RNAi can significantly increase the concentration of chemotherapeutic agents in tumor sites which can lead to the enhanced therapeutic efficacy (Patil et al. 2010; Jiang et al. 2010; Navarro et al. 2011; Yin et al. 2012). Notably, P-gp is also actively involved in the process of normal physiological detoxification and host protection throughout the human body (Szakács et al. 2006). Therefore, P-gp inhibition may occur in the normal tissue after treatment of P-gp targeting drugs, which could open sanctuary sites, such as the central nervous system to cytotoxic anti-cancer drugs, and result in severe side effects. For instance, P-gp modulators were found to significantly reduce the clearance capacity of chemotherapeutic drugs in tumor sites but meanwhile increase the severity or incidence of normal tissues to the adverse effects in the anti-cancer therapy. Due to the considerable overlap in their substrates, P-gp inhibition was also reported to interfere with drug metabolism in intestine and liver which are normally mediated by cytochrome P450-3A (CYP3A), thus giving rise to undesirable pharmacological activity in healthy organs and tissues (Benet et al. 2004). As such, cancer-tissue selectivity for P-gp with suppressive effects occur specifically in the cancer cells is highly required to minimize the undesirable side effects induced by potentially increased sensitivity of normal P-gp-expressing tissues to the anti-cancer drug.

In addition to the P-gp-associated MDR, other MDRs, including MRP1, MRP2, BCRP (ABCG2), and related cell signaling pathways, are also involved in the drug resistance during chemotherapy. For instance, the Notch signaling pathway is critical for tumor chemoresistance to the treatment agent platinum in ovarian cancer. Growing evidence has shown that the overexpression of Notch3 protein in cancer cells leads to the expansion of cancer stem cells (CSCs) and results in the elevated platinum chemoresistance (Capaccione and Pine 2013). Accordingly, application of either Notch-pathway inhibitors or Notch 3 RNAi strategies, such as γ-secretase inhibitor (GSI), could render tumors more sensitive to platinum chemotherapy, resulting in the more effective elimination of not only cancer cells but also CSCs by using the combination approach (McAuliffe et al. 2012).

4.3 Chemo-Immunotherapy

Chemotherapy that exploits chemical drugs has been widely identified as one of the gold-standard therapies in cancer therapy. Nevertheless, single chemotherapy is usually insufficient to overcome the complexity of cancers. Tumor recurrence is highly possible as oncogenic driver mutations and undetectable metastases are common in clinical cases. Moreover, MDR, which happens when tumors become resistant to specific chemodrugs with different targets and molecular structures, is also common in multiple cancer types. Therefore, combination strategy that can prevent cancer relapse and MDR is currently heavily evaluated to enhance therapeutic efficacy. In recent decades, immunotherapy combined with chemotherapy, named chemo-immunotherapy, has emerged as a novel and effective treatment against cancer. In this combination, chemotherapy is mainly applied to induce toxicity and death of tumor cells, while immunotherapy can guide prognostic and predictive mechanisms for patients under chemotherapy, thereby reducing the cancer relapse induced by the cancer invasion and metastasis (Rocco et al. 2019). Moreover, the chemotherapy could not solely kill the tumors by directly inducing apoptosis but also improve the antitumor immune responses which ICD primes. ICD is generally defined by the increased expression of certain molecules in cancer cells, such as the secretion of calreticulin (CRT), adenosine triphosphate (ATP), and high mobility group box 1 (HMGB1) (Wang et al. 2018b). These ICD-inducing therapeutics could not only eliminate Treg and MDSCs that result in immunosuppression but also render the cytotoxic effect of CTLs to cancer cells to a great extent, thus activating antitumor immunity to eradicate tumor cells (Asadzadeh et al. 2020). So far, the anti-cancer effect of both chemotherapy and immunotherapy largely relies on the efficiency of the delivery platform, indicating that the drug delivery system is a major unmet need in regard to achieving effective chemo-immunotherapy. In recent decades, the delivery platforms could offer shield to sensitive immunotherapeutic drugs incorporated and synergistically improved corresponding immunostimulating activity to kill tumor cells (Wang et al. 2018a). Meanwhile, nanomaterials-based drug delivery systems for chemotherapy could be beneficial to the administration of multi-drug release profiles and also reduce side effects of chemotherapeutic drugs to some extent, especially MDR (Madondo et al. 2016; Zitvogel et al. 2011).

Currently, various immunotherapeutic strategies such as immune checkpoint inhibitors (ICB)-based therapy, small immunostimulating agent molecules, and vaccine therapy have also been evaluated in the combination therapy with chemotherapeutic agents to synergistically overcome cancer through nanoplatforms. Among them, a combination of chemotherapeutic agents and ICBs in nanoplatforms is a common strategy and powerful option for improving antitumor immune effects. The translation of active T cells into exhausted T cells is considered as the most critical contributor to immune surveillance and escape on tumor cells. For the function recovery of T cells, ICBs, including programmed cell death protein 1 (PD-1), programmed cell death-ligand 1 (PD-L1), cytotoxic T lymphocyte antigen-4 (CTLA-4), and indoleamine 2,3-dioxygenase (IDO), have been extensively used to restore the antitumor immunity. Previously, multiple studies have demonstrated that neither monotreatment of ICBs nor single treatment of chemical drugs is sufficient enough to trigger desired antitumor responses. Therefore, harnessing nanoparticles that encapsulate both ICBs and chemotherapeutics have been developed, and these nanoplatforms can stabilize the delivery of cargos and render a controllable releasing profile of the encapsulated combined agents for prolonged therapeutic effects. As shown in Fig. 4.1, Kuai's group paired PD-1 and DOX for combinatory antitumor therapy through a synthetic lipoprotein-mimicking nanodiscs (Kuai et al. 2018). By virtue of this delivery system, DOX released from the nanodiscs triggered ICD of tumor cells and exerted pronounced antitumor efficacy, while PD-1 co-delivered induced complete tumor regression in around 90% of animals and prevented tumor recurrence in survivors. In addition to the application of PD-1 in chemo-immunotherapy, IDO inhibitor is also appealing in combinatory therapy as the incorporation of IDO inhibitors has been found to boost the efficacy of traditional chemotherapeutics dramatically. For example, Chen et al. developed a chemo-immunotherapeutic system where PTX was loaded in a nanocarrier with an IDO antagonist of NLG919 conjugated (Chen et al. 2016b). Results in multiple cancer mouse models showed that the co-administration of PTX and NLG919 could offer potent anti-cancer efficacy. These successful preclinical designs of dual delivery system provide generalizable frameworks and are believed to be able to extend to most of the chemotherapy and ICBs-based immunotherapy combination for future cancer treatment in the clinical stage.

Combining chemotherapeutic drugs with vaccination or immunostimulating agents is another promising strategy for improving antitumor efficiency. In general, introducing vaccination or immunostimulating agents results in a change of the tumor microenvironment and is considered to cooperate with chemodrugs for enhancing anti-cancer immune effects through improved APC and T cell activation. In a recent study, Lee et al. utilized potato virus X (PVX)-based vaccination with coadministration of DOX for combined antitumor therapy. As an immunotherapeutic for in situ vaccine, PVX could activate anti-cancer immunity and delay cancer progression (Lee et al. 2017). In the context of B16F10 melanoma, co-delivery of DOX with PVX presented enhanced antitumor effects when compared to a single treatment of PVX or DOX. Interestingly, coadministration of PVX with DOX showed a much better therapeutic effect than DOX linked with PVX (PVX – DOX), which, to great extent, overturn the conventional understanding that the most optimized therapeutic outcomes can be achieved through assembling multiple therapeutics with nanoplatforms. The paradigm shift in the combined chemo-immunotherapy indicated that the integration formation of immunotherapeutics with cytolytic chemotherapy still needs systematical investigation for maximizing synergy and efficacy. Besides, TLR agonists as an immunostimulating vaccination for activating DC maturation-mediated anti-tumor immunity have been reported to be able to cooperate with chemotherapeutics. Roy et al. developed a co-delivery system of PTX and TLR4 agonist based on the PLGA nanoparticle (TLNP) (Roy et al. 2013). In this chemo-immunotherapeutic modality (Fig. 4.2), the encapsulation rate of SP-LPS and PTX could reach over 60%. According to the in vivo outcomes, the combination

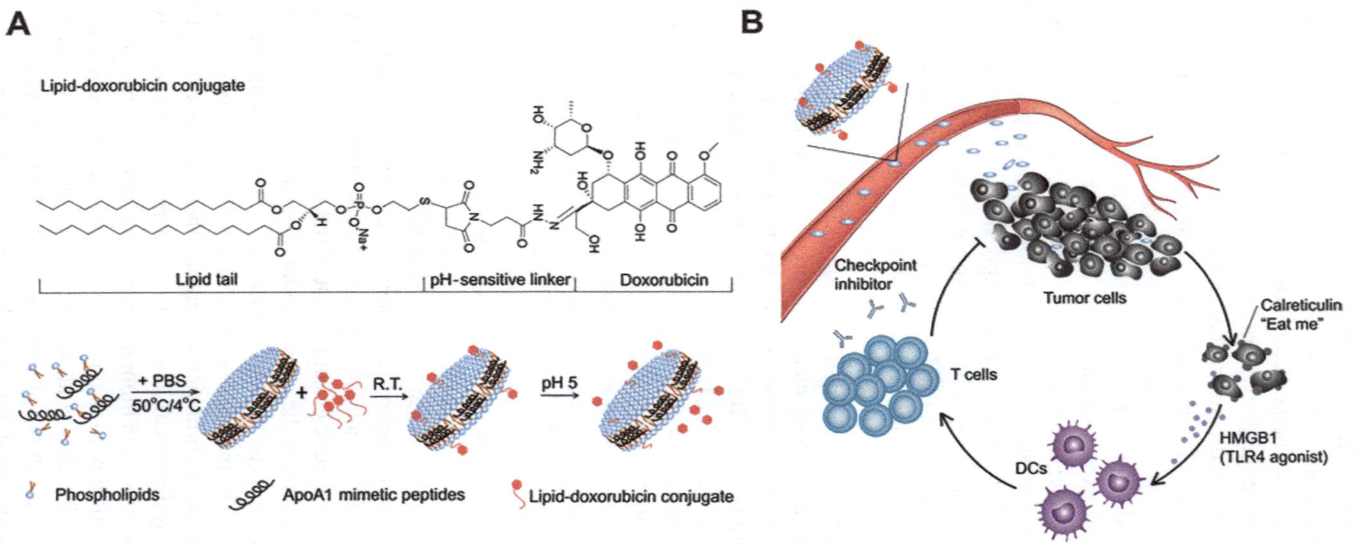

Fig. 4.1 (**a**) Schematic of synthetic lipoprotein-mimicking doxorubicin-loaded sHDL (sHDL-DOX) nanodiscs, which was formulated by incubation of lipid-DOX with preformed sHDL. (**b**) The assembled sHDL enabled intratumoral delivery of DOX, followed by internalization by tumor cells and pH-responsive release of DOX. With the synergy of sHDL-DOX and PD-1, an efficient elimination of established tumors and effective prevention of tumor relapse could be achieved. (Reused with permission from (Kuai et al. 2018) © 2022 American Association for the Advancement of Science)

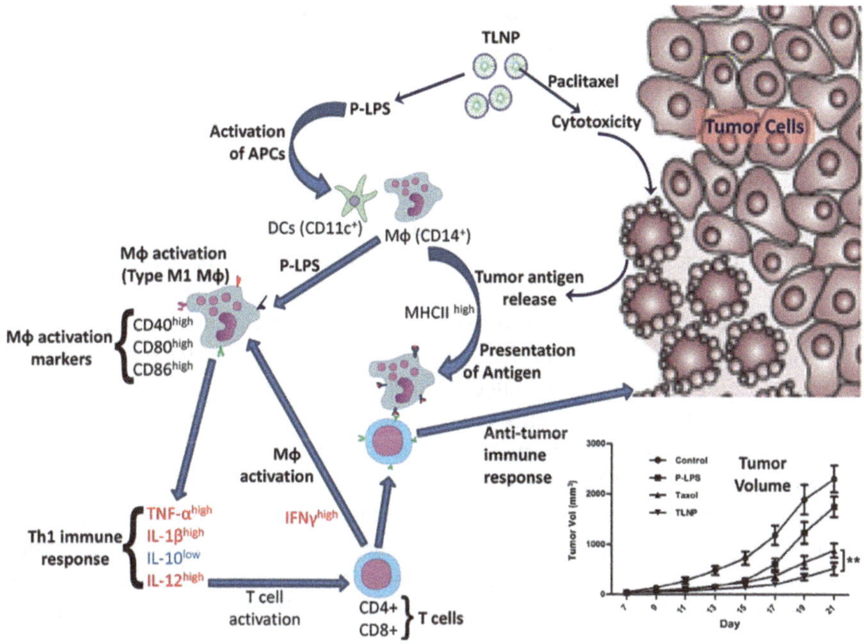

Fig. 4.2 Schematics of TLNP delivery system encapsulating PTX and TLR4 agonist, which led to significant tumor regression, as well as high infiltration and activation of APCs and T cells. (Reused with permission from (Roy et al. 2013) © 2013 Elsevier B.V)

could result in a greater than 40% cancer regression when compared to the single treatment of SP-LPS or PTX, as well as the enhanced APCs infiltration and T cells activation could be further confirmed by flow cytometric analysis of tumor-infiltrating immune cells. For an optimized synergy of incorporated immunostimulant agents and chemotherapeutics, Liu et al. further introduced dual pH-responsive nanocarrier to coadminister TLR7/8 agonist (R848) and DOX with controllable releasing profiles against breast cancer (Wilson et al. 2013). In this chemo-immunotherapeutic platform, a sequential and pH-dependent delivery was initiated when the assembled nanoparticles reached the tumor site, where the acid microenvironment (around pH 6.5) could drive the ionization of poly(L-histidine) (PHIS) and thus give rise to the release of drug to induce the maturation of DCs. Alternatively, the outer coating layer of HA-DOX conjugation was degraded in the acidic environment (pH $\approx$ 5.5), and then DOX was released to exert its anti-cancer effects inside the cells.

4.4 Combined Therapy of Immunotherapy with Photothermal Therapy (PTT) and Photodynamic Therapy (PDT)

Immunotherapy, as we discussed in Sect. 4.3, has shown tremendous progress in preclinical studies and clinical applications. However, mono-immunotherapy is far from adequate to achieve desirable therapeutic effectiveness for all kinds of cancers due to the intrinsic complexity and heterogeneity of some specific cancer types, low objective, and potential risk of immune-related side effects. Therefore, combined therapy has been widely involved to solve the current inefficiency of single immunotherapy. Other than chemical therapies as mentioned above, PTT and PDT are also widely used to improve the outcome in cancer therapy.

PTT refers to the methods using biomaterials which can absorb and then convert electromagnetic energy into thermal energy to trigger the tumor apoptosis or necrosis with hyperthermia (Zhou et al. 2012; Ju et al. 2017). Currently, substances termed photothermal agents (PTA) that can induce conversion of NIR to heat have been primarily applied in this therapy, and they can be classified into inorganic materials and organic materials. Due to the minimal invasiveness, small side effects, and good selectivity and controllability of these PTAs, PTT has been highly attractive in the recent decades. Introducing PTT into the combined therapy with immunotherapy not merely enhances the therapeutic efficacy of immunotherapy through several synergistic mechanisms, such as the promoted release of tumor-associated antigens (TAAs) and enhanced immune response by ICD, but also overcomes the challenges that single photothermal treatment cannot completely kill cancer cells with the metastasis or recurrence. As for increased production of TAAs, growing evidence has shown that hyperthermia triggered by PTT can directly induce the apoptosis of cancer cells, thereby largely produce TAAs that can be further uptaken and presented by APCs, which further contribute to the activation of a series of immune cells for anti-cancer immune responses. Currently, such anti-tumor immune responses via increased expression of TAAs have been widely combined with ICBs-based immunotherapy like PD-1/PD-L1 and CTLA-4 checkpoint blockades to achieve the repressive effect of tumor growth (Hu et al. 2020; Sun et al. 2019; Tian et al. 2017; Huang et al. 2020). For example, Liu et al. developed a synergistic immune-photothermal therapy, namely, SYMPHONY (Liu et al. 2017). In their study, anti-PD-L1 antibodies were combined with plasmonic gold nanostar (GNS) to realize a complete eradication of both primary tumor and distant untreated tumor sites in the mouse model studied. According to the in vivo results (Fig. 4.3), such a better anti-tumor effect of SYMPHONY as compared to mono-immunotherapy could be attributed to the mild hyperthermia triggered by GNS and thus resulted in a significantly improved proportion of CD4+ and CD8+ T cells to activate comprehensive anti-tumor responses in both local and distant sites. In another study, Wang et al. developed a single-walled nanotube (SWNT) to induce hyperthermia and combined this PTT with anti-CTLA-4 antibody therapy (Wang et al. 2014). As evaluated, incorporated SWNTs could result in the maturation of DCs and increase the

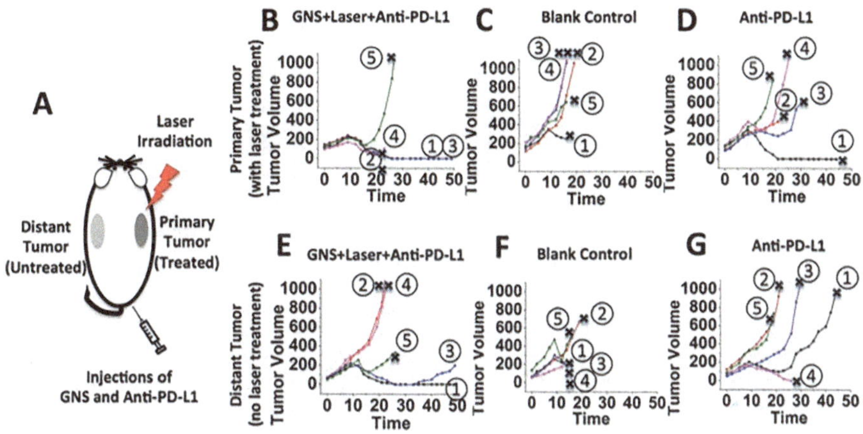

Fig. 4.3 (**a**) Schematics of SYMPHONY treatment. The right flank tumor was treated with PDT and the left flank tumor was not treated. (**b–d**) Tumor growth for primary tumors in the SYMPHONY group (GNS + Laser + Anti-PD-L1) (B), in the control group (**c**), and in the group with anti-PD-L1 alone (**d**). (**e–g**) Tumor growth for distant tumors in the SYMPHONY group (GNS + Laser + Anti-PD-L1) (**e**), in the control group (**f**), and in the group with anti-PD-L1 alone (**g**). The laser irradiation of photothermal therapy was performed only on the primary tumors. The line stopped (Signorell et al. 2018) if the mouse was sacrificed. (Reused with permission from (Liu et al. 2017) © 2017 Springer Nature)

secretion of IL-1 β and IL-12P. Moreover, the release of other anti-tumor cytokines such as TNF-α and IL-6 was also found to be upregulated. With the concomitant injection of anti-CTLA-4 antibodies, tumor size at distant sites was significantly decreased in the combination therapy group when compared to other single treatment. Their cytometry results (Fig. 4.4) further showed that the combination group could not only promote the infiltration of T cells and B cells in the distant tumor tissue but also increase the rate of active CD4+/CD8+ T cells to Treg cells. Interestingly, the combination therapy was also demonstrated to have high efficacy in a lung metastasis animal model as it reduced the number of metastasis nodes and promoted the animal survival rate, which indicated that SWNT-based PTT in combination with anti-CTLA-4 antibody worked effectively on not only primary tumor but also metastasis tumor.

Similar to PTT, PDT, which refers to methods using photosensitizers to activate a certain wavelength of excitation light (such as NIR) with molecular oxygen to produce singlet oxygen or other ROS for tumor apoptosis and necrosis, also holds great promise for treating multiple cancers (Chatterjee et al. 2008). Currently, nanotechnology-based PDT has emerged as a useful therapeutic approach, and it involves nanocarriers of photosensitizers to achieve high loading capacity and unprecedented optical performance. Several nanomaterials, like Au and carbon-related nanostructures that represent extraordinary PTT effect against the tumor through the production of singlet oxygen, are also promising in PDT. According to previous reports, PDT could be more immunostimulatory than PTT. For example,

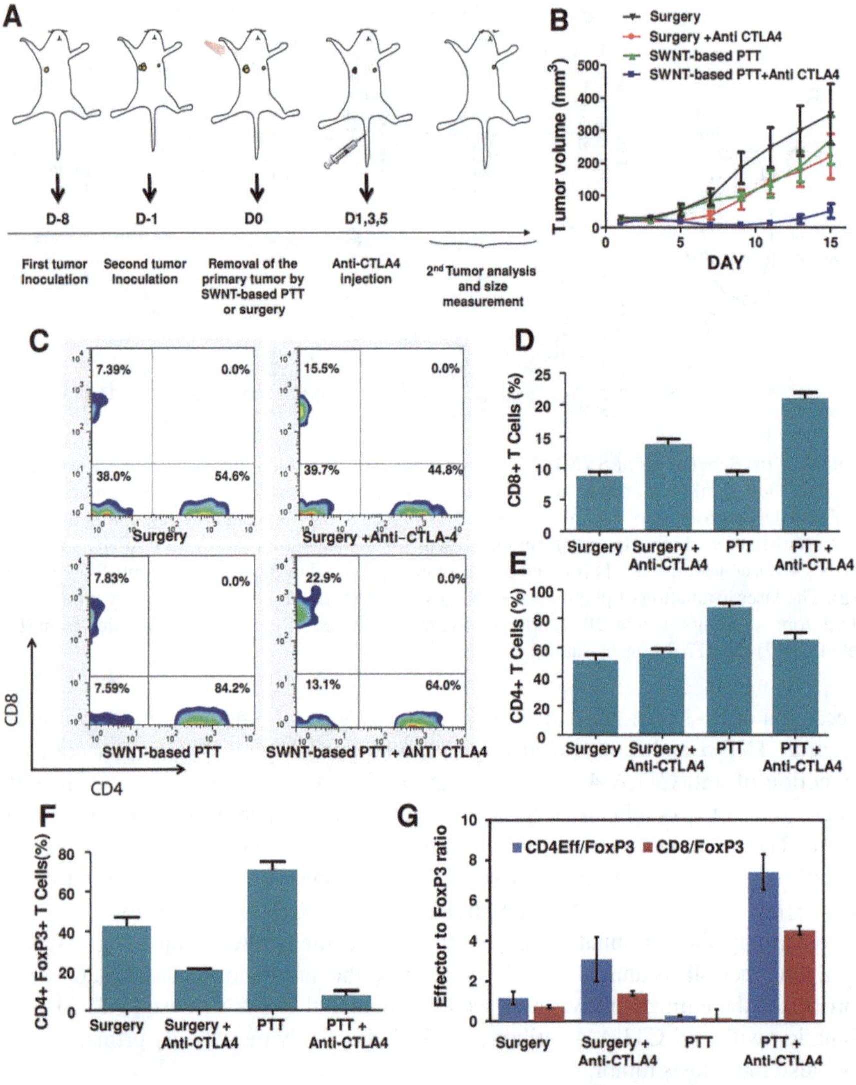

Fig. 4.4 Anti-tumor effect of SWNT-based PTT plus anti-CTLA-4 therapy. (**a**) Schematic illustration of SWNT-based PTT and anti-CTLA-4 combination therapy. (**b**) Tumor growth curves of different cohorts of mice after treatments indicated (n = 7–8 per group). (**c**) Quantification of T cells in secondary tumors. (**d, e**) Proportion of tumor-infiltrating CD8+ T cells (**d**) and CD4+ T cells (**e**). (**f**) Proportion of tumor-infiltrating CD4+ FoxP3+ T cells. (**g**) Ratios of tumor-infiltrating CD8+ T cells and effective CD4+ T cells over regulatory T cells in the secondary tumors upon various treatments to remove the first tumors. The combination treatment increased both the ratio of CD8+ T cells to Tregs and the ratio of effective CD4+ T cells to Tregs. (Reused with permission from (Wang et al. 2014) © 2014 WILEY-VCH Verlag GmbH & Co. KGaA, Weinheim)

Kabingu et al. found that PDT therapy could effectively inhibit the growth of primary tumors; meanwhile, the untreated tumors in the distant site were also killed because of the accompanied anti-tumor cytolytic activity triggered by increased CD8+ T cells (Kabingu et al. 2007). In another study, Yu et al. reported that using graphene oxide coated (HPPH)-PEG-HK as the nanocarrier for the PS HPPH delivery could induce strong anti-cancer effects through the activation of both DCs and CD8+ T cells, and its therapeutic efficacy was demonstrated in both primary tumors and lung metastasis models (Yu et al. 2017). Despite these exciting examples of single PDT design against tumors, PDT alone is facing the same challenges with PTT treatment as both have been identified to present limited stimulation and activation effects on either innate or adaptive immunity. Hence, combination of immunotherapy with PDT treatment was also evaluated to increase the treatment efficiency in recent years. As mentioned before, immune-modulating drugs, especially ICBs, could reverse the negative regulatory signals of the immune system when tumors occur. Thus, ICB agents have been utilized in the combined therapy to promote original anti-cancer efficiency induced by PDT. Recently, Xu et al. demonstrated that PDT/ICBs combined therapy, which was constructed by upconversion nanoparticles (UCNPs) with photosensitizer (chlorin e6(Ce6)) and imiquimod (R837), could effectively improve antitumor immunity and thus inhibit the progression of distant untreated tumors after PDT therapy (Xu et al. 2017). As shown in Fig. 4.5, this PDT/ICBs system was reported to remarkably improve DCs activation and maturation in PDT therapy with the involvement of CTLA-4 checkpoint blockade and presented efficient antitumor immunity that could inhibit both local and distant tumor growth. More interestingly, such a combined strategy could introduce immune memory cells during the therapy, and this immune memory could protect the body from tumor recurrence.

Although multiple preclinical applications of the combined immunotherapy/PTT/PDT have been developed and evaluated in antitumor research, there are still many concerns about such a strategy with only few cases have been applied clinically. The key challenge for the combined therapy is the safety issue of nanomaterials which is introduced for eliciting hyperthermia of PTT and oxygen or other ROS of PDT for tumor apoptosis and necrosis. As is known to all, the majority of these nanomaterials are synthetic inorganic nanoparticles, and there exist a high possibility of toxicity caused by undesirable degradability and long-term accumulation in the tissues. Previously, CYT-6091, a formulation of gold nanoparticles functionalized with recombinant human TNF-α (rhTNF-α), had been studied in a phase I clinical trial where more than 20 patients with advanced-stage cancer were enrolled (Libutti et al. 2010). The outcome demonstrated that rhTNF-α incorporated CYT-6091 could achieve the robust cancer targeting with minimum side effects after systemical administration. Though promising results are obtained from current clinical trials using some inorganic nanoparticles, their application in clinic is controversial and definitely needs more investigation. Moreover, a depth understanding of the relationship between tumor and immune systems, as well as systemic evaluation of every component in the combined therapy, is essential to support the

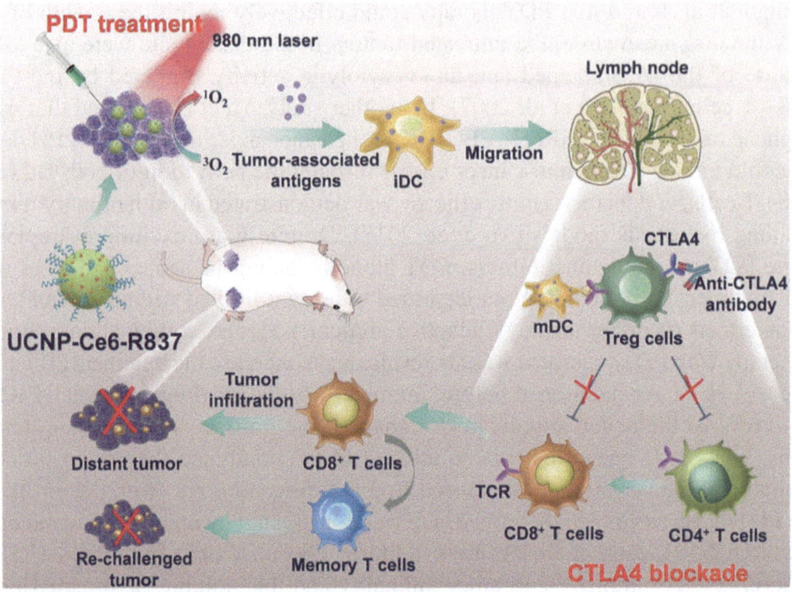

Fig. 4.5 Schematic of the mechanisms of combining NIR-mediated PDT with CTLA-4 checkpoint blockade for cancer immunotherapy. The generated tumor-associated antigens in the presence of those nanoparticles as the adjuvant effectively improved antitumor immunity. With CTLA-4 checkpoint blockade, primary local tumors were eliminated under direct NIR exposure, while distant tumors were also suppressed after the PDT treatment. (Reused with permission from (Xu et al. 2017) © 2017 American Chemical Society)

optimization of the synergetic approaches and establish safer and more effective multidisciplinary treatments for cancer.

4.5 Conclusion

Presently, combination therapy for cancer treatment is in full bloom, and a growing number of organizations of new or old anti-cancer therapeutics have been developed and screened in different combination regimens. Undoubtedly, combination therapies described in this chapter, including the combination of chemical and nucleic acid drugs, chem-immunotherapy, and combined therapy of immunotherapy with phototherapy, have demonstrated their feasibility, safety, and advantages in cancer treatment. Meanwhile, these combined therapies have shown the ability to address the treatment challenges caused by the intrinsic heterogeneity and complexity of cancer, especially MDR and high metastatic potential. Notably, several studies found not all combination strategies improved therapeutic effect when compared to the monotreatment, and efficiency of the combined therapies was highly relied on the delivery administration, such as optimum sequencing and timing. This requires more investigations on the mechanism of drug interaction involved in different combination regimens, as well as the development of appropriately designed drug delivery system.

Acknowledgments This work was supported by the National Natural Science Foundation of China (21HAA01203 (ZY)) and Department of Science and Technology of Jilin Province (20230402042GH (ZY)). The authors thank Jiayi Chen from School of Life Sciences, Jilin University for editorial assistance.

References

Albain KS, Nag SM, Calderillo-Ruiz G et al (2008) Gemcitabine plus paclitaxel versus paclitaxel mono-therapy in patients with metastatic breast cancer and prior Anthracycline treatment. J Clin Oncol 26:3950–3957. https://doi.org/10.1200/JCO.2007.11.9362

Asadzadeh Z, Safarzadeh E, Safaei S et al (2020) Current approaches for combination therapy of cancer: the role of immunogenic cell death. Cancers 12:1047. https://doi.org/10.3390/cancers12041047

Benet LZ, Cummins CL, Wu CY (2004) Unmasking the dynamic interplay between efflux transporters and metabolic enzymes. Int J Pharm 277:3–9. https://doi.org/10.1016/j.ijpharm.2002.12.002

Blagosklonny MV (2005) Overcoming limitations of natural anti-cancer drugs by combining with artificial agents. Trends Pharmacol Sci 26:77–81. https://doi.org/10.1016/j.tips.2004.12.002

Burrell RA, McGranahan N, Bartek J et al (2013) The causes and consequences of genetic heterogeneity in cancer evolution. Nature 501:338–345. https://doi.org/10.1038/nature12625

Capaccione KM, Pine SR (2013) The notch signaling pathway as a mediator of tumor survival. Carcinogenesis 34:1420–1430. https://doi.org/10.1093/carcin/bgt127

Chatterjee DK, Fong LS, Zhang Y (2008) Nanoparticles in photodynamic therapy: an emerging paradigm. Adv Drug Deliv Rev 60:1627–1637. https://doi.org/10.1016/j.addr.2008.08.003

Chen Y, Huang Y, Qin D et al (2016a) β-Cyclodextrin-based inclusion complexation bridged biodegradable self-assembly macromolecular micelle for the delivery of paclitaxel. PLoS One 11:e0150877. https://doi.org/10.1371/journal.pone.0150877

Chen Y, Xia R, Huang Y et al (2016b) An immunostimulatory dual-functional nanocarrier that improves cancer immunochemotherapy. Nat Commun 7:13443. https://doi.org/10.1038/ncomms13443

Fan WP, Yung B, Huang P et al (2017) Nanotechnology for multimodal synergistic cancer therapy. Chem Rev 117:13566–13638. https://doi.org/10.1021/acs.chemrev.7b00258

Fisher DE (2001) The p53 tumor suppressor: critical regulator of life & death in cancer. Apoptosis 6:7–15. https://doi.org/10.1023/A:1009659708549

Garattini S (2003) New approaches to cancer therapy. Ann Oncol 14:813–816. https://doi.org/10.1093/annonc/mdg261

Gilman A (1963) The initial clinical trial of nitrogen mustard. Am J Surg 105:574–578. https://doi.org/10.1016/0002-9610(63)90232-0

Hanahan D, Weinberg Robert A (2011) Hallmarks of cancer: the next generation. Cell 144:646–674. https://doi.org/10.1016/j.cell.2011.02.013

Hansen E, Woods RJ, Read AF (2017) How to use a chemotherapeutic agent when resistance to it threatens the patient. PLoS Biol 15:e2001110. https://doi.org/10.1371/journal.pbio.2001110

Hu C, Cai L, Liu S et al (2020) Copper-doped nanoscale covalent organic polymer for augmented photo/Chemodynamic synergistic therapy and immunotherapy. Bioconjug Chem 31:1661–1670. https://doi.org/10.1021/acs.bioconjchem.0c00209

Huang T-Y, Huang G-L, Zhang C-Y et al (2020) Supramolecular photothermal nanomedicine mediated distant tumor inhibition via PD-1 and TIM-3 blockage. Front Chem 8. https://doi.org/10.3389/fchem.2020.00001

Jiang J, Yang S-j, Wang J-c et al (2010) Sequential treatment of drug-resistant tumors with RGD-modified liposomes containing siRNA or doxorubicin. Eur J Pharm Biopharm 76:170–178. https://doi.org/10.1016/j.ejpb.2010.06.011

Ju YM, Zhang HL, Yu J et al (2017) Monodisperse au-Fe2C Janus nanoparticles: an attractive multifunctional material for triple-modal imaging-guided tumor photothermal therapy. ACS Nano 11:9239–9248. https://doi.org/10.1021/acsnano.7604461

Kabingu E, Vaughan L, Owczarczak B et al (2007) CD8+ T cell-mediated control of distant tumours following local photodynamic therapy is independent of CD4+ T cells and dependent on natural killer cells. Br J Cancer 96:1839–1848. https://doi.org/10.1038/sj.bjc.6603792

Kuai R, Yuan WM, Son S et al (2018) Elimination of established tumors with nanodisc-based combination chemoimmunotherapy. Sci Adv 4:eaao1736. https://doi.org/10.1126/sciadv.aao1736

Lebaron S, Zeltzer LK, Lebaron C et al (1988) Chemotherapy side effects in pediatric oncology patients: drugs, age, and sex as risk factors. Med Pediatr Oncol 16:263–268. https://doi.org/10.1002/mpo.2950160408

Lee KL, Murray AA, Le DHT et al (2017) Combination of plant virus nanoparticle-based in situ vaccination with chemotherapy potentiates antitumor response. Nano Lett 17:4019–4028. https://doi.org/10.1021/acs.nanolett.7600107

Li J, Wang Y, Zhu Y et al (2013) Recent advances in delivery of drug–nucleic acid combinations for cancer treatment. J Control Release 172:589–600. https://doi.org/10.1016/j.jconrel.2013.04.010

Libutti SK, Paciotti GF, Byrnes AA et al (2010) Phase I and pharmacokinetic studies of CYT-6091, a novel PEGylated colloidal gold-rhTNF nanomedicine. Clin Cancer Res 16:6139–6149. https://doi.org/10.1158/1078-0432.ccr-10-0978

Liu S, Guo Y, Huang R et al (2012) Gene and doxorubicin co-delivery system for targeting therapy of glioma. Biomaterials 33:4907–4916. https://doi.org/10.1016/j.biomaterials.2012.03.031

Liu Y, Maccarini P, Palmer GM et al (2017) Synergistic immuno photothermal nanotherapy (SYMPHONY) for the treatment of unresectable and metastatic cancers. Sci Rep 7:8606. https://doi.org/10.1038/s41598-017-09116-1

Ma Y, Yang Z, Huntoon K et al (2021) Advanced immunotherapy approaches for glioblastoma. Adv Ther 4:2100046. https://doi.org/10.1002/adtp.202100046

Madondo MT, Quinn M, Plebanski M (2016) Low dose cyclophosphamide: mechanisms of T cell modulation. Cancer Treat Rev 42:3–9. https://doi.org/10.1016/j.ctrv.2015.11.005

Marchetti S, Schellens JHM (2007) The impact of FDA and EMEA guidelines on drug development in relation to phase 0 trials. Br J Cancer 97:577–581. https://doi.org/10.1038/sj.bjc.6603925

McAuliffe SM, Morgan SL, Wyant GA et al (2012) Targeting Notch, a key pathway for ovarian cancer stem cells, sensitizes tumors to platinum therapy. Proc Natl Acad Sci USA 109:E2939. https://doi.org/10.1073/pnas.1206400109

Mokhtari RB, Kumar S, Islam SS et al (2013) Combination of carbonic anhydrase inhibitor, acetazolamide, and sulforaphane, reduces the viability and growth of bronchial carcinoid cell lines. BMC Cancer 13:378. https://doi.org/10.1186/1471-2407-13-378

Mokhtari RB, Homayouni TS, Baluch N et al (2017) Combination therapy in combating cancer. Oncotarget 8(23):38022–38043. https://doi.org/10.18632/oncotarget.16723

Nakamura K, Abu Lila AS, Matsunaga M et al (2011) A double-modulation strategy in cancer treatment with a chemotherapeutic agent and siRNA. Mol Ther 19:2040–2047. https://doi.org/10.1038/mt.2011.174

Navarro G, Sawant RR, Biswas S et al (2011) P-glycoprotein silencing with siRNA delivered by DOPE-modified PEI overcomes doxorubicin resistance in breast cancer cells. Nanomedicine 7:65–78. https://doi.org/10.2217/nnm.11.93

Ouyang X, Shi M, Jie F et al (2018) Phase III study of dulanermin (recombinant human tumor necrosis factor-related apoptosis-inducing ligand/Apo2 ligand) combined with vinorelbine and

cisplatin in patients with advanced non-small-cell lung cancer. Invest New Drugs 36:315–322. https://doi.org/10.1007/s10637-017-0536-y
Partridge AH, Burstein HJ, Winer EP (2001) Side effects of chemotherapy and combined Chemohormonal therapy in women with early-stage breast cancer. JNCI Monogr 2001:135–142. https://doi.org/10.1093/oxfordjournals.jncimonographs.a003451
Patil YB, Swaminathan SK, Sadhukha T et al (2010) The use of nanoparticle-mediated targeted gene silencing and drug delivery to overcome tumor drug resistance. Biomaterials 31:358–365. https://doi.org/10.1016/j.biomaterials.2009.09.048
Qin SY, Cheng YJ, Lei Q et al (2018) Combinational strategy for high-performance cancer chemotherapy. Biomaterials 171:178–197. https://doi.org/10.1016/j.biomaterials.2018.04.027
Rocco D, Della Gravara L, Battiloro C et al (2019) The role of combination chemo-immunotherapy in advanced non-small cell lung cancer. Expert Rev Anticancer 19:561–568. https://doi.org/10.1080/14737140.2019.1631800
Roy A, Singh MS, Upadhyay P et al (2013) Nanoparticle mediated co-delivery of paclitaxel and a TLR-4 agonist results in tumor regression and enhanced immune response in the tumor microenvironment of a mouse model. Int J Pharm 445:171–180. https://doi.org/10.1016/j.ijpharm.2013.01.045
Shang T, Yu X, Han S et al (2020) Nanomedicine-based tumor photothermal therapy synergized immunotherapy. Biomater Sci 8:5241–5259. https://doi.org/10.1039/D0BM01158D
Shi X, Li C, Gao S et al (2014) Combination of doxorubicin-based chemotherapy and polyethylenimine/p53 gene therapy for the treatment of lung cancer using porous PLGA microparticles. Colloids Surf B Biointerfaces 122:498–504. https://doi.org/10.1016/j.colsurfb.2014.07.020
Signorell RD, Luciani P, Brambilla D et al (2018) Pharmacokinetics of lipid-drug conjugates loaded into liposomes. Eur J Pharm Biopharm 128:188–199. https://doi.org/10.1016/j.ejpb.2018.04.003
Snajdauf M, Havlova K, Vachtenheim J et al (2021) The TRAIL in the treatment of human cancer: an update on clinical trials. Front Mol Biosci 8. https://doi.org/10.3389/fmolb.2021.628332
Sun W, Du Y, Liang X et al (2019) Synergistic triple-combination therapy with hyaluronic acid-shelled PPy/CPT nanoparticles results in tumor regression and prevents tumor recurrence and metastasis in 4T1 breast cancer. Biomaterials 217:119264. https://doi.org/10.1016/j.biomaterials.2019.119264
Szakács G, Paterson JK, Ludwig JA et al (2006) Targeting multidrug resistance in cancer. Nat Rev Drug Discov 5:219–234. https://doi.org/10.1038/nrd1984
Tian Y, Zhang Y, Teng Z et al (2017) pH-dependent transmembrane activity of peptide-functionalized gold Nanostars for computed tomography/photoacoustic imaging and Photothermal therapy. ACS Appl Mater Interfaces 9:2114–2122. https://doi.org/10.1021/acsami.6b13237
Waghray D, Zhang Q (2018) Inhibit or evade multidrug resistance P-glycoprotein in cancer treatment. J Med Chem 61:5108–5121. https://doi.org/10.1021/acs.jmedchem.7b01457
Wang C, Xu LG, Liang C et al (2014) Immunological responses triggered by Photothermal therapy with carbon nanotubes in combination with anti-CTLA-4 therapy to inhibit cancer metastasis. Adv Mater 26:8154–8162. https://doi.org/10.1002/adma.201402996
Wang Q, Ju X, Wang J et al (2018a) Immunogenic cell death in anti-cancer chemotherapy and its impact on clinical studies. Cancer Lett 438:17–23. https://doi.org/10.1016/j.canlet.2018.08.028
Wang Y-J, Fletcher R, Yu J et al (2018b) Immunogenic effects of chemotherapy-induced tumor cell death. Genes Dis 5:194–203. https://doi.org/10.1016/j.gendis.2018.05.003
Wilson JT, Keller S, Manganiello MJ et al (2013) pH-responsive nanoparticle vaccines for dual-delivery of antigens and Immunostimulatory oligonucleotides. ACS Nano 7:3912–3925. https://doi.org/10.1021/nn305466z
Xu Q, Xia Y, Wang C-H et al (2012) Monodisperse double-walled microspheres loaded with chitosan-p53 nanoparticles and doxorubicin for combined gene therapy and chemotherapy. J Control Release 163:130–135. https://doi.org/10.1016/j.jconrel.2012.08.032

Xu J, Xu LG, Wang CY et al (2017) Near-infrared-triggered photodynamic therapy with multitasking Upconversion nanoparticles in combination with checkpoint blockade for immunotherapy of colorectal cancer. ACS Nano 11:4463–4474. https://doi.org/10.1021/acsnano.7b00715

Yang Z, Ma Y, Zhao H et al (2020) Nanotechnology platforms for cancer immunotherapy. Wires Nanomed Nanotechnol 12:e1590. https://doi.org/10.1002/wnan.1590

Yin Q, Shen J, Chen L et al (2012) Overcoming multidrug resistance by co-delivery of Mdr-1 and survivin-targeting RNA with reduction-responsible cationic poly(β-amino esters). Biomaterials 33:6495–6506. https://doi.org/10.1016/j.biomaterials.2012.05.039

Yu XH, Gao D, Gao LQ et al (2017) Inhibiting metastasis and preventing tumor relapse by triggering host immunity with tumor-targeted photodynamic therapy using photosensitizer-loaded functional Nanographenes. ACS Nano 11:10147–10158. https://doi.org/10.1021/acsnano.7b04736

Zakir K, Ram PT, Noor K et al (2012) Induction of apoptosis and sensitization of head and neck squamous carcinoma cells to cisplatin by targeting Survivin gene expression. Curr Gene Ther 12:444–453. https://doi.org/10.2174/156652312803519805

Zhou F, Wu S, Song S et al (2012) Antitumor immunologically modified carbon nanotubes for photothermal therapy. Biomaterials 33:3235–3242. https://doi.org/10.1016/j.biomaterials.2011.12.029

Zitvogel L, Kepp O, Kroemer G (2011) Immune parameters affecting the efficacy of chemotherapeutic regimens. Nat Rev Clin Oncol 8:151–160. https://doi.org/10.1038/nrclinonc.2010.223

Chapter 5
Prodrug Nanocarrier

Kaicheng Tang, Shuang Wu, Xiaoyou Wang, and Chong Li

Abstract Despite the various types of nanocarriers, the widespread application of nanocarrier-based drug delivery systems is usually limited by problems such as drug burst release, low drug loading efficiency, and poor stability. The prodrug strategy specifically aims to modify the structures of the drugs themselves, and the majority of existing prodrugs have achieved clinical success, with new ones emerging. Meanwhile, the prodrug nanocarrier strategy integrates prodrugs with nanocarriers to enable effective drug loading and targeted delivery with controlled drug release both in vitro and in vivo, which is critical for improving drug efficacy. Numerous prodrug nanocarriers have been developed in response to the pathogenic features of various diseases, particularly in the treatment of tumors. This chapter primarily summarizes recent research progress in prodrug nanocarriers, including construction, stimuli-responsive strategies, conclusions, and future perspectives.

Keywords Prodrug nanocarrier · Construction strategy · Stimuli-responsive strategy · Cancer treatment

---

K. Tang
Medical Research Institute, College of Pharmaceutical Sciences, Southwest University, Chongqing, China

School of Pharmacy, Sichuan Industrial Institute of Antibiotics, Chengdu University, Chengdu, China

S. Wu · X. Wang · C. Li (✉)
Medical Research Institute, College of Pharmaceutical Sciences, Southwest University, Chongqing, China
e-mail: chongli@swu.edu.cn

L. Teng et al. (eds.), *Drug Delivery to Tumors*,
https://doi.org/10.1007/978-981-19-8930-8_5

5.1 Introduction

The therapeutic effectiveness of most traditional drug preparations is limited due to low tissue selectivity, poor cell membrane penetration, and uncontrolled drug release (Kamaly et al. 2016; Lin et al. 2019). Gradually, novel preparations have attracted a noticeable amount of attention from researchers, and nanocarriers are the most extensively studied. Nanocarriers are a type of drug carrier belonging to the nanolevel microscopic category, with a size between 10 ~ 1000 nm, and are generally composed of natural or synthetic polymer materials (He et al. 2019; Liu and Dong 2020; Li et al. 2019a; Jiao et al. 2019; Ariaeenejad et al. 2021). However, the broad application of nanocarrier drug delivery systems (DDS) in the clinic remains limited, and the reasons are mainly derived from the following three aspects: (1) Drug burst release. Some drug delivery systems may have a "burst release" behavior in the body that will produce high plasma concentrations beyond expectations and cause toxic side effects. (2) Low drug loading efficiency. Nanocarriers usually need a large amount of carrier materials and have low drug loading, which leads to a failure of adequate lesion concentration with insufficient curative effects. (3) Poor stability. After entering the complex body system, particularly those that are unfavorable to drug efficacy, some drugs are easily and quickly metabolized into inactive forms (Delplace et al. 2014; Vega-Villa et al. 2008; Patel and Misra 1999; Signorell et al. 2018; Plourde et al. 2017; Parmentier et al. 2011).

Prodrugs generally refer to drugs that are inactive or less active in vitro and are transformed into active drugs in vivo, which can increase the bioavailability and reduce the toxicity and side effects of drugs (Souza et al. 2019; Xiao et al. 2012; Zhang et al. 2017a, b; Hajnal et al. 2016; Li et al. 2014). This type of strategy is specifically aimed at modifying the structure of the drug itself; many prodrugs have enjoyed clinical success, and new ones are still emerging (Table 5.1). Prodrug nanocarriers have been evaluated by various researchers to further overcome the limitations of nanocarriers and improve the targeting ability of drug delivery systems based on the prodrug design concept (Lin et al. 2019). In short, the prodrug nanocarrier strategy integrates the prodrug strategy with the nanocarrier to achieve effective drug loading and accurate delivery. Researchers have developed a variety of prodrug nanocarrier systems, which have demonstrated promising prospects for use, especially in cancer treatment. The main types of nanocarriers are polymer nanoparticles (Kesharwani et al. 2014; He et al. 2019; Sur et al. 2019; Li et al. 2019a, b, c; Jiao et al. 2019; Xu et al. 2019), liposomes (Li et al. 2019a, b, c; Van Tran et al. 2019; Jain et al. 2019), micelles (Sun et al. 2016; Rezaei et al. 2020; Dai et al. 2017), nanoemulsions (Sánchez-López et al. 2019; Sheth et al. 2020; Zhang et al. 2020), which are made up primarily of materials that are well biocompatible and degradable and compatible with the clinical drugs development (Zaro 2014; Mura et al. 2015; Yin et al. 2013; Yu et al. 2018a, b; Shi et al. 2020; Hernández-Gil et al. 2015; AbouAitah et al. 2018). For example, poly(lactic-co-glycolic acid) (PLGA) is approved by the FDA as an organic material that meets the above conditions and is

Table 5.1 Some prodrugs were approved by the FDA during 2015–2018

No.	Tradename	Mechanism	Indication/use	Company	Approval date
1	Cresemba	Convert isavuconazonium to active isavuconazole by plasma lipase and inhibit ergosterol formation	Invasive aspergillosis and mucormycosis	Basilea and Astellas	2015
2	Aristada	Convert aripiprazole lauroxil to aripiprazole by hydrolysis	Adult schizophrenia	Alkermes	2015
3	Uptravi	Uptravi is hydrolyzed by carboxylesterase 1 to produce an active metabolite	Adult pulmonary hypertension	Actelion	2015
4	Xuriden	Xuriden can be metabolized in the body to produce ureidoside	Hereditary orotic aciduria	Wellstat	2015
5	Vemlidy	Vemlidy forms the intermediate product Ala-TFV by carboxylesterase 1, and Ala-TFV is converted into TFV, and then TFV is phosphorylated into TFV diphosphate	Hepatitis B virus (HBV)	Gilead	2016
6	Ingrezza	Inhibit vesicular monamine transporter-2 (VMAT 2)	Tardive dyskinesia	Neurocrine Biosciences	2017
7	Tavalisse	Tavalisse is converted into the active metabolite R406 in the intestine to inhibits spleen tyrosine kinase activity	Immune thrombocytopenia	Rigel	2018
8	Akynzeo	Dephosphorylation by plasma phosphatase	Nausea and vomiting	Helsinn	2018
9	Krintafel	Bioconversion by CYP2D6	Plasmodium vivax malaria	GSK	2018

widely applied in the preparation of prodrug nanocarriers (Danhier et al. 2012; Mir et al. 2017; Lu et al. 2020; Ding and Zhu 2018; Martins et al. 2018). In addition, there are various methods to construct prodrug nanocarriers, and drugs are expected to play different roles in the construction of prodrug nanocarriers. The types of endogenous or exogenous stimuli that can be selected to promote the transformation or release of active drugs are diverse. Therefore, prodrug nanocarriers can be designed to ensure optimal delivery of drugs according to the specific pathological and physiological characteristics of the disease, e.g., lower pH, hypoxia, higher interstitial pressure, and altered enzyme levels, are key features of the tumor microenvironment (TME) and are frequently devoted to designing responsive prodrug delivery systems (Delplace et al. 2014; Fang and Al-Suwayeh 2012; Luo et al. 2014; Taresco et al. 2018). This chapter will focus on the construction methods and stimulus-response types of prodrug nanocarriers.

5.2 Construction of Prodrug Nanocarriers

The construction of prodrug nanocarriers can be divided into three categories: (1) connecting two or more drugs using covalent or noncovalent interactions that can be termed "carrier-free" or "drug–drug" DDS (Sun et al. 2019a, b, c; Zhou et al. 2020; Zhong et al. 2016; AbouAitah et al. 2018). Generally, it includes hydrophobic and hydrophilic drugs to form an amphiphilic composite that can self-assemble to form nanoparticles for drug delivery; (2) covalently coupling the drug with the monomer/polymer to generate nanoparticles, which can be named "drug-monomer/polymer." Most of these conjugates are also amphiphilic, which can self-assemble for drug encapsulation; and (3) binding drugs to the matrix or surface of pre-prepared polymeric nanocarriers that can be named "drug-carrier." The combination was mainly based upon physical or chemical interactions, such as van der Waals forces or disulfide bonds. Irrespective of the form of prodrug nanocarriers, when they encounter a specific endogenous or exogenous stimulus, the prodrug can be responsively released and exert its efficacy.

5.2.1 *Drug–Drug*

Most of the current methods for the construction of polymer nanocarriers are self-assembly. Self-assembly is a natural process in which disordered monomers spontaneously form a certain structure or pattern. In the process of self-assembly, monomers spontaneously aggregate into a stable structure with a certain regular geometric morphology based on noncovalent bond interactions (Ragazzon and Prins 2018; Cummins et al. 2020; Woods et al. 2019; Penfold et al. 2019; Sun et al. 2019c). Taking advantage of different drug properties, two or more drugs are connected to form an amphiphilic composite drug, then the prodrug nanocarrier is constructed through self-assembly, and the drugs are loaded without any other materials (Shao et al. 2018; Li and Liu 2019; Li et al. 2020; Wang et al. 2018a, b; Zhong et al. 2016). In the field of nanomedicine, drug–drug nanocomposites are considered a promising strategy with great clinical transformation value. Zhu et al. (Wang et al. 2018a) proposed and constructed a supramolecular self-delivery prodrug nanocarrier system using a noncovalent molecular recognition strategy to link hydrophobic and hydrophilic drugs. The author selected the hydrophilic anticancer drug clofarabine (CA) and the hydrophobic anticancer drug raltitrexed (RT) as the research objects and obtained the new drug complex, and the complex monomers can self-assemble into spherical nanoparticles with high stability in an aqueous solution. Molecular dynamics simulations further demonstrated the feasibility of molecular recognition and self-assembly between CA and RT. In vitro and in vivo studies have shown that the system can significantly reduce the toxic side effects of free drugs on healthy tissues during treatment. Yan et al. (Sun et al. 2019a, b, c) also adopted an amphiphilic drug–drug conjugate strategy to bind chemotherapeutic agents with antiangiogenic drugs. An amphiphilic conjugate of hydrophilic floxuridine (FUDR) and hydrophobic pseudolaric acid B (PAB) was synthesized through a one-step esterification reaction, which can self-assemble to form nanoparticles in water. Further in vivo and in vitro experiments show that the FUDR-PAB conjugate not only has a high antitumor effect but also has excellent antiangiogenic activity. Luan et al. (Zhou et al. 2020) designed and synthesized the hydrophilic chemotherapeutic drug gemcitabine (GEM) and the hydrophobic indoleamine-2,3-dioxygenase inhibitor 1-methyl-tryptophan (1MT) amphipathic double prodrug (GEM-1MT) that can be self-assembled into nanoparticles in aqueous solution. The constructed drug delivery system GEM-1MT NPs can not only effectively inhibit the growth of tumors in situ but also have a significant inhibitory effect on lung metastasis of tumors (Fig. 5.1).

5.2.2 *Drug-Monomer/Polymer*

At present, the main strategy for constructing prodrug nanocarriers is to connect drugs and monomers/polymers with covalent bonds to form new composites and then to form prodrug nanocarriers by self-assembly. Generally, the properties of the drug can be optimized through this strategy. For instance, coupling with a hydrophilic polymer can improve hydrophobic drug water solubility to form an amphiphilic molecule. Wu et al. (Zhao et al. 2015) designed a multifunctional doxorubicin (DOX) prodrug nanocarrier system that can encapsulate the resistance reversal agent verapamil hydrochloride (VRP·HCl). Hydrophobic DOX is covalently linked to hydrophilic short-chain polyethylene glycol (mPEG) through a benzoic imine linkage to obtain a linear amphiphilic PEGylated prodrug, namely, mPEG-b-DOX. mPEG-b-DOX can self-assemble to form stable nanoparticles, and the resulting nanoparticles can be used as anticancer drugs and drug carriers at the same time. Zhu et al. (Guo et al. 2018) reported a platinum (IV) complex-based polyprodrug that can generate highly toxic platinum (II) species in situ as chemotherapeutics. Nanosized hydrogel-like polyprodrug could be synthesized by in situ polymerization of the platinum (IV) complex-based prodrug monomer (PPM) and 2-methacryloyloxy ethyl phosphorylcholine (MPC). The drug delivery system can achieve programmed drug release in vivo, exhibit excellent tumor accumulation, and reverse and prevent drug resistance (Fig. 5.2). Kundu et al. (Horo et al. 2019) synthesized 5-fluorouracil (5FU) and low molecular weight chitosan (LMWC) covalently coupled prodrugs. The conjugate is cleavable under 365 nm UV-A radiation and releases 5FU in a controlled manner. Compared with LMWC, the water solubility of the conjugate is improved, and in vitro release studies have shown that 5FU can be effectively released in a controlled manner.

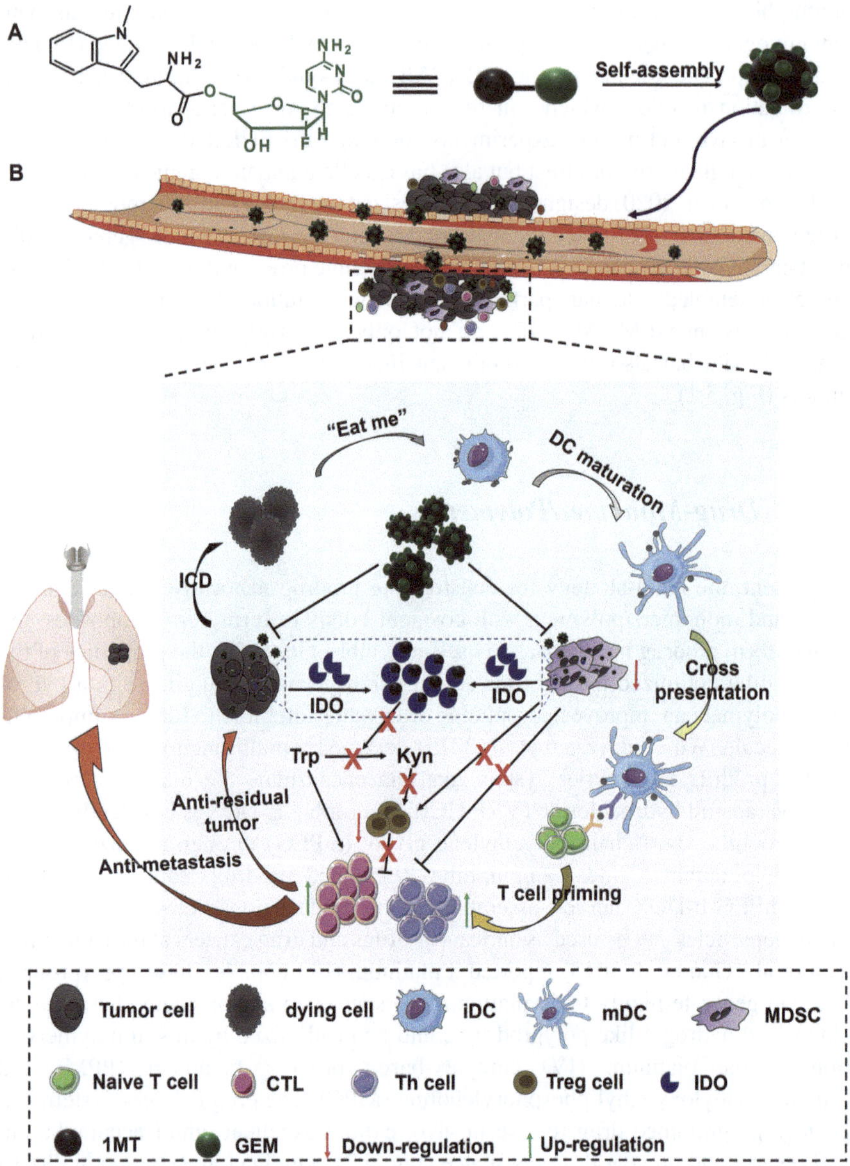

Fig. 5.1 (**a**) Schematic diagram of preparing GEM-1MT biprodrug NPs and (**b**) schematic diagram of GEM-1MT NPs for self-combining chemo-immunotherapy. (Reproduced with permission from Zhou et al. (2020))

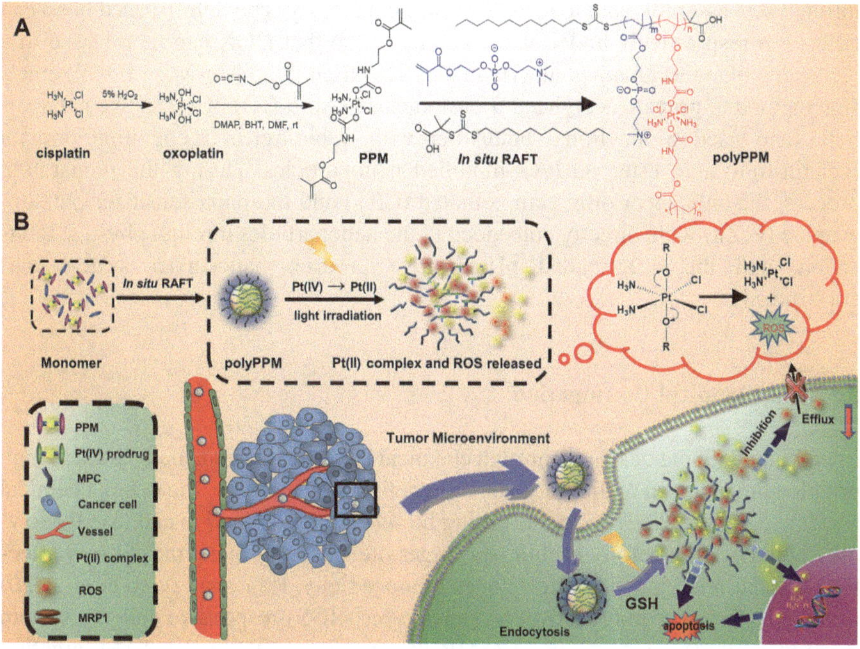

Fig. 5.2 Schematic diagram of the formation and antitumor mechanism of the photoactivable polyprodrug. (**a**) Synthetic route of PPM and polyPPM. (**b**) Illustrations of polyprodrug formation and photoreductive drug release. After internalization into cancer cells, the in situ generalization of ROS and cisplatin under light irradiation causes the death of cancer cells in O2-independent combinatorial chemo-photodynamic therapy. (Reproduced with permission from Guo et al. (2018))

5.2.3 Drug-Carrier Nanocomposites

Drugs can be conjugated with or entrapped in the nanocarriers during the preparation of nanoparticles, but the drugs do not affect or participate in the formation of nanocarriers. Most of the responsiveness is endowed to the nanocarrier rather than the drugs in this type of strategy, and the interaction between them can be physical or chemical.

5.2.3.1 Physical Entrapment

In this way, drugs are loaded inside or bind to the surface of the nanocarriers through physical interactions, and the constructed nanocarriers can disintegrate responsively to pathological characteristics or external stimuli to promote the release of the drugs. Huang et al. (Wei et al. 2020) established a new pH-sensitive prodrug nanocarrier delivery system using the physical interaction of drug and carrier materials for the codelivery of DOX and diosgenin to effectively inhibit tumor metastasis and proliferation. The mPEG-Schiff base diosgenin was used to construct nanocarriers through self-assembly and loaded DOX simultaneously. The Schiff-based bond can be broken responsively under acidic conditions so that DOX can be released in a controlled manner. Zhang et al. (Hu et al. 2019) used the oil-in-water (O/W) emulsion method to prepare acetylated β-cyclodextrin (Ac-β-CD) pH-sensitive nanoparticles and used a functional emulsifier with a biological recognition domain (gelatin-folic acid ester (G-FA))-modified nanoparticles. During the preparation process, the anticancer drug camptothecin (CPT) and the photosensitizer phthalocyanine (PcZn) were directly embedded in the nanoparticles through physical interaction, and the drug-loaded pH-sensitive prodrug nanocarrier system was constructed.

5.2.3.2 Chemical Conjugation

Interacting with nanocarriers through chemical bonding, DDSs usually possess high drug loading and can realize targeted drug delivery. In addition, the strong chemical bonding can keep the DDS stable during blood circulation in vivo and lose the lowest amount of drug before reaching the target site, thus improving the delivery effectiveness and reducing side effects or immunogenicity. Han et al. (Guo et al. 2020) designed a hypoxia and reactive oxygen species (ROS)-responsive delivery system *p*-(aminomethyl) benzoic acid (PAMB)/doxorubicin (DOX) for TAM-targeting depletion. Under a circumstance of hypoxic, the overlayer PEG would be rapidly shed off via azobenzene degradation, exposing mannan. On the other hand, phenylboronic acid pinacol ester serves as the hydrophobic unit to form the ROS-responsive core, embedding with DOX. The hydrophobic groups collapse via intracellular ROS oxidization to release DOX. The dissociation of this prodrug nanocarrier was successfully inhibited extracellular matrix (ECM) degradation and neovascularization and tumor invasion contributed by TAM depletion (Fig. 5.3).

5.3 Stimuli-Responsive Strategies for Prodrug Nanocarriers

The principal advantage of prodrug nanocarriers is to respond to specific stimuli in vivo and trigger the release or conversion of active drugs to maximize their efficacy compared with traditional nanocarriers.

The stimuli can be classified into two types: (1) endogenous and (2) exogenous. Most endogenous stimuli originate from the body's lesions, for example, changes in the tumor microenvironment (TME). Compared with normal tissues, tumor tissues generally have lower pH, hypoxia, higher interstitial pressure, and altered enzyme levels, which can be perceived by prodrug nanocarriers (Fang and Al-Suwayeh 2012; Delplace et al. 2014; Luo et al. 2014; Taresco et al. 2018). On the other hand, exogenous stimuli mostly refer to external interventions, such as the use of light, ultrasound, heat, and magnetism, interfering with prodrug nanocarriers to induce targeted and controlled drug delivery.

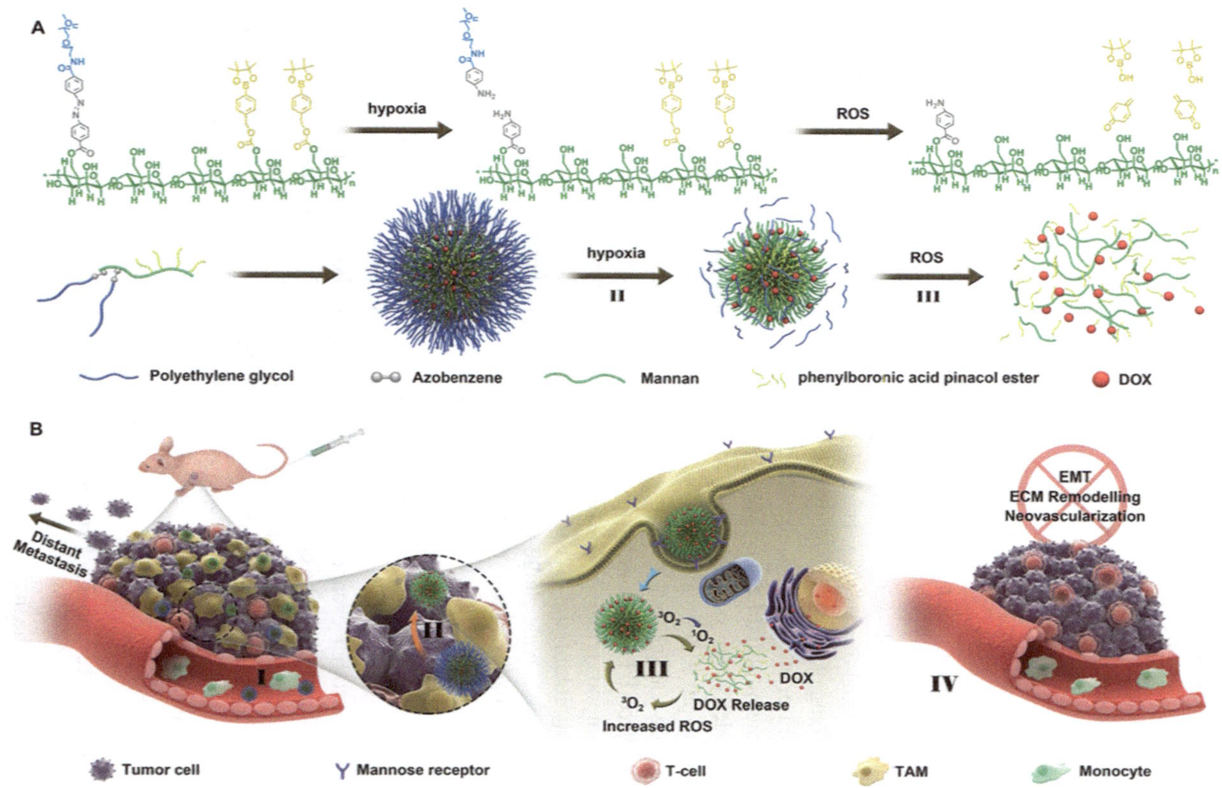

Fig. 5.3 Illustration of the hypoxia and reactive oxygen species (ROS)-responsive release of PAMB/DOX in vivo. Specifically eliminating tumor-associated macrophages with an extra- and intracellular stepwise-responsive nanocarrier for inhibiting metastasis (Guo et al. 2020)

5.3.1 Endogenous Stimuli

Low selectivity and severe side effects have been challenging obstacles to successful chemotherapy for tumor treatment. Based on the difference between lesions and normal tissues or cells, a variety of prodrug nanocarrier DDSs responsive to the endogenous pathological microenvironment were designed and showed promising prospects for clinical applications.

5.3.1.1 pH-Sensitive Prodrug Nanocarriers

The TME displays a lower pH relative to normal tissues. The pH of tumor tissue is approximately 6.5–7.0 and 4.5–6.5 for tumor cells, which lays the groundwork for the development of pH-sensitive delivery systems for effectively targeting tumors rather than normal tissues (Wei et al. 2020; Noh et al. 2015; Song et al. 2016; Zhu et al. 2013; Jiang et al. 2017; Zhang et al. 2016, 2019a; Sauraj et al. 2018; Zhai et al. 2018; Câmara et al. 2017). Since boric acid can form an acid-cleavable covalent bond with mannitol, Pan et al. (Ma et al. 2019b) designed a pH-sensitive prodrug nanocarrier delivery system loaded with the antitumor drug bortezomib (BTZ). Under neutral physiological conditions, the drug delivery system is relatively stable to reduce drug leakage, but it will trigger the rapid dissociation of catechol/BA ester and release free BTZ after entering tumor cells in a weakly acidic environment to achieve continuous release in the tumor (Fig. 5.4). Platinum (Pt)-based chemotherapeutics mainly inhibit the division of cancer cells by causing DNA replication disorders and have a wide range of clinical applications, but they can cause serious toxic side effects throughout the body due to their lack of selectivity to tumors and normal tissues or cells (Ding et al. 2020; Hu and Steinmetz 2020). Hu et al. (Chen et al. 2020b) reported a strategy to develop tumor extracellular pH-sensitive nanoparticle (esNP)-loaded Pt(IV) prodrugs (esNP-Pt). When the pH $\leq$ 6.8, the prodrug nanocarrier could achieve charge reversal for specific cellular uptake and endosomal/lysosomal escape by breaking the amide bond. In vitro cell uptake experiments showed that the cellular uptake rate of esNPs was similar to that of pH-insensitive succinic acid nanoparticles (isNPs) at pH 7.4 and was 6 times that at pH 6.8. The results demonstrated that esNP-Pt showed selective toxicity to tumors and weak systemic toxicity, reduced the side effects of platinum drugs, and had certain clinical research value.

5.3.1.2 Redox-Responsive Prodrug Nanocarrier

Hypoxia is one of the most salient characteristics of tumors and a hotspot for researchers relative to normal tissues or cells. Chen et al. (Zeng et al. 2018) systematically summarized the design and research of prodrugs sensitive to hypoxia

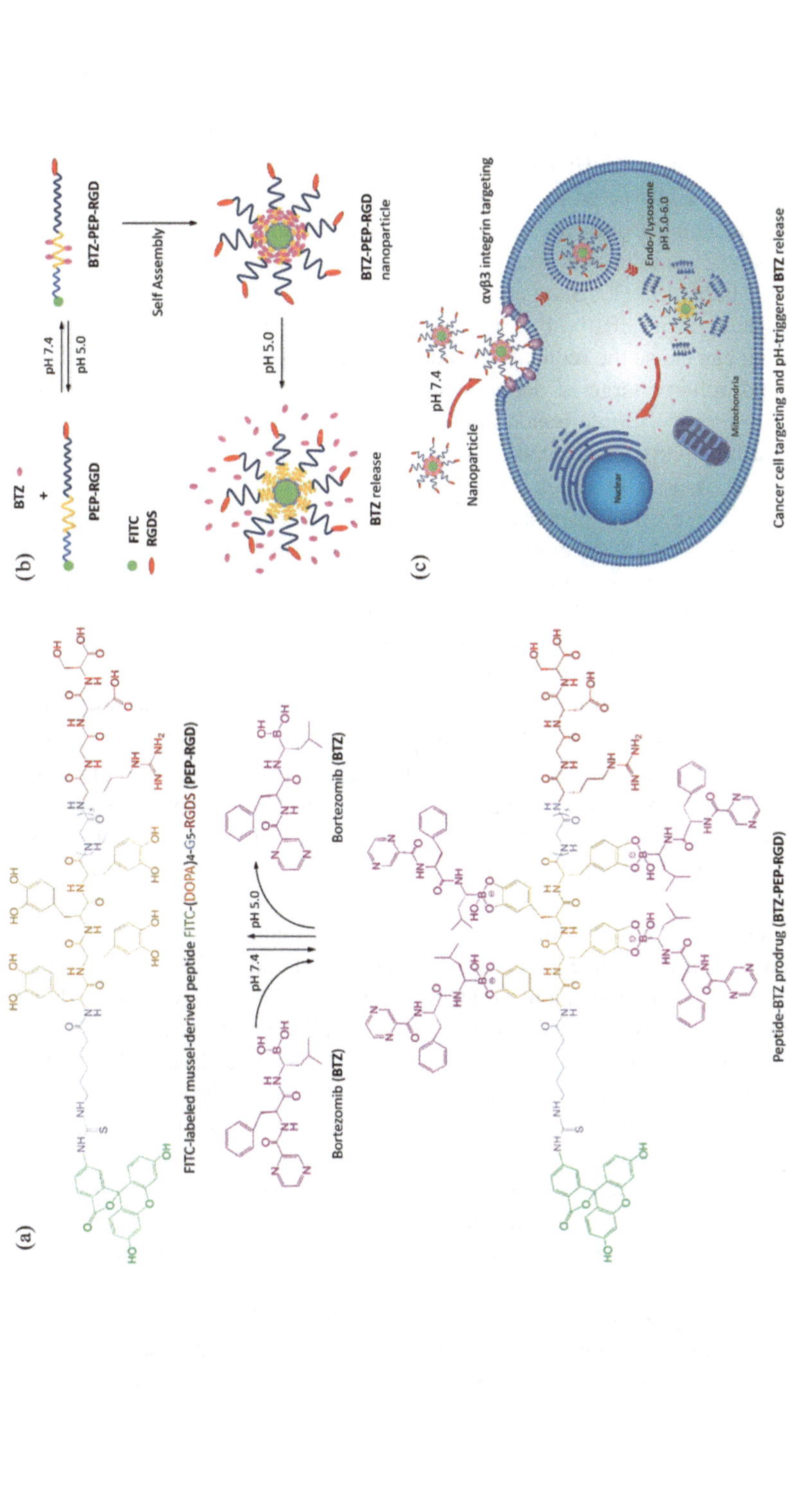

Fig. 5.4 **(a)** Structural formula of the mussel-derived cancer cell-targeting peptide and the dynamic conjugation of the antitumor drug BTZ with pH responsiveness. **(b, c)** Schematic illustration of the peptide-prodrug-based nanoparticle with cancer cell targeting and pH-sensitive drug release properties. (Reproduced with permission from Ma et al. (2019a, b))

characteristics; therefore, we will focus on introducing the design of prodrug nanocarriers responsive to hypoxia.

Responsive design for low-oxygen environments mostly relies on the cleavage of chemical bonds, involving nitroimidazole groups, azo groups, and disulfide bonds, among which disulfide bonds have been extensively studied (Fu et al. 2015; Wang et al. 2014; Sun et al. 2018; Huang et al. 2010; Ma et al. 2019a, b; Yang et al. 2020a). Nanocarriers are constructed by disulfide bonds to load drugs or directly connect the drugs to the nanocarriers. Meanwhile, the concentration of glutathione (GSH) in the cytoplasm of some tumors is approximately 7 times that in normal cells (Raza et al. 2018; Ma et al. 2020; Shen et al. 2018; Hernández-Gil et al. 2015; Luo et al. 2016a, b; Zhang et al. 2019a, b; Liu et al. 2011, 2017; Jin et al. 2018; Ma et al. 2010), which can cause the reductive reaction to break the disulfide bond to release the drug and achieve targeted delivery into tumors. Shi et al. (Yu et al. 2018b) directly introduced disulfide bonds into the framework of traditional mesoporous organosilica nanoparticles (MONs). Once the drug-loading system was taken up into the tumor tissue with a high concentration of GSH, the disulfide bonds in the MON skeleton cleaved in response to the reductive microenvironment, leading to subsequent structural disintegration, and the DOX loaded was quickly released from the nanoparticles. In addition, directly connecting the drug and the carrier through disulfide bonds is also an effective construction strategy (Fig. 5.5). Yang et al. (Li et al. 2017) reported α-amylase- and redox-responsive nanoparticles based on hydroxyethyl starch (HES) for tumor-targeted delivery. Paclitaxel (PTX) is bound to HES through disulfide bonds to furnish HES-SS-PTX, and the conjugates self-assemble to form stable and monodisperse nanoparticles (NPs). After NPs entered the tumor tissue, they could quickly disintegrate under reductive conditions to release PTX. Compared with free paclitaxel, HES-SS-PTX nanoparticles showed higher in vivo antitumor efficiency (63.6% vs 52.4%) and lower toxicity in 4T1 tumor-bearing mice. Sulfur-based chemical bonds have been widely studied in redox-responsive prodrug nanocarrier DDSs. Selenium (Se) belongs to the same group of elements as sulfur in the periodic table, and selenium-based polymers containing selenide or diselenide have also been developed and studied for the design and construction of redox-responsive DDSs. Sun et al. (2019a, b, c) designed and synthesized four redox-responsive paclitaxel-citronellol prodrugs based on thioether bonds, disulfide bonds, selenoether bonds, and diselenide bonds. The nonresponsive carbon chain conjugate was used as a negative control. Several hydrophobic prodrugs can self-assemble into nanoparticles with uniform size and high drug loading. Among them, prodrug nanoassemblies constructed using diselenium bonds have displayed many therapeutic advantages, such as good colloidal stability, long circulation time, and high tumor accumulation. In addition, Sun et al. (Yang et al. 2020a, b) developed a trisulfide bond-based tumor drug delivery system based on previous disulfide bond research. Compared with disulfide bonds and thioether bonds, trisulfide bonds effectively promoted the self-assembly ability of DOX dimer prodrugs, thereby improving the colloidal stability of the prodrug nanoassembly and the metabolic process in vivo. Compared with traditional disulfide bonds,

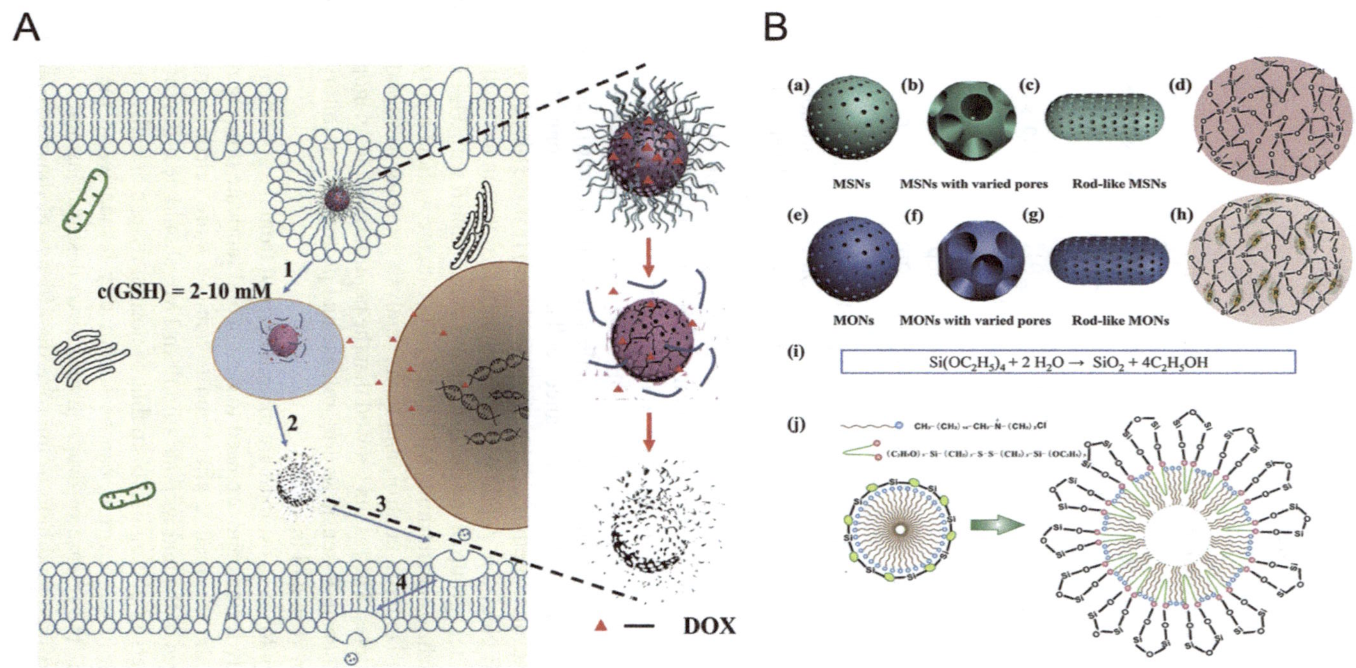

Fig. 5.5 (**a**) Schematic illustration of the transport of DOX@PEG-MONs in blood vessels and the subsequent intracellular delivery, biodegradation, and excretion of PEG-MONs. "1" indicates the reducing-responsive degradation in lysosomes, "2" indicates the degraded products suffered from the lysosomes, and "3, 4" indicate the excretion process of degraded fragments from cells. (**b**) Schematic diagrams of typical (a) MSNs and (e) MONs, (b) MSNs and (f) MONs with varied pores, (c) rod-like MSNs and (g) MONs and framework of (d) MSNs and (h) MONs. (i) Chemical reaction equations for the preparation of typical MSNs via the sol-gel method and (j) diagrammatic drawing of the MON synthetic process. (Reproduced with permission from Yu et al. (2018a, b))

trisulfide bonds also showed higher glutathione sensitivity. By simply inserting the trisulfide bond, the DOX dimer prodrug self-assembled into nanoparticles, which featured high drug loading, high self-assembly stability, and high tumor selectivity and had profound research significance.

5.3.1.3 Enzyme-Responsive Prodrug Nanocarriers

Enzymes are proteins or RNAs produced by living cells that possess substrates with high specificity and high catalytic efficacy. They play a critical role in catalyzing almost all biological processes and can be employed as a typical endogenous stimulus (Scomparin et al. 2017; Mu et al. 2018; Gordhan et al. 2020; Linderoth et al. 2009). Many enzymes are being used to design and build new prodrug nanocarriers, particularly some representative cancer-related enzymes, including matrix metalloproteinases (MMPs), cathepsin, hyaluronidase, and phospholipase. MMP2 has been extensively studied from the MMP family, which is overexpressed in human multidrug-resistant tumors and provides a possibility for the construction of an enzyme-responsive prodrug DDS. Gao et al. (Li et al. 2018) reported a new multifunctional nanoformulation based on mesoporous silica-coated iron oxide nanoparticles (Fe_3O_4@MSNs) for the delivery of enzyme-responsive anticancer drugs and magnetic resonance imaging (MRI) in vivo. The anticancer drug DOX was wrapped in the porous cavities, and the MMP2 enzyme-reactive peptide was covalently conjugated to the surface of the nanoparticles. The results of in vitro experiments showed that the system had high specificity for controlled release to cell lines overexpressing MMP2. In addition, cathepsin B is a lysosomal cysteine protease that can degrade extracellular matrix components during the invasion and metastasis of cancer cells. Lee et al. (Jeong et al. 2015) synthesized a dendrimer-methoxy poly(ethylene glycol) (MPEG)-DOX conjugate using cathepsin B cleavable peptides. The polypeptide was coupled with the terminal carboxylic acid group of the dendrimer and then coupled with MPEG amine and DOX. Meanwhile, a dendrimer-MPEG-DOX conjugate without a polypeptide chain was synthesized as a control, and then a dialysis strategy was used to prepare the nanoparticles. The results showed that the peptide-modified nanoparticles demonstrated cathepsin B-sensitive drug release and preferably inhibitory effects on tumors both in vitro and in vivo.

In summary, endogenous stimuli are mainly derived from the specific microenvironment of body lesions and are used to design and fabricate prodrug nanocarriers to enhance the therapeutic effect and to achieve more selective treatment. The subtle changes of the endogenous stimuli in TME should be different enough for the nanocarriers to distinguish from that of normal tissues or cells, so that the constructed prodrug nanocarriers can be specifically responsive to the lesions, which also avoids the release of drugs in undesired destinations and minimizes side effects and offers great help to improve the efficacy of diagnosis and treatment of diseases.

5.3.2 *Exogenous Stimuli*

Compared with endogenous microenvironmental stimuli, the response of exogenous stimuli does not depend on the characteristics of the lesion in the body, and the release and delivery of drugs are controlled by external conditioned stimuli (mainly including light, ultrasound, heat, magnetism, etc.), usually in cooperation with endogenous stimuli to construct prodrug nanocarriers to achieve high-efficiency delivery.

5.3.2.1 Light-Responsive Prodrug Nanocarriers

Photodynamic therapy (PDT) is an emerging method of treating tumors with photosensitive drugs and laser activation. Irradiating the tumor with lasers of specific wavelengths can activate the photosensitive drugs that selectively accumulate in the tumor tissues and trigger photochemical reactions to destroy the tumor (Jiménez-Amezcua et al. 2018; Song et al. 2016). Jiang et al. (Ihsanullah et al. 2020) reported a DDS based on the tumor microenvironment and external stimuli to improve PDT-induced chemotherapy of different tumor cells at the proximal and distal ends of blood vessels. The nanoparticles modified with the TAT peptide promoted tumor cell uptake and penetration due to their responsiveness to low pH. The photosensitizer chlorine e6 (Ce6) and hypoxia-activated prodrug tirapazamine (TPZ) were retained in tumor cells and accumulated near and distal to blood vessels. Under extracorporeal laser irradiation, high-efficiency photodynamic therapy was achieved, and TPZ was released and activated, thereby improving the combined therapeutic effect against tumor cells at the distal end of the blood vessel (Fig. 5.6). On the other hand, general PDT has been widely used in the diagnosis and treatment of various cancers using a wide range of light (375–732 nm) and photosensitizers. Therefore, the construction of a light-sensitive prodrug nanocarrier can also directly optimize the drug itself. Bilgicer et al. (Xiao et al. 2014) reported the design and characterization of a series of novel photosensitive Pt(IV)-azide prodrugs and micellar nanoformulations. One advantage of this prodrug nanocarrier system compared to general PDT is that it does not require a photosensitive catalyst or an oxygen-rich environment. Under mild UVA irradiation, free Pt(IV) complexes and micellar nanoparticles can quickly release biologically active Pt(II) while maintaining extremely high stability in the dark. Experimental results showed that compared with its free prodrug analogs (cisplatin and oxaliplatin), micellar nanoparticles significantly enhanced the uptake of photosensitive Pt(IV) prodrugs by ovarian cancer SKOV-3 cells.

5.3.2.2 Ultrasound-Responsive Prodrug Nanocarriers

Ultrasound (US) is widely used clinically because of its advantages of noninvasiveness, no ionizing radiation, low cost, and the ability to regulate the tissue penetration depth by adjusting the frequency, period, and exposure time. It has performed

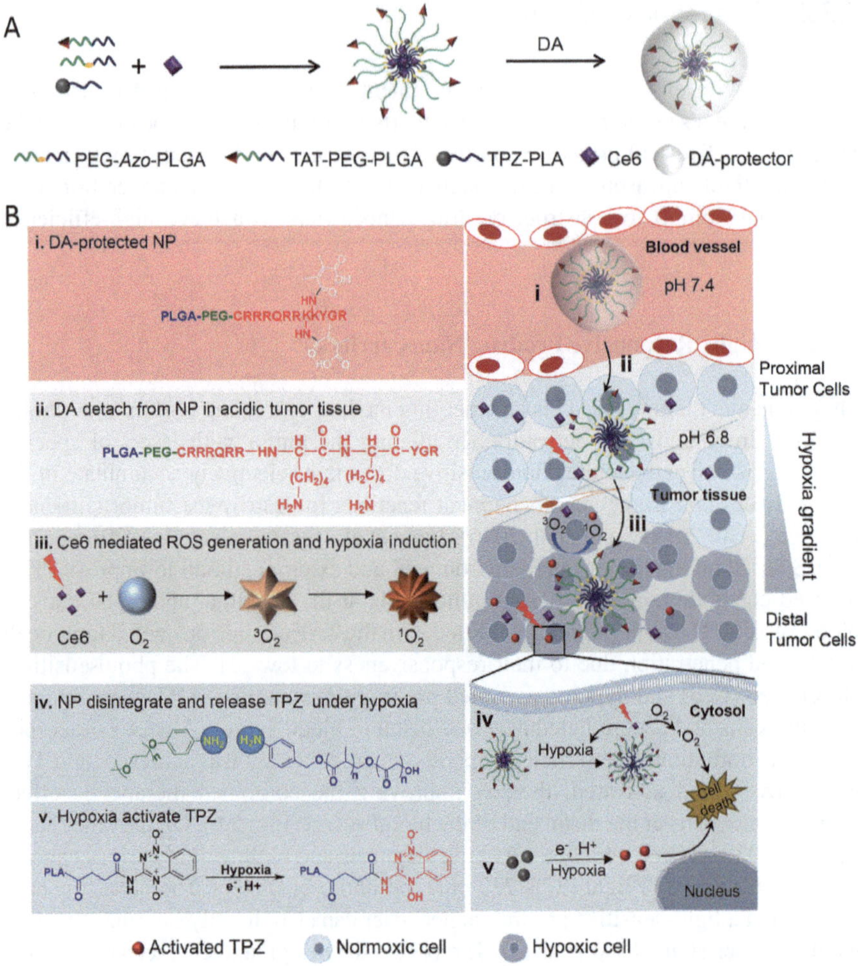

Fig. 5.6 **(a, b)** Schematic illustration of tumoral pH-activatable and hypoxia-responsive NPs for enhancing PDT-induced hypoxia-activated chemotherapy. (Reproduced with permission from Ihsanullah et al. (2020))

well in disease diagnosis, treatment, and drug delivery. Researchers have constructed numerous prodrug nanocarrier delivery systems based on ultrasound strategies (Wu et al. 2017a, b; Gao et al. 2019; Lea-Banks et al. 2019; Paris et al. 2015). Vallet-Regı´ et al. (Paris et al. 2015) grafted a copolymer on the surface of mesoporous silica nanoparticles as a pore gatekeeper, which loaded drugs in an open conformation at low temperature (4 °C) and collapsed at 37 °C while closing the pore entrance. Therefore, drug leakage was prevented at physiological temperature. Upon ultrasound irradiation, the polymer changed its hydrophobicity and opened the gate like a coil to release the drug. In vitro experiments showed that this strategy can release DOX and cause LNCaP cell death with sole exposure to ultrasound.

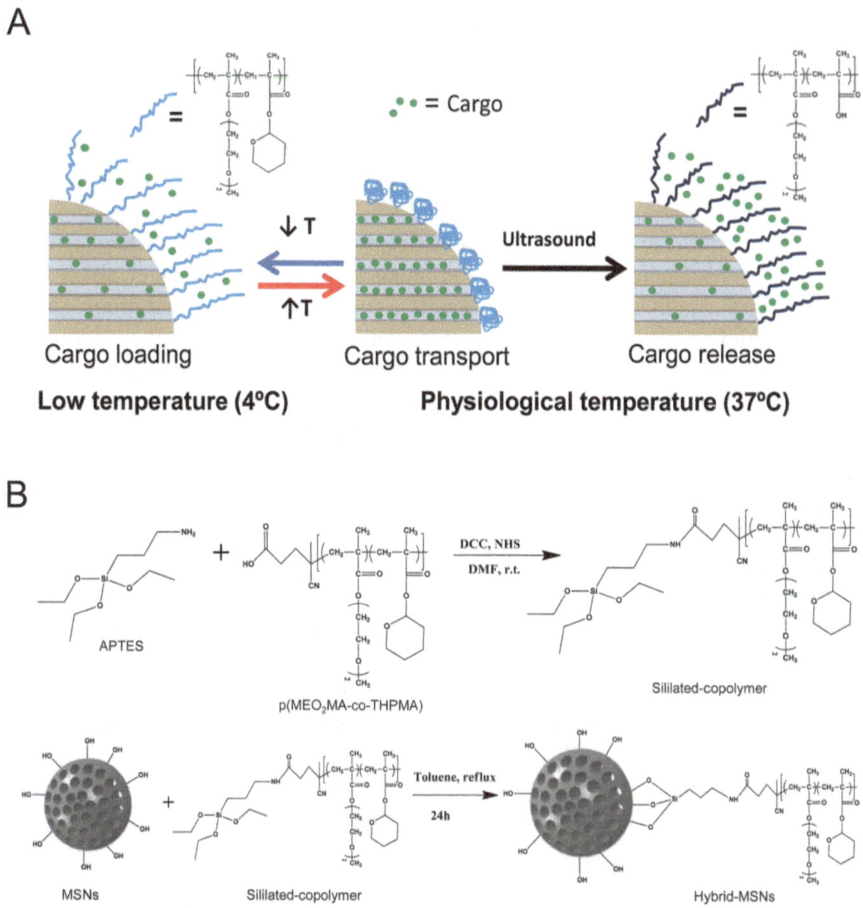

Fig. 5.7 (**a**) Schematic illustration of the behavior of the dual responsive release system in aqueous medium. (**b**) Preparation of p(MEO_2MA-*co*-THPMA)-SiO_2 nanoparticles (Hybrid-MSNs): APTES conjugation to p(MEO_2MA-*co*-THPMA) (top) and grafting of the silylated copolymer to mesoporous SiO_2 nanoparticles (bottom). (Reproduced with permission from Paris et al. (2015))

However, the most conventional ultrasound-sensitive strategies are mainly to change the behavior of the nanocarriers after exposure to stimuli to release the drug, and few focus on the regulation of the drug itself (Fig. 5.7). Recently, Herrmann et al. (Huo et al. 2021) used ultrasound mechanochemistry to achieve selective cleavage of weak covalent bonds or noncovalent bonds and achieved precise activation of small-molecule chemotherapeutics in polymers or nanostructures for the first time. The study relied on a disulfide polymer centered on the anticancer drug camptothecin (CPT). The ultrasound mechanical force selectively tore apart the disulfide bond, converted it into thiol, and promoted intramolecular 5-exo-trig cyclization, thereby releasing and activating camptothecin. This result is the first report of the use of ultrasound mechanical force to achieve precise regulation of drug activity at the molecular level and provides an extremely precise treatment strategy.

5.3.2.3 Thermoresponsive Prodrug Nanocarriers

Extensive research has been carried out to control the response of nanocarriers through temperature changes. In particular, polymers with lower critical solution temperature (LCST) or upper critical solution temperature (UCST) are frequently used as temperature-sensitive polymers. LCST polymers are soluble below the critical temperature and insoluble above, while UCST polymers are the opposite, which is the key to establishing temperature-sensitive prodrug nanocarrier systems (Bordat et al. 2019; McDaniel et al. 2013; Farjadian et al. 2020; Zhang et al. 2017a). Temperature-sensitive responses are usually a supplement to targeted drug delivery strategies and result in a synergistic effect with conventional or other responsive systems to enhance the efficacy of drugs. Chattopadhyay et al. (Panja et al. 2016) used a novel architecture of four-arm star block copolymers to prepare smart temperature-sensitive micelles. Temperature-induced micelles had a higher DOX release rate in the temperature range above the LCST, and the release rate was controlled by temperature changes. The results showed that the micelles were selectively internalized into cancer cells (C6 glioma) instead of normal cells (HaCaT). The tumor volume was significantly suppressed, and there was no obvious systemic toxicity when DOX-loaded micelles were injected in vivo into a C6 glioma rat tumor model (Fig. 5.8). Qian et al. (Qu et al. 2015) constructed a thermoresponsive hybrid hydrogel DDS based on the characteristics of the hydrogel. Briefly, gold nanorods (GNRs) were doped into thermally responsive hydrogels, and the shrinkage of thermally responsive hydrogels was induced by the photothermal effect of

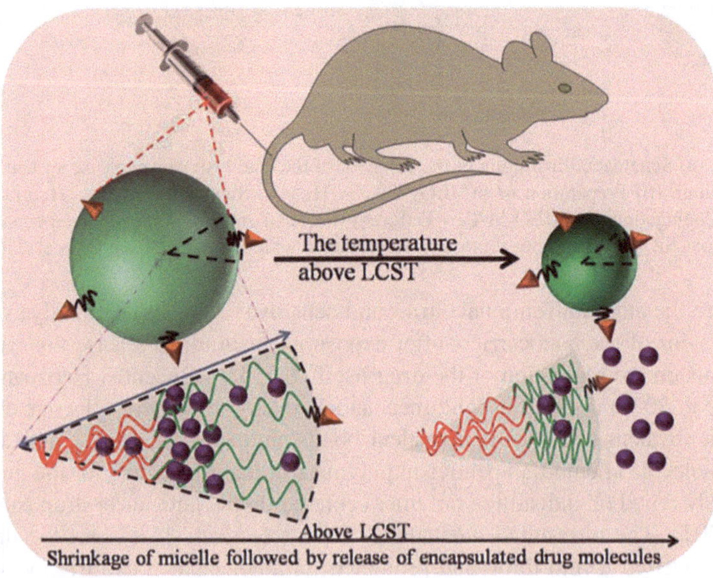

Fig. 5.8 Illustration of the thermoresponsive release of DOX-loaded micelles. (Reproduced with permission from Panja et al. (2016))

gold nanorods under the action of near-infrared (NIR) lasers, thereby releasing loaded DOX. In a 4T1 breast cancer model of the in vivo locoregional prevention of postoperative recurrence, this strategy has shown an excellent effect of inhibiting tumor recurrence to 16.7% compared with 83.3% for control GNRs with NIR irradiation, which has shown great potential in preventing postoperative recurrence of cancer.

5.3.2.4 Magnetism-Responsive Prodrug Nanocarriers

Regulation of drug delivery by magnetism has become a crucial supplement to the development of pharmaceutics in recent decades (Liu et al. 2019; Mansouri et al. 2017). Research on magnetism-responsive drug delivery systems has benefited from the development of magnetic nanoparticles (MNPs). Due to the inherent magnetic properties of MNPs, they have been widely studied in targeted drug delivery and magnetic resonance imaging (MRI). Through the control of an external magnetic field, MNPs can be guided to the target site to achieve efficient drug delivery and make a great contribution to disease diagnosis. In research on MNP drug delivery, mesoporous silica nanoparticles (MSNs) have been employed to construct magnetism-responsive prodrug nanocarriers. María et al. (Guisasola et al. 2018) proposed a strategy for in vivo tumor treatment mediated by MSNs. The system consisted of magnetic iron oxide nanoparticles embedded in a mesoporous silica matrix and coated with a thermosensitive polymer. Magnetic nanoparticles acted as an internal heat source under an alternating magnetic field (AMF), which raised the temperature of the surrounding environment and triggered the transformation of polymers, thereby releasing the drug contained in the pores of silica and realizing a controllable program for drug release (Fig. 5.9). Lesieur et al. (Marie et al. 2015) prepared magnetic-fluid-loaded liposomes (MFLs) by embedding superparamagnetic maghemite nanocrystals in PEGylated phospholipid microcapsules. The

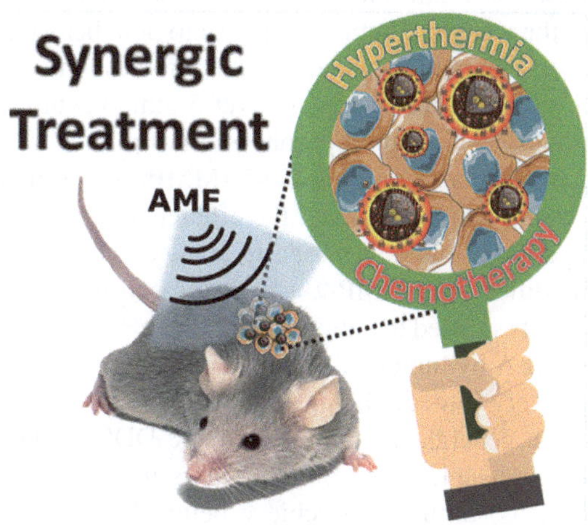

Fig. 5.9 Illustration of the procedure for the thermoresponsive treatment. (Reproduced with permission from Guisasola et al. (2018). Copyright 2018, American Chemical Society)

nanocarrier provided the possibility for selective magnetic targeted therapy of malignant tumors in the brain and noninvasive tracking of MRI intravascular drug delivery. Controlled by an external magnetic field, MFLs effectively aggregated into the striatum of a mouse model of human glioma cells (U87). Experimental results showed that the accumulation of MFLs in malignant tissues was significantly increased, and endothelial cells and glioblastoma cells effectively internalized MFLs. Therefore, magnetism-responsive prodrug nanocarriers have promising prospects in drug delivery, cancer treatment, and even imaging.

In summary, prodrug nanocarrier delivery systems responsive to exogenous stimuli have been a supplement to delivery systems sensitive to endogenous stimuli and an essential part of prodrug nanocarrier DDSs, which greatly enriches the design and development of prodrug nanocarriers and extends the possibility of their clinical applications.

5.3.3 Multiresponsive Prodrug Nanocarrier

Prodrug nanocarriers responsive to endogenous or exogenous stimuli introduced above are all single-responsive systems. In addition, many researchers have proposed a multiresponsive prodrug delivery strategy, namely, a single nanocarrier that responds to two or more stimuli (Zhao et al. 2018; Ashjari et al. 2020; Chen et al. 2018, 2020a, b; Kazempour et al. 2019; Su et al. 2020). The different stimuli can be endogenous stimuli, exogenous stimuli, or a combination of both (Li et al. 2019a, b, c; Cao et al. 2017). Chen et al. (Song et al. 2018) used PEG and 3,3′-dithiodipropionic acid (DTPA) to synthesize a new type of reduction-heat dual responsive polymer through direct polycondensation. The prepared PEG-DTPA copolymer was soluble in water at room temperature and exhibited a reversible thermoresponsive property. Meanwhile, due to the existence of disulfide bonds on the main chain, the PEG-DTPA copolymer exhibited reduction-sensitive properties. The resulting copolymer can self-assemble into spherical micelles, exhibiting structural changes in the water medium in response to both heat and reduction, and was expected to be used as a reduction and thermal dual responsive prodrug nanocarrier in tumor therapy. In another case, to deliver 5-aminolevulinic acid (ALA), the precursor of the photosensitizer protoporphyrin IX (PpIX), Ji et al. (Wu et al. 2017a, b) constructed a matrix metalloproteinase-2 (MMP-2) and pH dual-sensitive ALA prodrug nanocarrier, which improved the efficiency of ALA targeted delivery. The pH-sensitive hydrazone bond was used to connect ALA, and gold nanoparticles (AuNPs) were modified with MMP-2 activatable cell-penetrating peptides (CPPs). Experimental results showed that this enzyme and pH dual responsive system enhanced the uptake of ALA by tumor cells and significantly improved the efficiency of tumor-targeted delivery (Fig. 5.10).

In general, the multiresponsive DDS of prodrug nanocarriers is able to better distinguish lesions from normal tissues, release drugs more precisely in response to various stimuli, and achieve better therapeutic effects than DDS responsive to a single stimulus.

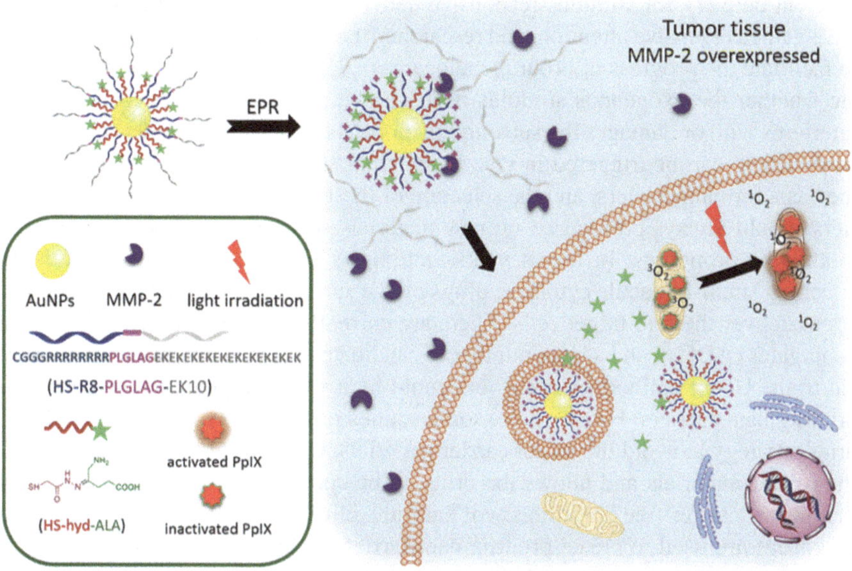

Fig. 5.10 Schematic illustration of enzyme and pH dual-sensitive ALA-conjugated AuNPs for targeted PDT. (Reproduced with permission from Wu et al. (2017a, b))

5.4 Conclusion and Future Perspectives

Cancer is one of the most damaging malignant diseases in human health. The precise delivery of drugs to tumors is a significant principle for the construction of drug delivery systems. In recent years, many researchers have proposed various strategies combining prodrugs and nanocarriers to achieve precise and controlled drug release. This chapter introduced the construction methods and stimulus-responsive types of prodrug nanocarriers and shed light on the main research directions and strategies currently involved.

The clinical applications of nanocarriers have proven their immense development potential and value (such as paclitaxel liposomes for injection, doxorubicin hydrochloride liposomes for injections, etc.), but research on the preparation of prodrug nanocarriers is still mainly in the clinical trial stage or even at the animal level, suggesting that there are still stumbling challenges in developing prodrug nanocarrier delivery systems for the clinic. (1) The sensitivity of prodrug nanocarriers to endogenous stimuli needs to be evaluated. For example, the extracellular pH of tumor cells is 6.5–6.9, approximate to that in the normal tissue environment (pH $\approx$ 7.4), which is a severe challenge for the sensitivity of pH-responsive systems. Wang et al. (2017) reported a hybrid ultra-pH-sensitive (HyUPS) nanotransistor probe to image and quantify the acidification process of endocytic organelles in living cells, which can achieve an ultra-pH-sensitive response (<0.25 pH). Ultrasensitive strategies are a promising way to improve the sensitivity of nanocarriers to achieve efficient delivery for stimulus-responsive delivery systems. (2) Exogenous stimuli have gained extensive attention, and researchers have carried out a variety of studies to facilitate the progress of prodrug nanocarriers. However, it is necessary to evaluate whether the exogenous stimulus itself can be precisely controlled, whether it interferes with or damages the surrounding normal tissues, and whether the prodrug nanocarriers can be triggered in vivo by external stimuli. (3) Carrier materials are the basis for nanocarriers, and the selection of these materials for prodrug nanocarriers should always be focused on new materials with decent responsiveness and modifiable properties. In recent years, antibodies have been used as carriers by attaching small-molecule cytotoxic drugs on the surface of antibodies through linkers to deliver them to target cells. There are currently six approved antibody–drug conjugates (ADCs), and a large number of candidates are in different stages of clinical trials. Using antibodies as carriers could be a valuable guide to the design of prodrug nanocarriers. Furthermore, some studies have directly adopted the "drug–drug" strategy to avoid the use of carrier materials, which minimizes the toxicity of the carrier materials and allows the drugs to be completely loaded. It is expected that more "carrier-free" prodrugs will enter the clinic in the future. (4) At present, few drugs are used to create prodrug nanocarriers, primarily paclitaxel, doxorubicin, and platinum, due to the abundant modifiable properties found in these drugs. For example, many functional groups in paclitaxel can be modified into esters or amides. However, other drugs lack modifiable groups and sites, which restricts their further study; thus, it is also necessary to continue to explore a wider range of modification methods, such as oxidizing the hydrogen of cisplatin to hydroxyl groups by an oxidant for subsequent research. Taken together, the drug library needs to be broadened to include molecular targeted drugs, immunomodulatory drugs, and even biomacromolecule drugs. (5) It is also worth paying attention to another well-known phenomenon of tumor-related complications. At present, the majority of cancer patients have already entered the middle or late stages when diagnosed. Tumor complications such as cachexia, local hemorrhage, infection, and destruction of organ physiological functions are the main causes of patient suffering, disability, and death. These complications were directly or indirectly caused by the development or treatment of malignant tumors. For instance, postoperative breast cancer is often accompanied by complications such as lymphedema of the upper limbs, subcutaneous fluid, and skin flap necrosis (McNeely et al. 2012; Ho et al. 2018). Therefore, prodrug nanocarriers are expected to enhance the therapeutic potential for tumor treatment while simultaneously managing tumor-related complications.

In summary, the prodrug nanocarrier strategy shows unique advantages in the treatment of tumors, especially targeted delivery based on the response to prominent pathological characteristics or external stimuli. It also provides many design ideas for the construction of drug delivery systems. While some of the mentioned challenges are still outlined, it is undeniable that this strategy has become a significant part of drug delivery for tailored precision therapy with tremendous potential for the diagnosis and treatment of disease.

References

AbouAitah K, Swiderska-Sroda A, Farghali AA et al (2018) Folic acid-conjugated mesoporous silica particles as nanocarriers of natural prodrugs for cancer targeting and antioxidant action. Oncotarget 9(41):26466–26490. https://doi.org/10.18632/oncotarget.25470

Ariaeenejad S, Motamedi E, Hosseini Salekdeh G (2021) Application of the immobilized enzyme on magnetic graphene oxide nano-carrier as a versatile bi-functional tool for efficient removal of dye from water. Bioresour Technol 319:124228. https://doi.org/10.1016/j.biortech.2020.124228

Ashjari M, Kazemi M, Abi MN et al (2020) Poly (isopropyl-oxazoline) micelle nanocarrier as dual-responsive prodrug for targeted doxorubicin delivery. J Drug Deliv Sci Technol 59. https://doi.org/10.1016/j.jddst.2020.101914

Bordat A, Boissenot T, Nicolas J et al (2019) Thermoresponsive polymer nanocarriers for biomedical applications. Adv Drug Deliv Rev 138:167–192. https://doi.org/10.1016/j.addr.2018.10.005

Câmara ALS, Nagel G, Tschiche HR et al (2017) Acid-sensitive lipidated doxorubicin prodrug entrapped in nanoemulsion impairs lung tumor metastasis in a breast cancer model. Nanomedicine 12(15):1751–1765. https://doi.org/10.2217/nnm-2017-0091

Cao XT, Le CMQ, Thi HHP et al (2017) Redox-responsive core cross-linked prodrug micelles prepared by click chemistry for pH-triggered doxorubicin delivery. Express Polym Lett 11(10):832–845. https://doi.org/10.3144/expresspolymlett.2017.79

Chen M, Zhang W-G, Li J-W et al (2018) Preparation of pH- and reductive-responsive prodrug nanoparticles via polymerization-induced self-assembly. Sci China Chem 61(9):1159–1166. https://doi.org/10.1007/s11426-018-9268-y

Chen L, Zhuang W, Hu C et al (2020a) pH and singlet oxygen dual-responsive GEM prodrug micelles for efficient combination therapy of chemotherapy and photodynamic therapy. J Mater Chem B 8(26):5645–5654. https://doi.org/10.1039/d0tb00622j

Chen S-Q, Song G, He C et al (2020b) Tumor extracellular pH-sensitive polymeric nanocarrier-grafted platinum(iv) prodrugs for improved intracellular delivery and cytosolic reductive-triggered release. Polym Chem-UK 11(12):2212–2221. https://doi.org/10.1039/c9py01838g

Cummins C, Lundy R, Walsh JJ et al (2020) Enabling future nanomanufacturing through block copolymer self-assembly: a review. Nano Today 35. https://doi.org/10.1016/j.nantod.2020.100936

Dai L, Cai R, Li M et al (2017) Dual-targeted cascade-responsive prodrug micelle system for tumor therapy in vivo. Chem Mater 29(16):6976–6992. https://doi.org/10.1021/acs.chemmater.7b02513

Danhier F, Ansorena E, Silva JM et al (2012) PLGA-based nanoparticles: an overview of biomedical applications. J Control Release 161(2):505–522. https://doi.org/10.1016/j.jconrel.2012.01.043

Delplace V, Couvreur P, Nicolas J (2014) Recent trends in the design of anticancer polymer prodrug nanocarriers. Polym Chem 5(5):1529–1544. https://doi.org/10.1039/c3py01384g

Ding D, Zhu Q (2018) Recent advances of PLGA micro/nanoparticles for the delivery of biomacromolecular therapeutics. Mater Sci Eng C 92:1041–1060. https://doi.org/10.1016/j.msec.2017.12.036

Ding F, Zhang L, Chen H et al (2020) Enhancing the chemotherapeutic efficacy of platinum prodrug nanoparticles and inhibiting cancer metastasis by targeting iron homeostasis. Nanoscale Horiz 5(6):999–1015. https://doi.org/10.1039/d0nh00148a

Fang J-Y, Al-Suwayeh SA (2012) Nanoparticles as delivery carriers for anticancer prodrugs. Expert Opin Drug Deliv 9(6):657–669. https://doi.org/10.1517/17425247.2012.679927

Farjadian F, Ghasemi S, Andami Z et al (2020) Thermo-responsive nanocarrier based on poly(N-isopropylacrylamide) serving as a smart doxorubicin delivery system. Iran Polym J 29(3):197–207. https://doi.org/10.1007/s13726-020-00785-w

Fu Q, Wang Y, Ma Y et al (2015) Programmed hydrolysis in designing paclitaxel prodrug for nanocarrier assembly. Sci Rep-UK 5(1). https://doi.org/10.1038/srep12023

Gao J, Yu B, Li C et al (2019) Ultrasound triggered phase-change nanodroplets for doxorubicin prodrug delivery and ultrasound diagnosis: an in vitro study. Colloids Surf B Biointerfaces 174:416–425. https://doi.org/10.1016/j.colsurfb.2018.11.046

Gordhan D, Swainson SME, Pearce AK et al (2020) Poly (glycerol adipate): from a functionalized nanocarrier to a polymeric-prodrug matrix to create amorphous solid dispersions. J Pharm Sci-US 109(3):1347–1355. https://doi.org/10.1016/j.xphs.2019.12.004

Guisasola E, Asín L, Beola L et al (2018) Beyond traditional hyperthermia: in vivo cancer treatment with magnetic-responsive mesoporous silica nanocarriers. ACS Appl Mater & Interfaces 10(15):12518–12525. https://doi.org/10.1021/acsami.8b02398

Guo D, Xu S, Huang Y et al (2018) Platinum(IV) complex-based two-in-one polyprodrug for a combinatorial chemo-photodynamic therapy. Biomaterials 177:67–77. https://doi.org/10.1016/j.biomaterials.2018.05.052

Guo N, Zhou Y, Wang T et al (2020) Specifically eliminating tumor-associated macrophages with an extra- and intracellular stepwise-responsive nanocarrier for inhibiting metastasis. ACS Appl Mater Interfaces:57798–57809. https://doi.org/10.1021/acsami.0c19301

Hajnal K, Gabriel H, Aura R et al (2016) Prodrug strategy in drug development. Acta Med Marisiensis 62(3):356–362. https://doi.org/10.1515/amma-2016-0032

He C, Zhang X, Yan R et al (2019) Enhancement of cisplatin efficacy by lipid–CaO2 nanocarrier-mediated comprehensive modulation of the tumor microenvironment. Biomater Sci-UK 7(10):4260–4272. https://doi.org/10.1039/c9bm00797k

Hernández-Gil J, Cobaleda-Siles M, Zabaleta A et al (2015) An iron oxide nanocarrier loaded with a Pt(IV) prodrug and immunostimulatory dsRNA for combining complementary cancer killing effects. Adv Healthc Mater 4(7):1034–1042. https://doi.org/10.1002/adhm.201500080

Ho OA, Lin C-Y, Pappalardo M et al (2018) Comparisons of submental and groin vascularized lymph node flaps transfer for breast cancer-related lymphedema. Plast Reconstr Surg Glob Open 6(12):e1923–e1923. https://doi.org/10.1097/GOX.0000000000001923

Horo H, Das S, Mandal B et al (2019) Development of a photoresponsive chitosan conjugated prodrug nano-carrier for controlled delivery of antitumor drug 5-fluorouracil. Int J Biol Macromol 121:1070–1076. https://doi.org/10.1016/j.ijbiomac.2018.10.095

Hu H, Steinmetz NF (2020) Cisplatin prodrug-loaded nanoparticles based on physalis mottle virus for cancer therapy. Mol Pharm 17(12):4629–4636. https://doi.org/10.1021/acs.molpharmaceut.0c00834

Hu X, Gao Z, Tan H et al (2019) A pH-responsive multifunctional nanocarrier in the application of chemo-photodynamic therapy. J Nanomater 2019:1–12. https://doi.org/10.1155/2019/3898564

Huang B, Desai A, Tang S et al (2010) The synthesis of a c(RGDyK) targeted SN38 prodrug with an indolequinone structure for bioreductive drug release. Org Lett 12(7):1384–1387. https://doi.org/10.1021/ol1002626

Huo S, Zhao P, Shi Z et al (2021) Mechanochemical bond scission for the activation of drugs. Nat Chem. https://doi.org/10.1038/s41557-020-00624-8

Ihsanullah KM, Kumar BN, Zhao Y et al (2020) Stepwise-activatable hypoxia triggered nanocarrier-based photodynamic therapy for effective synergistic bioreductive chemotherapy. Biomaterials 245. https://doi.org/10.1016/j.biomaterials.2020.119982

Jain A, Hurkat P, Jain SK (2019) Development of liposomes using formulation by design: basics to recent advances. Chem Phys Lipids 224. https://doi.org/10.1016/j.chemphyslip.2019.03.017

Jeong Y-I, Park H-K, Lee SJ et al (2015) Enzyme-responsive doxorubicin release from dendrimer nanoparticles for anticancer drug delivery. Int J Nanomedicine. https://doi.org/10.2147/ijn.S87145

Jiang K, Chi T, Li T et al (2017) A smart pH-responsive nano-carrier as a drug delivery system for the targeted delivery of ursolic acid: suppresses cancer growth and metastasis by modulating P53/MMP-9/PTEN/CD44 mediated multiple signaling pathways. Nanoscale 9(27):9428–9439. https://doi.org/10.1039/c7nr01677h

Jiao L, Zhang X, Cui J et al (2019) Three-in-one functional silica nanocarrier with singlet oxygen generation, storage/release, and self-monitoring for enhanced fractional photodynamic therapy. ACS Appl Mater Interfaces 11(29):25750–25757. https://doi.org/10.1021/acsami.9b08371

Jiménez-Amezcua I, Carmona FJ, Romero-García I et al (2018) Silk fibroin nanoparticles as biocompatible nanocarriers of a novel light-responsive CO-prodrug. Dalton Trans 47(31):10434–10438. https://doi.org/10.1039/c8dt02125b

Jin H, Sun M, Shi L et al (2018) Reduction-responsive amphiphilic polymeric prodrugs of camptothecin–polyphosphoester for cancer chemotherapy. Biomater Sci-UK 6(6):1403–1413. https://doi.org/10.1039/c8bm00162f

Kamaly N, Yameen B, Wu J et al (2016) Degradable controlled-release polymers and polymeric nanoparticles: mechanisms of controlling drug release. Chem Rev 116(4):2602–2663. https://doi.org/10.1021/acs.chemrev.5b00346

Kazempour M, Edjlali L, Akbarzadeh A et al (2019) Synthesis and characterization of dual pH-and thermo-responsive graphene-based nanocarrier for effective anticancer drug delivery. J Drug Deliv Sci Technol 54. https://doi.org/10.1016/j.jddst.2019.101158

Kesharwani P, Jain K, Jain NK (2014) Dendrimer as nanocarrier for drug delivery. Prog Polym Sci 39(2):268–307. https://doi.org/10.1016/j.progpolymsci.2013.07.005

Lea-Banks H, O'Reilly MA, Hynynen K (2019) Ultrasound-responsive droplets for therapy: a review. J Control Release 293:144–154. https://doi.org/10.1016/j.jconrel.2018.11.028

Li J, Liu P (2019) Self-assembly of drug–drug conjugates as drug self-delivery system for tumor-specific pH-triggered release. Part Part Syst Charact 36(7). https://doi.org/10.1002/ppsc.201900113

Li N-N, Lin J, Gao D et al (2014) A macromolecular prodrug strategy for combinatorial drug delivery. J Colloid Interface Sci 417:301–309. https://doi.org/10.1016/j.jcis.2013.11.061

Li Y, Hu H, Zhou Q et al (2017) α-amylase- and redox-responsive nanoparticles for tumor-targeted drug delivery. ACS Appl Mater Interfaces 9(22):19215–19230. https://doi.org/10.1021/acsami.7b04066

Li E, Yang Y, Hao G et al (2018) Multifunctional magnetic mesoporous silica nanoagents for in vivo enzyme-responsive drug delivery and MR imaging. Nanotheranostics 2(3):233–242. https://doi.org/10.7150/ntno.25565

Li H, Zhao Y, Jia Y et al (2019a) Covalently assembled dopamine nanoparticle as an intrinsic photosensitizer and pH-responsive nanocarrier for potential application in anticancer therapy. Chem Commun 55(100):15057–15060. https://doi.org/10.1039/c9cc08294h

Li J, Hu Z-E, Yang X-L et al (2019b) GSH/pH dual-responsive biodegradable camptothecin polymeric prodrugs combined with doxorubicin for synergistic anticancer efficiency. Biomater Sci-UK 7(8):3277–3286. https://doi.org/10.1039/c9bm00425d

Li M, Du C, Guo N et al (2019c) Composition design and medical application of liposomes. Eur J Med Chem 164:640–653. https://doi.org/10.1016/j.ejmech.2019.01.007

Li J, Li X, Liu P (2020) Doxorubicin-doxorubicin conjugate prodrug as drug self-delivery system for intracellular pH-triggered slow release. Colloids Surf B Biointerfaces 185. https://doi.org/10.1016/j.colsurfb.2019.110608

Lin M-H, Hung C-F, Hsu C-Y et al (2019) Prodrugs in combination with nanocarriers as a strategy for promoting antitumoral efficiency. Future Med Chem 11(16):2131–2150. https://doi.org/10.4155/fmc-2018-0388

Linderoth L, Peters GH, Madsen R et al (2009) Drug delivery by an enzyme-mediated cyclization of a lipid prodrug with unique bilayer-formation properties. Angew Chem Int Edit 48(10):1823–1826. https://doi.org/10.1002/anie.200805241

Liu D-M, Dong C (2020) Recent advances in nano-carrier immobilized enzymes and their applications. Process Biochem 92:464–475. https://doi.org/10.1016/j.procbio.2020.02.005

Liu J, Huang W, Pang Y et al (2011) Molecular self-assembly of a homopolymer: an alternative to fabricate drug-delivery platforms for cancer therapy. Angew Chem Int Edit 50(39):9162–9166. https://doi.org/10.1002/anie.201102280

Liu Y, Zhu S, Gu K et al (2017) GSH-activated NIR fluorescent prodrug for podophyllotoxin delivery. ACS Appl Mater Interfaces 9(35):29496–29504. https://doi.org/10.1021/acsami.7b07091

Liu JF, Jang B, Issadore D et al (2019) Use of magnetic fields and nanoparticles to trigger drug release and improve tumor targeting. Wiley Interdiscip Rev Nanomed Nanobiotechnol 11(6). https://doi.org/10.1002/wnan.1571

Lu X, Miao L, Gao W et al (2020) Engineered PLGA microparticles for long-term, pulsatile release of STING agonist for cancer immunotherapy. Sci Transl Med 12(556):eaaz6606. https://doi.org/10.1126/scitranslmed.aaz6606

Luo C, Sun J, Sun B et al (2014) Prodrug-based nanoparticulate drug delivery strategies for cancer therapy. Trends Pharmacol Sci 35(11):556–566. https://doi.org/10.1016/j.tips.2014.09.008

Luo C, Sun J, Liu D et al (2016a) Self-assembled redox dual-responsive prodrug-nanosystem formed by single thioether-bridged paclitaxel-fatty acid conjugate for cancer chemotherapy. Nano Lett 16(9):5401–5408. https://doi.org/10.1021/acs.nanolett.6b01632

Luo C, Sun J, Sun B et al (2016b) Facile fabrication of tumor redox-sensitive nanoassemblies of small-molecule oleate prodrug as potent chemotherapeutic nanomedicine. Small 12(46):6353–6362. https://doi.org/10.1002/smll.201601597

Ma N, Li Y, Xu H et al (2010) Dual redox responsive assemblies formed from diselenide block copolymers. J Am Chem Soc 132(2):442–443. https://doi.org/10.1021/ja908124g

Ma X, Özliseli E, Zhang Y et al (2019a) Fabrication of redox-responsive doxorubicin and paclitaxel prodrug nanoparticles with microfluidics for selective cancer therapy. Biomater Sci-UK 7(2):634–644. https://doi.org/10.1039/c8bm01333k

Ma Y, He P, Tian X et al (2019b) Mussel-derived, cancer-targeting peptide as pH-sensitive prodrug nanocarrier. ACS Appl Mater Interfaces 11(27):23948–23956. https://doi.org/10.1021/acsami.9b09031

Ma W, Sun J, Xu J et al (2020) Sensitizing triple negative breast cancer to tamoxifen chemotherapy via a redox-responsive vorinostat-containing polymeric prodrug nanocarrier. Theranostics 10(6):2463–2478. https://doi.org/10.7150/thno.38973

Mansouri M, Nazarpak MH, Solouk A et al (2017) Magnetic responsive of paclitaxel delivery system based on SPION and palmitoyl chitosan. J Magn Magn Mater 421:316–325. https://doi.org/10.1016/j.jmmm.2016.07.066

Marie H, Lemaire L, Franconi F et al (2015) Superparamagnetic liposomes for MRI monitoring and external magnetic field-induced selective targeting of malignant brain tumors. Adv Funct Mater 25(8):1258–1269. https://doi.org/10.1002/adfm.201402289

Martins C, Sousa F, Araújo F et al (2018) Functionalizing PLGA and PLGA derivatives for drug delivery and tissue regeneration applications. Adv Healthc Mater 7(1):1701035. https://doi.org/10.1002/adhm.201701035

McDaniel JR, Bhattacharyya J, Vargo KB et al (2013) Self-assembly of thermally responsive nanoparticles of a genetically encoded peptide polymer by drug conjugation. Angew Chem Int Edit 52(6):1683–1687. https://doi.org/10.1002/anie.201200899

McNeely ML, Binkley JM, Pusic AL et al (2012) A prospective model of care for breast cancer rehabilitation: postoperative and postreconstructive issues. Cancer 118(S8):2226–2236. https://doi.org/10.1002/cncr.27468

Mir M, Ahmed N, Rehman A (2017) Recent applications of PLGA based nanostructures in drug delivery. Colloids Surf B Biointerfaces 159:217–231. https://doi.org/10.1016/j.colsurfb.2017.07.038

Mu J, Lin J, Huang P et al (2018) Development of endogenous enzyme-responsive nanomaterials for theranostics. Chem Soc Rev 47(15):5554–5573. https://doi.org/10.1039/c7cs00663b

Mura S, Bui DT, Couvreur P et al (2015) Lipid prodrug nanocarriers in cancer therapy. J Control Release 208:25–41. https://doi.org/10.1016/j.jconrel.2015.01.021

Noh I, Kim H-O, Choi J et al (2015) Co-delivery of paclitaxel and gemcitabine via CD44-targeting nanocarriers as a prodrug with synergistic antitumor activity against human biliary cancer. Biomaterials 53:763–774. https://doi.org/10.1016/j.biomaterials.2015.03.006

Panja S, Dey G, Bharti R et al (2016) Tailor-made temperature-sensitive micelle for targeted and on-demand release of anticancer drugs. ACS Appl Mater Interfaces 8(19):12063–12074. https://doi.org/10.1021/acsami.6b03820

Paris JL, Cabañas MV, Manzano M et al (2015) Polymer-grafted mesoporous silica nanoparticles as ultrasound-responsive drug carriers. ACS Nano 9(11):11023–11033. https://doi.org/10.1021/acsnano.5b04378

Parmentier J, Becker MMM, Heintz U et al (2011) Stability of liposomes containing bio-enhancers and tetraether lipids in simulated gastro-intestinal fluids. Int J Pharm 405(1):210–217. https://doi.org/10.1016/j.ijpharm.2010.12.005

Patel VB, Misra AN (1999) Encapsulation and stability of clofazimine liposomes. J Microencapsul 16(3):357–367. https://doi.org/10.1080/026520499289077

Penfold NJW, Yeow J, Boyer C et al (2019) Emerging trends in polymerization-induced self-assembly. ACS Macro Lett 8(8):1029–1054. https://doi.org/10.1021/acsmacrolett.9b00464

Plourde K, Derbali RM, Desrosiers A et al (2017) Aptamer-based liposomes improve specific drug loading and release. J Control Release 251:82–91. https://doi.org/10.1016/j.jconrel.2017.02.026

Qu Y, Chu BY, Peng JR et al (2015) A biodegradable thermo-responsive hybrid hydrogel: therapeutic applications in preventing the post-operative recurrence of breast cancer. NPG Asia Mater 7(8):e207. https://doi.org/10.1038/am.2015.83

Ragazzon G, Prins LJ (2018) Energy consumption in chemical fuel-driven self-assembly. Nat Nanotechnol 13(10):882–889. https://doi.org/10.1038/s41565-018-0250-8

Raza A, Hayat U, Rasheed T et al (2018) Redox-responsive nano-carriers as tumor-targeted drug delivery systems. Eur J Med Chem 157:705–715. https://doi.org/10.1016/j.ejmech.2018.08.03

Rezaei SJT, Sarijloo E, Rashidzadeh H et al (2020) pH-triggered prodrug micelles for cisplatin delivery: preparation and in vitro/vivo evaluation. React Funct Polym 146. https://doi.org/10.1016/j.reactfunctpolym.2019.104399

Sánchez-López E, Guerra M, Dias-Ferreira J et al (2019) Current applications of nanoemulsions in cancer therapeutics. Nanomaterials 9(6). https://doi.org/10.3390/nano9060821

Sauraj, Kumar SU, Kumar V et al (2018) pH-responsive prodrug nanoparticles based on xylan-curcumin conjugate for the efficient delivery of curcumin in cancer therapy. Carbohydr Polym 188:252–259. https://doi.org/10.1016/j.carbpol.2018.02.006

Scomparin A, Florindo HF, Tiram G et al (2017) Two-step polymer- and liposome-enzyme prodrug therapies for cancer: PDEPT and PELT concepts and future perspectives. Adv Drug Deliv Rev 118:52–64. https://doi.org/10.1016/j.addr.2017.09.011

Shao W, Liu X, Sun G et al (2018) Construction of drug–drug conjugate supramolecular nanocarriers based on water-soluble pillar[6]arene for combination chemotherapy. Chem Commun 54(68):9462–9465. https://doi.org/10.1039/c8cc05180a

Shen W, Liu W, Yang H et al (2018) A glutathione-responsive sulfur dioxide polymer prodrug as a nanocarrier for combating drug-resistance in cancer chemotherapy. Biomaterials 178:706–719. https://doi.org/10.1016/j.biomaterials.2018.02.011

Sheth T, Seshadri S, Prileszky T et al (2020) Multiple nanoemulsions. Nat Rev Mater 5(3):214–228. https://doi.org/10.1038/s41578-019-0161-9

Shi L, Wang Y, Wang Q et al (2020) Transforming a toxic drug into an efficacious nanomedicine using a lipoprodrug strategy for the treatment of patient-derived melanoma xenografts. J Control Release 324:289–302. https://doi.org/10.1016/j.jconrel.2020.05.025

Signorell RD, Luciani P, Brambilla D et al (2018) Pharmacokinetics of lipid-drug conjugates loaded into liposomes. Eur J Pharm Biopharm 128:188–199. https://doi.org/10.1016/j.ejpb.2018.04.003

Song H, Kang X, Sun J et al (2016) Nanoparticle delivery of sterically hindered platinum(iv) prodrugs shows 100 times higher potency than that of cisplatin upon light activation. Chem Commun 52(11):2281–2283. https://doi.org/10.1039/c5cc09534d

Song L, Zhang B, Jin E et al (2018) A reduction-sensitive thermo-responsive polymer: synthesis, characterization, and application in controlled drug release. Eur Polym J 101:183–189. https://doi.org/10.1016/j.eurpolymj.2018.02.022

Souza C, Pellosi DS, Tedesco AC (2019) Prodrugs for targeted cancer therapy. Expert Rev Anticancer Ther 19(6):483–502. https://doi.org/10.1080/14737140.2019.1615890

Su X, Zhuang W, Yu T et al (2020) ROS and GSH dual-responsive GEM prodrug micelles for ROS-triggered fluorescence turn on bioimaging and cancer therapy. Adv Mater Interfaces 7(13). https://doi.org/10.1002/admi.202000294

Sun J, Chen Y, Li K et al (2016) A prodrug micellar carrier assembled from polymers with pendant farnesyl thiosalicylic acid moieties for improved delivery of paclitaxel. Acta Biomater 43:282–291. https://doi.org/10.1016/j.actbio.2016.07.014

Sun B, Luo C, Yu H et al (2018) Disulfide bond-driven oxidation- and reduction-responsive prodrug nanoassemblies for cancer therapy. Nano Lett 18(6):3643–3650. https://doi.org/10.1021/acs.nanolett.8b00737

Sun B, Luo C, Zhang X et al (2019a) Probing the impact of sulfur/selenium/carbon linkages on prodrug nanoassemblies for cancer therapy. Nat Commun 10(1). https://doi.org/10.1038/s41467-019-11193-x

Sun M, Qian Q, Shi L et al (2019b) Amphiphilic drug-drug conjugate for cancer therapy with combination of chemotherapeutic and antiangiogenesis drugs. Sci China Chem 63(1):35–41. https://doi.org/10.1007/s11426-019-9602-4

Sun Y, Chen C, Stang PJ (2019c) Soft materials with diverse suprastructures via the self-assembly of metal–organic complexes. Acc Chem Res 52(3):802–817. https://doi.org/10.1021/acs.accounts.8b00663

Sur S, Rathore A, Dave V et al (2019) Recent developments in functionalized polymer nanoparticles for efficient drug delivery system. Nano-Struct Nano-Objects 20. https://doi.org/10.1016/j.nanoso.2019.100397

Taresco V, Alexander C, Singh N et al (2018) Stimuli-responsive prodrug chemistries for drug delivery. Adv Ther 1(4). https://doi.org/10.1002/adtp.201800030

Van Tran V, Moon J-Y, Lee Y-C (2019) Liposomes for delivery of antioxidants in cosmeceuticals: challenges and development strategies. J Control Release 300:114–140. https://doi.org/10.1016/j.jconrel.2019.03.003

Vega-Villa KR, Takemoto JK, Yáñez JA et al (2008) Clinical toxicities of nanocarrier systems. Adv Drug Deliv Rev 60(8):929–938. https://doi.org/10.1016/j.addr.2007.11.007

Wang Y, Liu D, Zheng Q et al (2014) Disulfide bond bridge insertion turns hydrophobic anticancer prodrugs into self-assembled nanomedicines. Nano Lett 14(10):5577–5583. https://doi.org/10.1021/nl502044x

Wang Y, Wang C, Li Y et al (2017) Digitization of endocytic pH by hybrid ultra-pH-sensitive nanoprobes at single-organelle resolution. Adv Mater 29(1):1603794. https://doi.org/10.1002/adma.201603794

Wang D, Yu C, Xu L et al (2018a) Nucleoside analogue-based supramolecular nanodrugs driven by molecular recognition for synergistic cancer therapy. J Am Chem Soc 140(28):8797–8806. https://doi.org/10.1021/jacs.8b04556

Wang Y, Wang X, Deng F et al (2018b) The effect of linkers on the self-assembling and anti-tumor efficacy of disulfide-linked doxorubicin drug-drug conjugate nanoparticles. J Control Release 279:136–146. https://doi.org/10.1016/j.jconrel.2018.04.019

Wei Z, Wang H, Xin G et al (2020) A pH-sensitive prodrug nanocarrier based on diosgenin for doxorubicin delivery to efficiently inhibit tumor metastasis. Int J Nanomedicine 15:6545–6560. https://doi.org/10.2147/ijn.S250549

Woods D, Doty D, Myhrvold C et al (2019) Diverse and robust molecular algorithms using reprogrammable DNA self-assembly. Nature 567(7748):366–372. https://doi.org/10.1038/s41586-019-1014-9

Wu J, Han H, Jin Q et al (2017a) Design and proof of programmed 5-aminolevulinic acid prodrug nanocarriers for targeted photodynamic cancer therapy. ACS Appl Mater Interfaces 9(17):14596–14605. https://doi.org/10.1021/acsami.6b15853

Wu P, Jia Y, Qu F et al (2017b) Ultrasound-responsive polymeric micelles for sonoporation-assisted site-specific therapeutic action. ACS Appl Mater Interfaces 9(31):25706–25716. https://doi.org/10.1021/acsami.7b05469

Xiao H, Song H, Yang Q et al (2012) A prodrug strategy to deliver cisplatin(IV) and paclitaxel in nanomicelles to improve efficacy and tolerance. Biomaterials 33(27):6507–6519. https://doi.org/10.1016/j.biomaterials.2012.05.049

Xiao H, Noble GT, Stefanick JF et al (2014) Photosensitive Pt(IV)–azide prodrug-loaded nanoparticles exhibit controlled drug release and enhanced efficacy in vivo. J Control Release 173:11–17. https://doi.org/10.1016/j.jconrel.2013.10.020
Xu J, Merlier F, Avalle B et al (2019) Molecularly imprinted polymer nanoparticles as potential synthetic antibodies for immunoprotection against HIV. ACS Appl Mater Interfaces 11(10):9824–9831. https://doi.org/10.1021/acsami.8b22732
Yang B, Wei L, Wang Y et al (2020a) Oxidation-strengthened disulfide-bridged prodrug nanoplatforms with cascade facilitated drug release for synergetic photochemotherapy. Asian J Pharm Sci 15(5):637–645. https://doi.org/10.1016/j.ajps.2019.09.001
Yang Y, Sun B, Zuo S et al (2020b) Trisulfide bond–mediated doxorubicin dimeric prodrug nanoassemblies with high drug loading, high self-assembly stability, and high tumor selectivity. Sci Adv 6(45):eabc1725. https://doi.org/10.1126/sciadv.abc1725
Yin Q, Shen J, Zhang Z et al (2013) Reversal of multidrug resistance by stimuli-responsive drug delivery systems for therapy of tumor. Adv Drug Deliv Rev 65(13–14):1699–1715. https://doi.org/10.1016/j.addr.2013.04.011
Yu J, Li W, Yu D (2018a) Atrial natriuretic peptide modified oleate adenosine prodrug lipid nanocarriers for the treatment of myocardial infarction: in vitro and in vivo evaluation. Drug Des Devel Ther 12:1697–1706. https://doi.org/10.2147/dddt.S166749
Yu L, Chen Y, Lin H et al (2018b) Ultrasmall mesoporous organosilica nanoparticles: morphology modulations and redox-responsive biodegradability for tumor-specific drug delivery. Biomaterials 161:292–305. https://doi.org/10.1016/j.biomaterials.2018.01.046
Zaro JL (2014) Lipid-based drug carriers for prodrugs to enhance drug delivery. AAPS J 17(1):83–92. https://doi.org/10.1208/s12248-014-9670-z
Zeng Y, Ma J, Zhan Y et al (2018) Hypoxia-activated prodrugs and redox-responsive nanocarriers. Int J Nanomedicine 13:6551–6574. https://doi.org/10.2147/ijn.S173431
Zhai Y, Zhou X, Zhang Z et al (2018) Design, synthesis, and characterization of Schiff base bond-linked pH-responsive doxorubicin prodrug based on functionalized mPEG-PCL for targeted cancer therapy. Polymers 10(10). https://doi.org/10.3390/polym10101127
Zhang Y, Yang C, Wang W et al (2016) Co-delivery of doxorubicin and curcumin by pH-sensitive prodrug nanoparticle for combination therapy of cancer. Sci Rep-UK 6(1). https://doi.org/10.1038/srep21225
Zhang Q, Chen X, Shi H et al (2017a) Thermo-responsive mesoporous silica/lipid bilayer hybrid nanoparticles for doxorubicin on-demand delivery and reduced premature release. Colloids Surf B Biointerfaces 160:527–534. https://doi.org/10.1016/j.colsurfb.2017.10.005
Zhang X, Li X, You Q et al (2017b) Prodrug strategy for cancer cell-specific targeting: a recent overview. Eur J Med Chem 139:542–563. https://doi.org/10.1016/j.ejmech.2017.08.010
Zhang H, Zhou T, Yu Q et al (2019a) pH-Sensitive betulinic acid polymer prodrug nanoparticles for efficient and targeted cancer cells treatment. Int J Polym Mater Po 69(10):659–668. https://doi.org/10.1080/00914037.2019.1596916
Zhang J, Guo Y, Ding F et al (2019b) A camptothecin-grafted DNA tetrahedron as a precise nanomedicine to inhibit tumor growth. Angew Chem Int Edit 58(39):13794–13798. https://doi.org/10.1002/anie.201907380
Zhang R, Zhang Z, McClements DJ (2020) Nanoemulsions: an emerging platform for increasing the efficacy of nutraceuticals in foods. Colloids Surf B Biointerfaces 194. https://doi.org/10.1016/j.colsurfb.2020.111202
Zhao D, Liu N, Shi K et al (2015) Preparation of a multifunctional verapamil-loaded nano-carrier based on a self-assembling PEGylated prodrug. Colloids Surf B Biointerfaces 135:682–688. https://doi.org/10.1016/j.colsurfb.2015.08.018
Zhao Y, Deng Y, Tang Z et al (2018) Zwitterionic reduction-activated supramolecular prodrug nanocarriers for photodynamic ablation of cancer cells. Langmuir 35(5):1919–1926. https://doi.org/10.1021/acs.langmuir.8b02745
Zhong T, Huang R, Tan L (2016) Amphiphilic drug–drug assembly via dual-responsive linkages for small-molecule anticancer drug delivery. RSC Adv 6(71):66420–66430. https://doi.org/10.1039/c6ra15675d

Zhou S, Shang Q, Wang N et al (2020) Rational design of a minimalist nanoplatform to maximize immunotherapeutic efficacy: four birds with one stone. J Control Release 328:617–630. https://doi.org/10.1016/j.jconrel.2020.09.035
Zhu L, Wang D, Wei X et al (2013) Multifunctional pH-sensitive superparamagnetic iron-oxide nanocomposites for targeted drug delivery and MR imaging. J Control Release 169(3):228–238. https://doi.org/10.1016/j.jconrel.2013.02.015

Chapter 6
Smart Drug Delivery Vehicle

Zhaohua Miao and Zhengbao Zha

Abstract Smart drug delivery vehicles, which are responsive to endogenous or exogenous stimuli, are appealing therapeutic platforms with less side effects and enhanced treatment efficacy for the aim to echo the increased requirements of precision medications (Fig. 6.1). On the basis of the stimuli-responsive property, drugs could be controllably released and accumulated in the region of interest by designing appropriate delivery vehicles that are sensitive to endogenous physiological signals or extrinsic stimulations. Here, various endogenous and exogenous stimulations-responsive drug vehicles are presented and discussed.

Keywords pH-responsive · Redox · ATP · Enzyme · Hypoxia · Glucose

6.1 Endogenous Stimulation

The endogenous stimulations mainly include pH, redox, ATP, enzymes, hypoxia, and so on (Fig. 6.1).

6.1.1 pH-Responsive Vehicles

The heterogeneous distribution of pH in living organisms is the basic principle for pH-responsive vehicles. The normal pH value of the majority of body fluids is 7.4. However, at the cellular level, lysosomes maintain acid environment with a pH of ~5 through the special transporter (H+ pump) on the membrane. In addition, for tumor microenvironment or inflammatory sections, their pH ranges from 5.5 to 6.8.

Z. Miao · Z. Zha (✉)
School of Food and Biological Engineering, Hefei University of Technology, Hefei, China
e-mail: zbzha@hfut.edu.cn

L. Teng et al. (eds.), *Drug Delivery to Tumors*,
https://doi.org/10.1007/978-981-19-8930-8_6

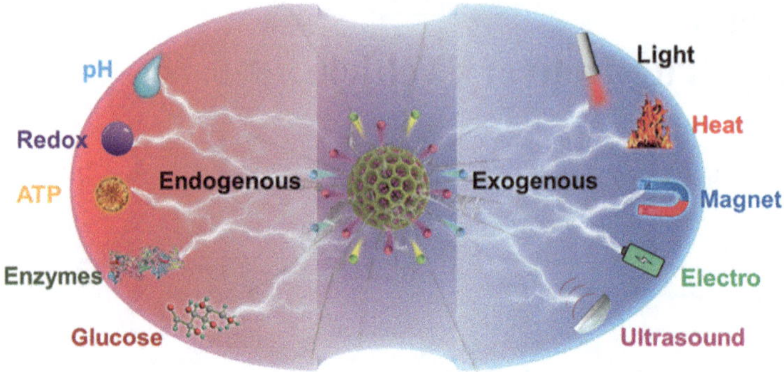

Fig. 6.1 Schematic illustration of smart drug delivery vehicles with endogenous or exogenous stimulation-responsive capabilities

These pH differences in various regions encourage researchers to design a series of pH-sensitive vehicles.

Nanoformulations with acid-responsive-hydrolysis chemical bonds (e.g., amide, acetal) were firstly designed and explored as pH-sensitive vehicles. Liu et al. described a biomimetic "platesome" drug delivery system (PEOz-platesome-DOX) in response to the acidic microenvironment of lysosomes (Liu et al. 2019). As shown in Fig. 6.2, the platesome nanovehicles were composed of platelet membranes and synthetic liposomes modified by pH-sensitive lipid DSPE-PEOz with amide bonds. Upon intravenous injection into mouse tumor models, platesome could spontaneously accumulate in tumor region, followed by the disassembly process of platesome in the acidic environment (tumor microenvironment and lysosomal compartments), thus enabling the selective release of DOX cargo. After the treatment, excellent tumor treatment efficiency was achieved using the pH-sensitive platesome, and the therapeutic results were much better than that of conventional liposomes.

To design pH-sensitive vehicles, physical swelling is another widely used strategy. For instance, Li et al. designed pH-induced charge reversal photodynamic nanoagents (PPNs) consisting of upconversion nanoparticles and pH-responsive polymeric ligands (Li et al. 2018). As shown in Fig. 6.3, because of the imidazole group ionization (pKa $\approx$ 6.8) in acidic environment, negatively charged PPNs began to swell and partially dissociated in $6.5 < pH < 7.4$ and further achieved the complete dissociation in $pH < 6.5$, thus depicting excellent antitumor efficacy in vitro and in vivo.

6.1.2 Redox-Responsive Vehicles

Hydroxyl radicals (•OH), superoxide anions (O^{2-}), hydrogen peroxide (H_2O_2), and other reactive oxygen species (ROS) molecules play important roles in signal transduction and homeostasis in vivo. Mitochondrial oxidative phosphorylation, drug

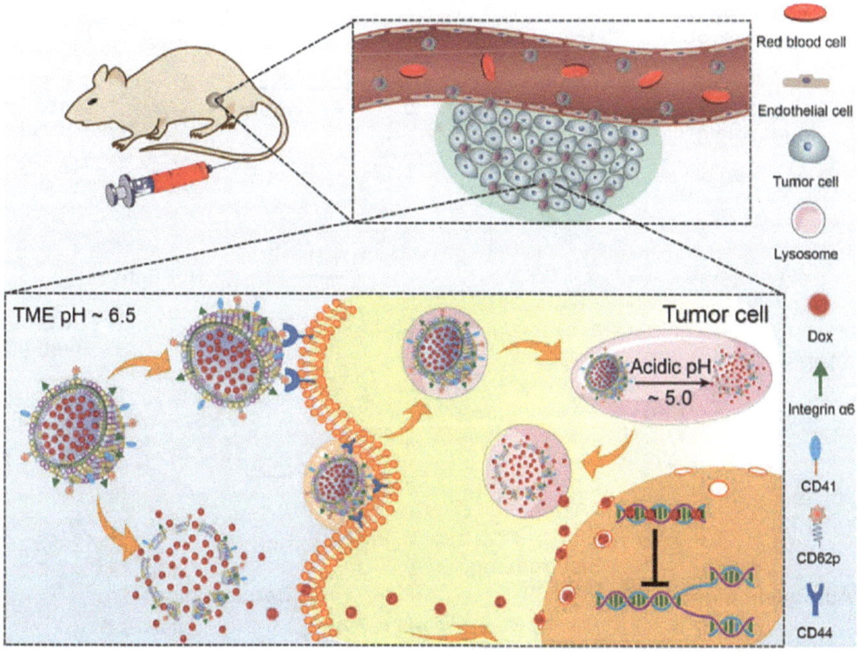

Fig. 6.2 Schematic illustration of the antitumor mechanism of platesome under the acidic pH of the tumor microenvironment/lysosomes. (Reused with permission from Liu et al. (2019))

uptake, microbial infection, and ionizing radiation would increase the ROS generation. The produced ROS could damage the DNA and proteins in cells, causing cell apoptosis. Thus, multiple ROS-responsive drug vehicles have been exploited for cancer treatments. These ROS-responsive drug vehicles usually contain thioether, diselenide, thioketal, aminoacrylate, and arylboronic ester-based moieties.

Thioketal is one of the most common moieties used to fabricate ROS-responsive drug vehicle. The thioketal can be cleaved into two thiols by various ROS, leading to the vehicle crush and drug release (Yang et al. 2021). For example, Wang and his team developed a ROS-responsive polymeric nanoparticle system with the capability to self-destruct for chemo-photodynamic therapy (Wang et al. 2019). By coupling the ROS-sensitive pendant thioketal to aliphatic polycarbonates, DOX could be released from polymer-conjugated DOX(PEG-PBC-TKDOX) since large-scale of ROS was produced by the photodynamic effect of the photosensitizer chlorin e6 (Ce6) under laser irradiation (Fig. 6.4). Such self-destructive polymeric system enhanced the efficiency of chemo-photodynamic therapy and decreased the off-target toxicity.

The other ROS-responsive moieties include diselenide, thioether, aminoacrylate, and arylboronic ester, etc. Diselenide bonds had similar design scheme with thioketal but were more sensitive than thioketal under mild stimulating conditions because of their lower bond energy (172 kJ mol^{-1}) (Saravanakumar et al. 2017). For instance,

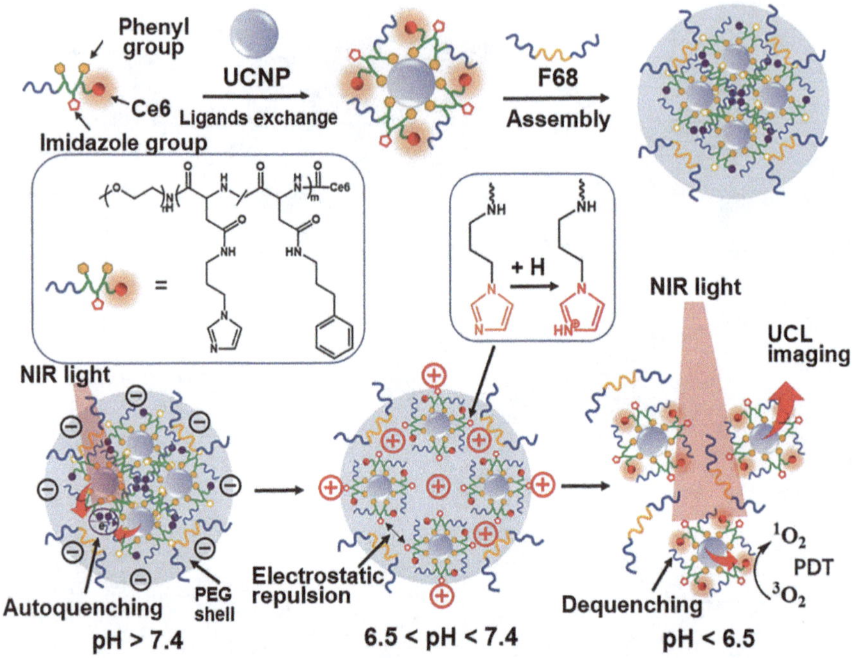

Fig. 6.3 Schematic illustration of pH-induced charge reversal and disassemble of PPNs under acid environment. (Reused with permission from Li et al. (2018))

Zhang et al. firstly reported diselenide bond-based ROS-responsive micelles (Ma et al. 2011). They designed a triblock copolymer of poly (ethylene glycol)-b-polyurethane-b-poly (ethylene glycol) (PEG-PUSeSe-PEG) containing diselenide bonds that could self-assemble into micelles. The obtained micelles were responsive to not only H_2O_2 stimuli but also γ-radiation (5 Gy) in low doses, suggesting high potential application for combined radiotherapy and chemotherapy.

6.1.3 ATP-Responsive Vehicles

Adenosine-5′-triphosphate (ATP), the intracellular energy transfer molecule, has been successfully applied as an effective and responsive agent release molecule due to striking ATP concentration difference between extracellular (<0.4 mM) and intracellular (1–10 mM) environments. Interaction of ATP with diverse molecules is a significant prerequisite for achieving controlled release of drugs. During the design of the ATP-responsive drug release system, the competitive coordination of ATP with biomolecules or metal ions was commonly employed.

DNA aptamers, in the presence of ATP, have a tendency to form a stable tertiary structure due to its high affinity with ATP. Therefore, they have been extensively

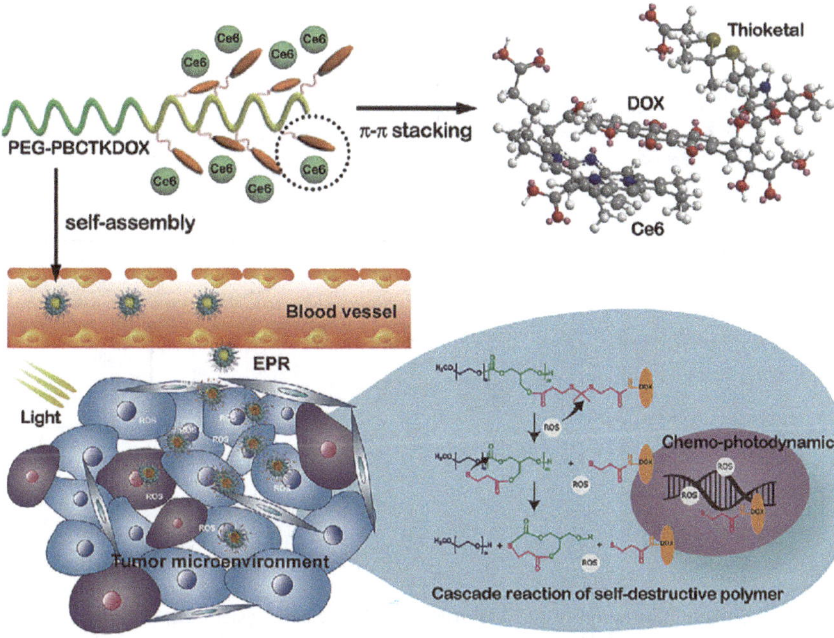

Fig. 6.4 Schematic illustration of ROS-sensitive self-destructive polymeric nanomicelles containing thioketal moiety for chemo-photodynamic therapy. (Reused with permission from Wang et al. (2019))

applied for controlled release of drugs. Mo et al. constructed an ATP-responsive nanogels encapsulating DOX (DOX/NG) (Fig. 6.5a) (Mo et al. 2014). As described in Fig. 6.4, HA on the nanocarrier could specifically target to the CD44 receptor on tumor cells through active targeting, thereby achieving intratumor aggregation of nanoparticles. Once the DNA aptamers combined with high concentration of ATP in cytosol to form a stable structure, drugs were released from the site of GC pairs (Fig. 6.5b, c). Eventually, the accumulation of DOX in the nucleus led to tumor cell death. Compared with non-ATP-responsive nanovehicles, ATP-responsive nanovehicles could significantly inhibit tumor growth. In addition, Liao et al. successfully fabricated a DNA microcapsule using $CaCO_3$ as a template by layer-by-layer assemble (Song et al. 2018). When exposed to ATP, ATP aptamer complex broke the DNA shell, resulting in the release of encapsulated macromolecules, biomolecules, or nanoparticle cargos.

Competitive coordination between ATP and metal ions is the second type of ATP-responsive drug release system (Yang et al. 2019). For example, Song et al. synthesized self-assembled quantum dots (QDs)-phenolic nanoclusters (NCs) by assembling QDs, tannic acid (TA), and DOX drug (QDs@TA-PEG/DOX) (Yang et al. 2019). The zinc-phenol ligands between the ZnS shell and various phenolic groups of TA caused the assembly of QDs. ATP was bounded to QDs by

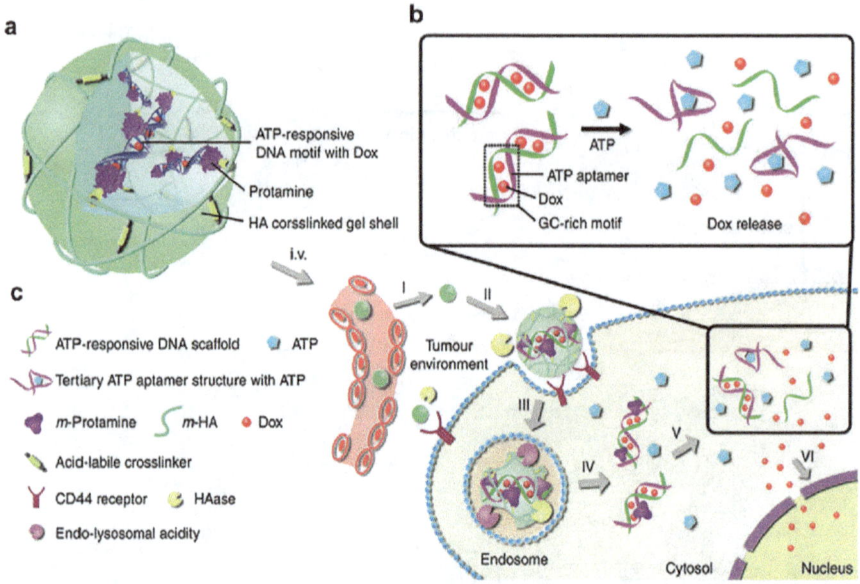

Fig. 6.5 Schematic illustration of the ATP-triggered Dox release system. (**a**) The main components of ATP-responsive DNA motif with Dox, protamine and a HA-crosslinked gel shell. (**b**) Mechanism of ATP-triggered release of Dox. (**c**) Delivery process of ATP-responsive DNA nanogels. (Reused with permission from Mo et al. (2014))

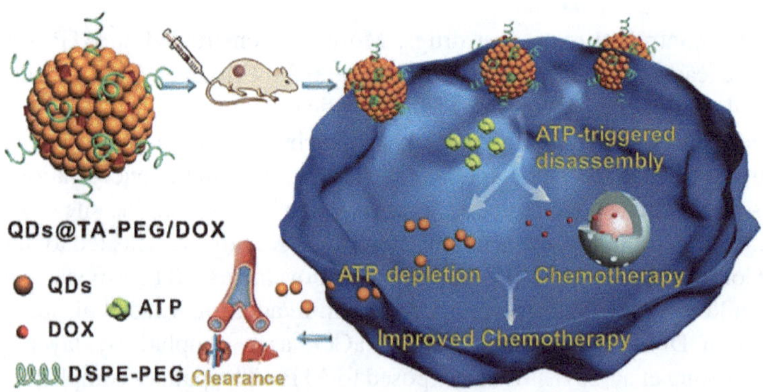

Fig. 6.6 Schematic illustration of QDs@TA-PEG/DOX enhancing chemotherapy by simultaneous ATP-responsive chemical drug release and cancer cell sensitization. (Reused with permission from Song et al. (2018))

metal-triphosphate coordination, which resulted in decomposition of QDs by substituting TA with ATP on the surface of QDs. On one hand, this process of ATP-responsive release of DOX enabled chemotherapy at specific sites; on the other hand, nanocarriers-induced ATP depletion enhanced chemosensitivity (Fig. 6.6).

6.1.4 Enzyme-Responsive Vehicles

Enzyme-responsive reactions have high catalytic efficiency and specificity to substrates. A large number of enzymes exist in the biological environment, and it is possible to use these enzymes as endogenous stimulus.

Many specifically and highly expressed enzymes exist in the tumor microenvironment, and integrating them into the nanoparticle system can give the material enzyme-responsive release characteristics. As shown in Fig. 6.7, Han et al. used the matrix metalloproteinase-9 (MMP-9) and cathepsin B-specific cleavage peptides as the corresponding switches and covalently connected them with poly (ethylene glycol) (PEG) and gemcitabine, respectively (Han et al. 2017). The two enzymes were located in different positions in tumor microenvironment. The specific peptides in nanoparticles were cleaved by MMP-9 existing in the tumor extracellular environment, leading to the removal of PEG shell to expose the internal targeting RGD peptides, which promoted the endocytosis of nanoparticles by tumor cells. When nanoparticles entered into the lysosome, drugs were released because cathepsin B cleaved the corresponding specific peptides.

The enzyme stimulation response not only directly cleaves specific peptides to release drugs but also changes the properties of nanoparticles by anchoring specific groups to change the chemical structure, resulting in the disassembly of the nanoparticles to release drugs indirectly. As shown in Fig. 6.8, Cai et al. prepared smart nanoclusters-based perylene monoimide (PMI) dyes, which were disassembled by enzyme response (Cai et al. 2020). The tetrachloroperylene monoimide core was modified by hydrophilic galactose and hydrophobic β-alanine methyl ester which both could be degraded by carboxylesterase (CE). After modification, the CE-responsive amphiphilic molecular precursor was synthesized and assembled with the folate-modified protein into a nanocluster (FHP) with a diameter of ~100 nm. The precursor of the amphiphilic molecule was enzymatically hydrolyzed to produce carboxylic acid hydrolysates when accumulated in tumor state. Then, its hydrophobic end was changed to a hydrophilic end, leading to the nanoparticle's disassembly. Meanwhile, the extra electrostatic repulsion of the deprotonated carboxyl group also promoted the disintegration of nanoclusters. As a result, the small molecule photosensitizers were successfully released into the deep section of tumor.

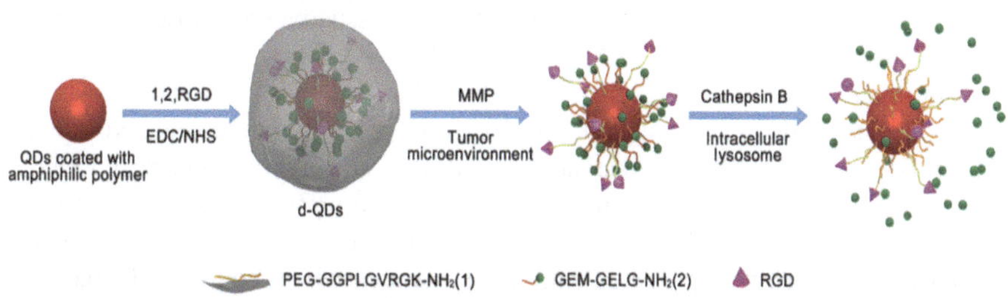

Fig. 6.7 Schematic illustration of dual enzymatic reactive gemcitabine delivery systems. (Reused with permission from Han et al. (2017))

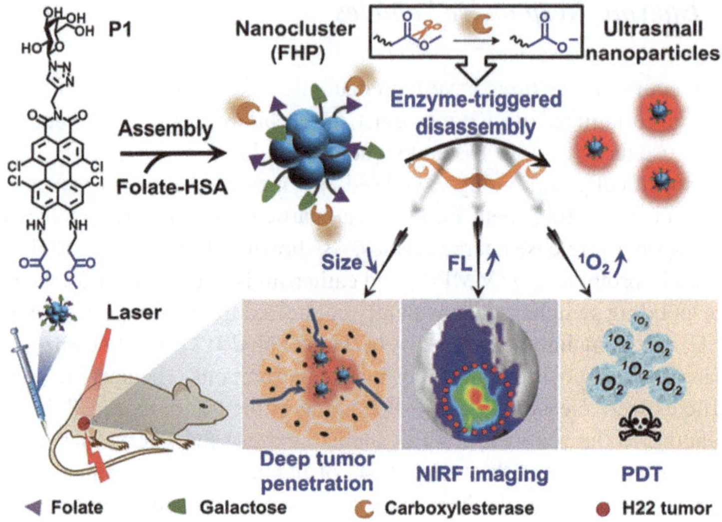

Fig. 6.8 Schematic illustration of enzyme-triggered disassembly of perylene monoimide-based nanoclusters for photodynamic therapy. (Reused with permission from Cai et al. (2020))

6.1.5 Hypoxia-Responsive Vehicles

Hypoxia is caused by the imbalance between oxygen supply and oxygen consumption in rapidly proliferating cells, and it is one of main features of many solid tumors. When the tumor cells are far away from the blood vessels, the oxygen concentration may drop to a relatively low level. Under the condition of hypoxia, tumor cells will change their own microenvironment and become resistant to many anti-cancer drugs, especially chemotherapy, radiotherapy, and photodynamic therapy (Zhang et al. 2020b). Therefore, hypoxia targeting drug delivery system has become a promising strategy for precise tumor treatment. The most commonly used hypoxia targeting therapy strategy is based on the covalent combination of hypoxia sensitive adhesive and drug to form the prodrug of hypoxia response. Hypoxia-responsive chemical bonds (such as azo/nitro) have been unitized for the preparation of nanomaterials which can be decomposed or self-assembled to change their size to improve the infiltration and accumulation of tumor tissue when exposed to hypoxia tumor regions.

Nano-formulation with azo chemical bonds that can be hydrolyzed at low oxygen concentrations was designed and explored as a hypoxic-sensitive carrier (Ihsanullah et al. 2020). For instance, Kai Zhang and his co-authors reported a hypoxic reactive albumin nanoparticle system for full tumor infiltration with good treatment effect (Zhang et al. 2020a). They designed and synthesized metal–organic frameworks (MOF) nanoparticles with azo structures which could be degraded in the hypoxic tumor microenvironment (TME) rapidly. In this way, Ce6 and Cu^{2+} were quickly released and enhanced tumor penetration was achieved. As shown in Fig. 6.9, a large quantity of Cu-MOF NPs were accumulated validly in tumors by

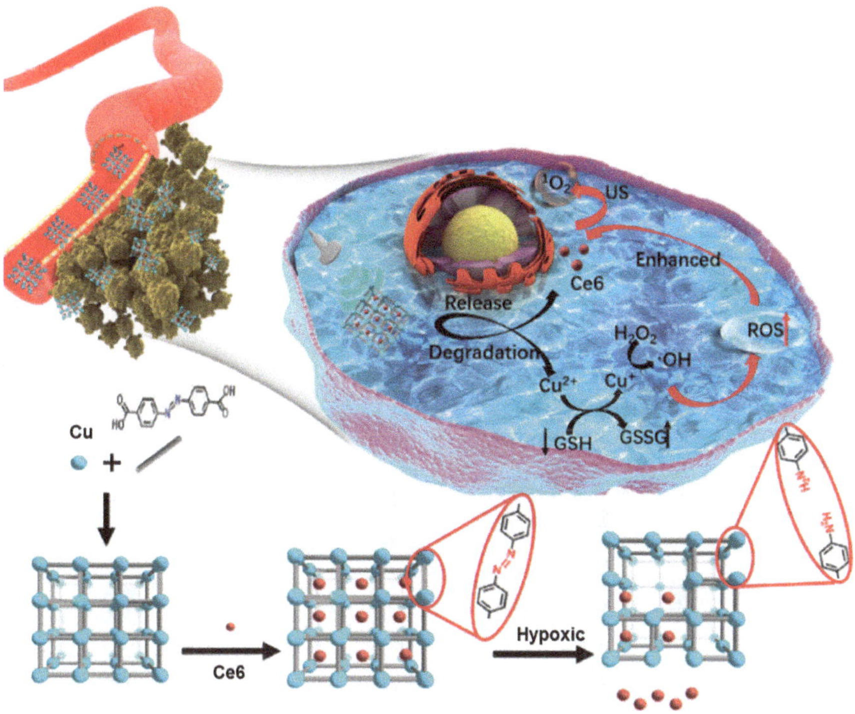

Fig. 6.9 Schematic illustration of synthesis of hypoxia-responsive Cu-MOF nanosystem for enhanced cancer therapy. (Reused with permission from Zhang et al. (2020a))

enhanced permeability and retention effect (EPR), and the hypoxia TME triggered their degradation, resulting in the Cu^{2+} and Ce6 release deeply into tumors. Moreover, GSH exhaustion was induced by redox between significant intracellular GSH and free Cu^{2+}, accompanied by the reduction of Cu^{2+} to Cu^{+}.

To design hypoxia-sensitive vehicles, nitro group is another widely used strategy. For instance, through self-assembly of 2-nitroimidazole-modified polypeptides and cationic lipids, Yujing Li and his co-authors prepared the hypoxic-responsive nanoparticles (HRNP) with siRNA targeting the hypoxic-associated tumor-promoting gene CDC20 in the treatment of breast cancer (Li et al. 2020). As depicted in Fig. 6.10, HRNP-mediated siCDC20 targeting vector could effectively silence the expression of CDC20 and showed strong anti-tumor effect.

6.1.6 Glucose-Responsive Vehicles

Materials responding to the changes of glucose concentration is the basic principle for designing a glucose response system. The initial glucose-sensitive substance was prepared by incorporating glucose oxidase (GOx) into pH-sensitive hydrogel.

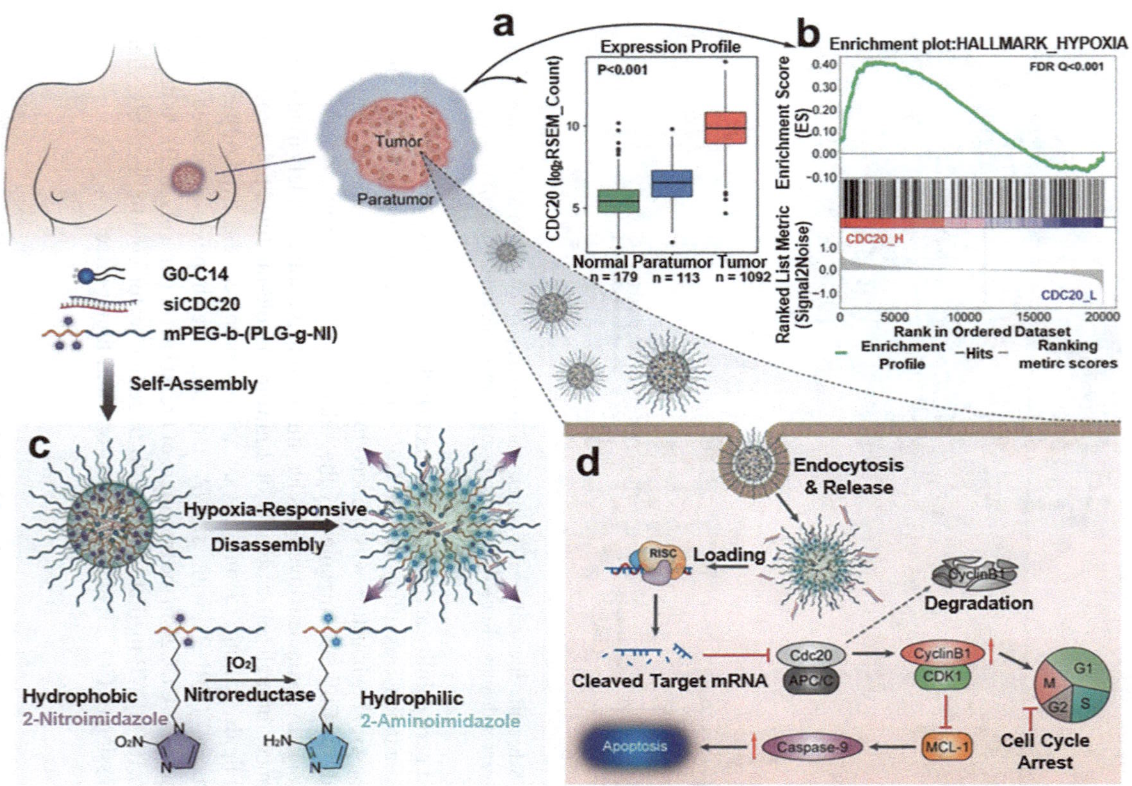

Fig. 6.10 Hypoxia-responsive nanoparticle (HRNP) for hypoxia-correlated pro-tumorigenic gene (CDC20) in targeting breast cancer therapy. (**a**) CDC20 expression levels in normal, paratumor, and tumor tissues. (**b**) The positive correlation of CDC20 with tumor hypoxia. (**c**) Hypoxia-responsive disassembly of nanoparticles. (**d**) Schematic of HRNP/siCDC20 delivery into tumor cell cytoplasm to induce G2/M arrest and cell apoptosis. (Reused with permission from Li et al. (2020))

The glucose-response characteristics of pH-sensitive substances were conferred by the immobilization of GOx, based on that GOx could convert glucose into gluconic acid when oxygen was present. The production of gluconic acid caused the reduction in pH, which further initiated the swelling of pH-sensitive polymer chains (Kashyap et al. 2007). Most of insulin transport and distribution systems were based on the swelling of hydrogels or membranes (Li and Luo 2009).

Hydrogel is a polymer network with hydrophilicity but remains indissoluble due to crosslinking. Glucose-reactive hydrogels usually contain pH-sensitive polymers and GOx as the glucose-reactive transducer, respectively. When GOx is trapped and covalently attached to the polymer, the hydrogel will swell or collapse after PH alteration, caused by the production of gluconic acid from glucose. Polymers sensitive to pH could be classified into cationic and anionic polymers. The response to solubilization consists of the charge properties of the main chain monomer. In more detail, when the pKa of the cationic group is lower than the surrounding pH, the copolymer exhibits hydrophobicity and discharges water; in contrast, when the pKa is higher than the surrounding pH, the cationic group becomes protonated and exhibits hydrophilicity, which in turn absorbs more water. Anionic polymers follow a similar mechanism to swell according to pH changes in biological environment (Wu et al. 2011).

Polymer composite membrane is another system for drug release. As shown in Fig. 6.11, a composite membrane was construct by GOx, catalase (CAT), and pH-responsive nanoparticles in the hydrophobic polymer solution (Gordijo et al. 2010). With the glucose concentration increased, the pH decreased in the system, causing the pH smart nanoparticles to collapse and releasing the insulin. In addition, MnO_2 nanoparticles were added to construct a bio-inorganic hybrid membrane, which both enhanced the mechanical intensity and the permanent stability of enzyme, causing the designed glucose-responsive system close to practical applications.

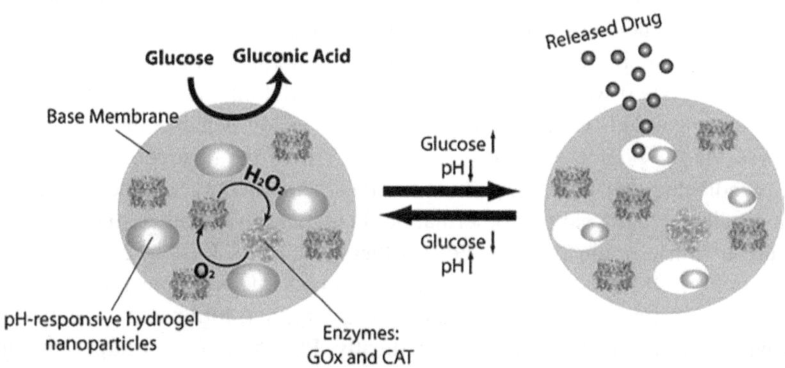

Fig. 6.11 The glucose-responsive process of the composite membrane. (Reused with permission from Gordijo et al. (2010))

6.2 Exogenous Stimuli

The exogenous stimulations include light, heat, magnet, electro and ultrasound, and so on.

6.2.1 *Light-Responsive Vehicles*

Light-responsive drug delivery is widely used for cancer therapy because it can achieve efficient and accurate targeted drug delivery under non-contact conditions. The light sources with various wavelengths include ultraviolet (UV) (200–400 nm), visible (Vis) (400–700 nm), and near infrared (NIR) (700–1100 nm) lights. Among them, NIR light is the most useful excitation source for drug delivery systems due to its low toxicity and deep tissue penetration.

One effective strategy of light-triggered drug delivery is to break specific chemical bonds in the nano-drug carrier through light to induce drug release. These chemical bonds include nitrobenzyl (Cao et al. 2013; Wong et al. 2015), coumarin, azobenzene (Croissant et al. 2014), and anthracene (Xie et al. 2016), etc. For example, the hydrophobic NIR absorbing dye methyl cyanine was loaded into the hydrophobic core of chitosan-based micelles with o-nitrobenzyl groups. Under NIR (765 nm) irradiation, the dye molecule displayed 808-nm fluorescence, which induced the break of o-nitrobenzyl part through a two-photon process, further promoting the drug release. Compared with the direct irradiation of 808-nm light source, the local luminescence of two-photon wavelength (808 nm) accelerated the photolysis of o-nitrobenzyl.

The combination of photothermal or photodynamic therapy (PDT) with other treatment is another effective strategy for light-responsive drug delivery. For instance, Gao et al. designed a light-responsive biomimetic DOX-cisplatin-loaded gold nanorods (GNRs) nanogel system in which the surface of GNRs was modified with hyaluronic acid. The DOX and cisplatin were loaded through electrostatic interaction and chelation reaction, respectively (Fig. 6.12a) (Gao et al. 2020). Due to the dual stabilizing effects of hyaluronic acid nanogel and 4T1 cell membrane, the delivery system was maintained in blood circulation and thus premature drug release was avoided. Under NIR laser irradiation, on-demand drug release was achieved via the photothermal effect of the GNRs in the nanogel reservoir. These results indicated that combined photothermal/chemotherapy depicted better treatment efficiency than sole photothermal or chemotherapy. Another study reported a mixture of nitric oxide (NO)-based gas therapy and PDT (Wan et al. 2018). As shown in Fig. 6.12b, LArg@PCN@Mem nanomedicine was constructed by incorporating L-Arg into the MOF material PCN as a carrier, followed by the modification with cancer cell membranes. Specifically, PCN produced a large amount of singlet oxygen after exposure to NIR radiation. The optically responsive ROS in

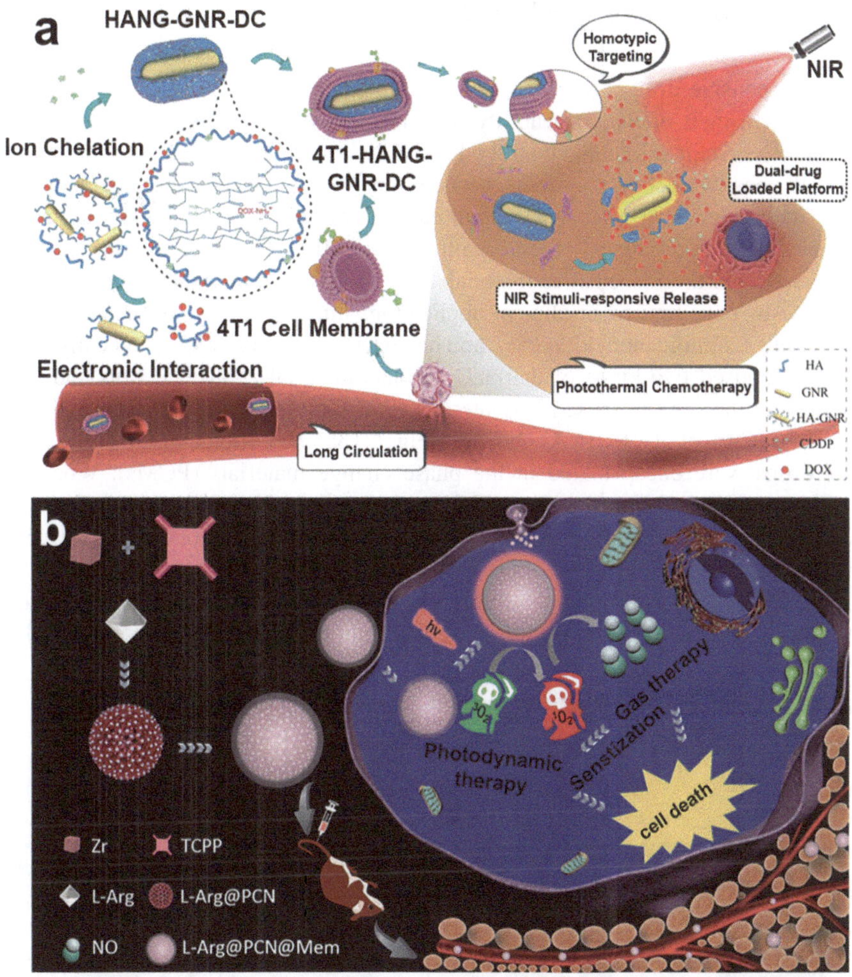

Fig. 6.12 (**a**) Schematic illustration of the synthesis of biomimetic DOX-cisplatin-loaded GNRs for combined chemotherapy and homotypic targeted photothermal therapy of cancer cells. (Reproduced from Gao et al. (2020)) (**b**) Schematic illustration of the synthesis of L-Arg@PCN@Mem and lethal mechanism of combined gas therapy and photodynamic therapy of tumor cells. (Reused with permission from Wan et al. (2018))

L-Arg@PCN@Mem showed a double-effect. From one point of view, it killed tumor cells directly as high-performance PDT. On the other hand, ROS could also induce the change of L-Arg into NO through oxidation, which was used for gas therapy. Due to the sensitization effect of NO, it could enhance the PDT effect under hypoxic environment.

6.2.2 Heat-Responsive Vehicles

Heat-responsive drug carriers are sensitive to temperature changes, and the drug carrier will melt or its permeability will increase when the external temperature exceeds a certain value, and then the loaded drug will be released. Among various stimuli-sensitive delivery systems, heat-responsive vehicles provide great potential over their counterparts because of their ease of control (e.g., microwave, radio frequency, light, high intensity focused ultrasound (HIFU)) and preparation.

Traditional thermosensitive materials (e.g., poly(N-isopropylacrylamide) (poly-NIP AAm)) are limited in clinical applications due to their unsatisfactory biodegradability, significant cytotoxicity, and harsh triggering conditions. Recently, Choi et al. proposed two kinds of thermal response materials: fatty alcohols and fatty acids, and the state of these two materials changed along with the change of temperature (Choi et al. 2010). As shown in Fig. 6.13a, the colloidal particles containing drugs were encapsulated in the phase change materials (PCMs). When the temperature rose, the particles leaked from the melting phase change materials, and the drug was subsequently released from the particles. This work effectively promoted the on-demand release of drugs.

Liposomes are widely used as drug carriers because they can avoid multidrug resistance and reduce host immune response, and they have also been explored as

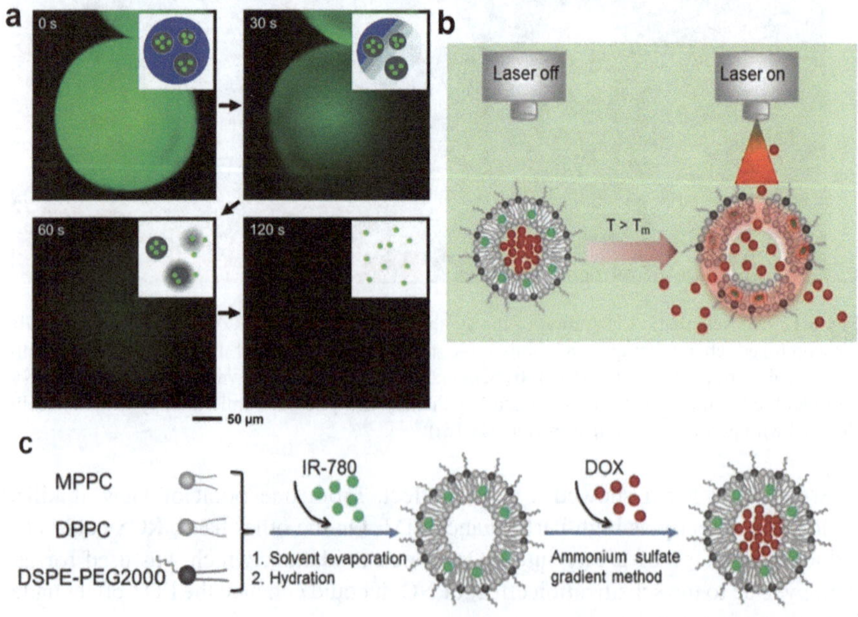

Fig. 6.13 (**a**) The schematic diagram shows the process of drug release from PCM with the increase of temperature. (**b**) Schematic illustration of DOX release from DITSL under NIR-laser irradiation. (**c**) Schematic diagram of preparation of DITSL. (Reused with permission from Yan et al. (2016))

thermo-sensitive drug delivery system. For example, Yan et al. reported a smart DOX/IR-780-loaded thermo-responsive liposome (DITSL) (Yan et al. 2016). This thermo-responsive drug delivery system was composed of liposomes containing IR-780 and DOX by using DPPC, MPPC, and DSPE-PEG2000 (Fig. 6.13c). Under the irradiation of NIR light, photothermal conversion agent IR-780 converted light into heat. When the temperature rose to 42 °C, the bilayer structure of liposomes was destroyed, and the permeability increased, thus resulting in the rapid drug release (Fig. 6.13b). This thermo-response drug delivery system not only reduced the side effects and improved the curative effect through controlled drug release but also had high drug loading and biocompatibility, exhibiting excellent antitumor efficiency in vitro and in vivo.

6.2.3 Magnet-Responsive Vehicles

Magnets have attracted extensive attentions in the past decades for the potential applications in new motors, remote controlled drug release, novel cancer thermal ablation therapy, and micro-machines (Wang et al. 2016; Mirvakili et al. 2020). Especially, the transport capacity and magnetic induction heating of magnets have been applied extensively for drug transportation. Wu et al. developed a γ-Fe_2O_3 microactuator with better magnetization properties than Fe_3O_4 nanoparticles and revealed that the release mechanism was mainly attributed to the expanding of liberal scope in the polymer network under magnetic induction heating (Wu et al. 2014). Wu et al. also reported superparamagnetic Fe_2O_3@MOF with high crystallite and porosity and excellent magnetic properties for controlled magnetic separation and drug release behavior.

In recent years, microbots and nanobots have been the potential solutions for targeted therapeutic intervention and manipulating drug delivery due to their precise targeting capabilities (Tang et al. 2020). The greatest advantage is their excellent controllability. Xiong et al. reported a shape-adaptive DNA robot that could perform magnetically driven navigation in a confined and unstructured space (Xiong et al. 2016). The results showed that the DNA robot could be used as a carrier to transport cells in limited space due to its three-dimensional porous network structure and good biocompatibility. In addition, Chen et al. constructed FeGa@P(VDF-TrFE) core-shell magnetoelectric nano-wires loaded with Taxol (Chen et al. 2017). Figure 6.14a shows that when the delivery system was surrounded by an alternating magnetic field, the combination of drug molecules with polydopamine was disrupted by the instantaneous change in polarization mode caused by the magnetoelectric effect, resulting in the release of the drug. As shown in Fig. 6.14b, the survival rate of cells incubated with drug-loaded nanowires under AC magnetic field was decreased by 40%. This was significantly higher than that in other groups, indicating the importance of magnet in killing cancer cells.

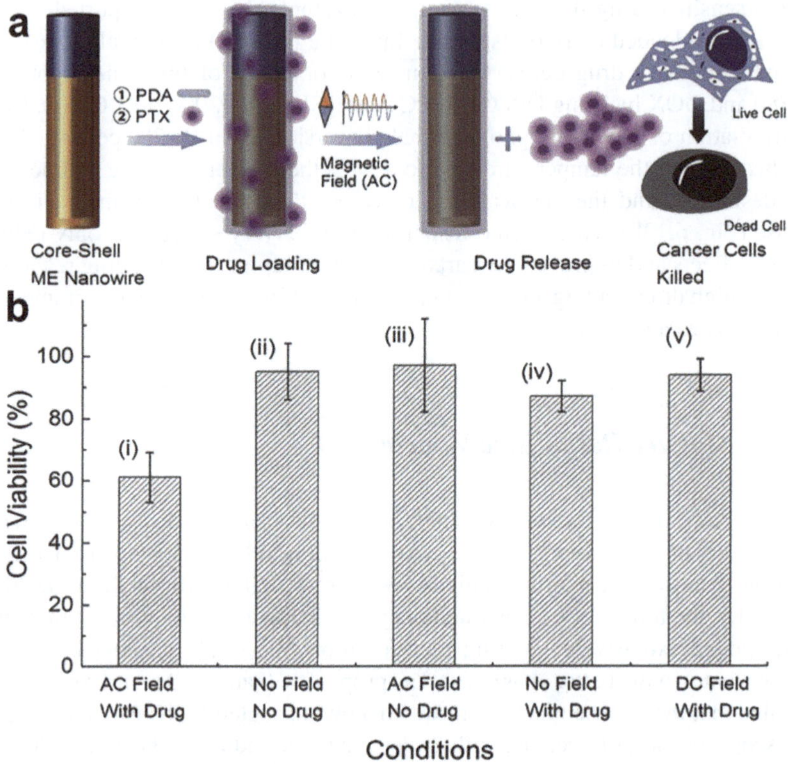

Fig. 6.14 (**a**) Scheme of FeGa@P(VDF-TrFE) core–shell nanowires for on-demand drug delivery experiment. (**b**) The killing efficacy of on-demand drug delivery with/without AC field. (Reused with permission from Chen et al. (2017))

6.2.4 Electric-Responsive Vehicles

Electric field is one of the physical triggers for biotherapy. In a variety of applications, electric fields can trigger the release of drugs from delivery systems. Different from other stimulus, the electrical response system enables spatiotemporal control of drug release. Therefore, the active ingredient is expected to use a low intensity external electric field and released by a continuous or pulsed mode. High intensity extraneous electric fields are reported to directly affect cell membrane permeability. Thus, electric field not only triggers the controlled release of drugs but also improves the healing and restores tissue integrity. Electric field response systems, especially these based on conductive polymers, represent a class of materials which can control the release of active ingredients, such as small molecules, peptides, and proteins.

Conductive polymers and hydrogel-based materials are the promising substrates for electronically controlled release of drugs or as bioactive scaffolds. For example, Servant et al. designed an electro-responsive polymer hydrogel containing pristine multiwalled carbon nanotubes (pMWNT) by in situ radical polymerization for

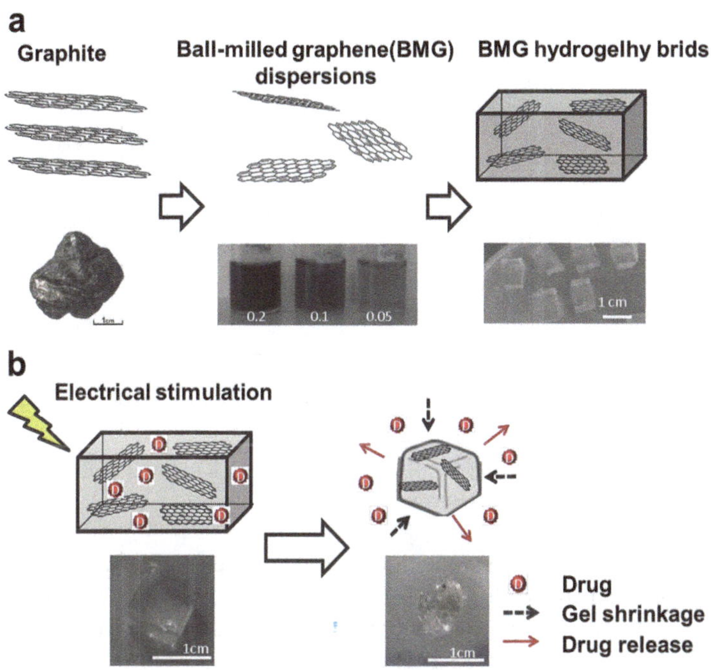

Fig. 6.15 Electro-responsive graphene/PMAA hydrogel hybrids for "on-demand" drug delivery. (**a**) A scheme of the synthesis of BMG hydrogel hybrids. (**b**) Pulsaltile drug release upon ON/OFF application of a DC electric field. (Reused with permission from Servant et al. (2013))

pulsatile drug release (Servant et al. 2013). Then, radio-labeled sucrose was applied to a hydrophilic small drug molecule model to determine the release characteristics. The system combined the electrical sensitivity of conductivity and improved the responding ability to electric fields, thus achieving encouraging results for pulsed drug release in vivo. Besides, their group also developed an electro-responsive polymer hydrogel device using graphene sheets which were well dispersed and could be used for pulsatile drug delivery in vivo as shown in Fig. 6.15 (Servant et al. 2014). By using ball milling, graphene aqueous suspensions were obtained via liquid exfoliation of graphite by means of its molecular interactions with melamine. Then, graphene hydrogel hybrids were synthesized by in situ radical polymerization when the methacrylic acid (MAA) and N, N′-methylenebisacrylamide (BIS) exist. This hybrid hydrogel has strengthened mechanical, electrical, and thermal properties. It could maintain its structural integrity at low concentrations and controlled substance release in a pulsed manner in the ON/OFF application at low voltage. In the meantime, this system showed depicted drug release in the body without any resistance to heating in vivo.

Electroporation is a newly developed technology for drug delivery. However, it has the problems of low-throughput and severe cell damage. To solve these problems, Liu et al. designed a high-throughput electroporation system, in which

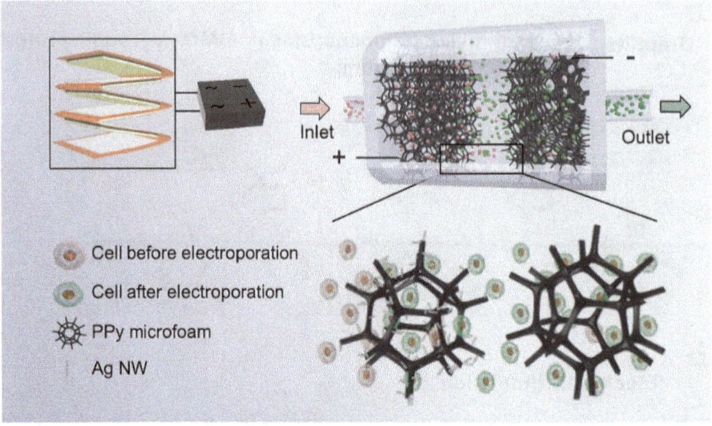

Fig. 6.16 Schematic illustration of the triboelectric nanogenerator-driven high-throughput electroporation system. (Reused with permission from Liu et al. (2020))

flexible polypyrrole (PPy) microfoam was used as the electrode in the flow channel due to its biocompatibility (Fig. 6.16) (Liu et al. 2020). In order to achieve high transfection efficiency while not compromising cell vitality, 1D silver nanorods (Ag NWs) were modified on microbubble electrodes to create a locally enhanced electric field and decrease cell destruction based on the "lightening rod effect." The pulsed voltage was produced by the triboelectric nanogenerator (TENG) because an electricity supply provided a stable DC pulse for self-powered electroporation. When cells passed through the Ag NWs-modified PPy microbubble electrode, the continuous external electric field destroyed the plasma membrane, further driving the foreign molecules into the cytoplasm. Through the technology, various molecules are efficiently transported to the intracellular matrix with high-throughput, low-damage, and high delivery efficiency.

6.2.5 Ultrasound-Responsive Vehicles

Ultrasound (US) has been widely explored for drug delivery. The different mechanisms of interaction between US and the drug carrier, including cavitation, pressure variation, and hyperthermia, can be classified into thermal and nonthermal. These US-based strategies could greatly promote drug uptake and release in the region of interest.

Nonthermal interaction referred to molecule vibration and high pressure that was induced by the interaction between the US (20 kHz to 1 MHz frequency) and the medium (Bushberg et al. 2003). Microbubbles could enhance the interaction because gas core allowed the microbubble volumetrically oscillate in response to US, namely, cavitation (Qin et al. 2009). For instance, Dai et al. designed a microbubble drug carrier (PCF-MBs) for targeted drug- delivery in response to US (Chen et al.

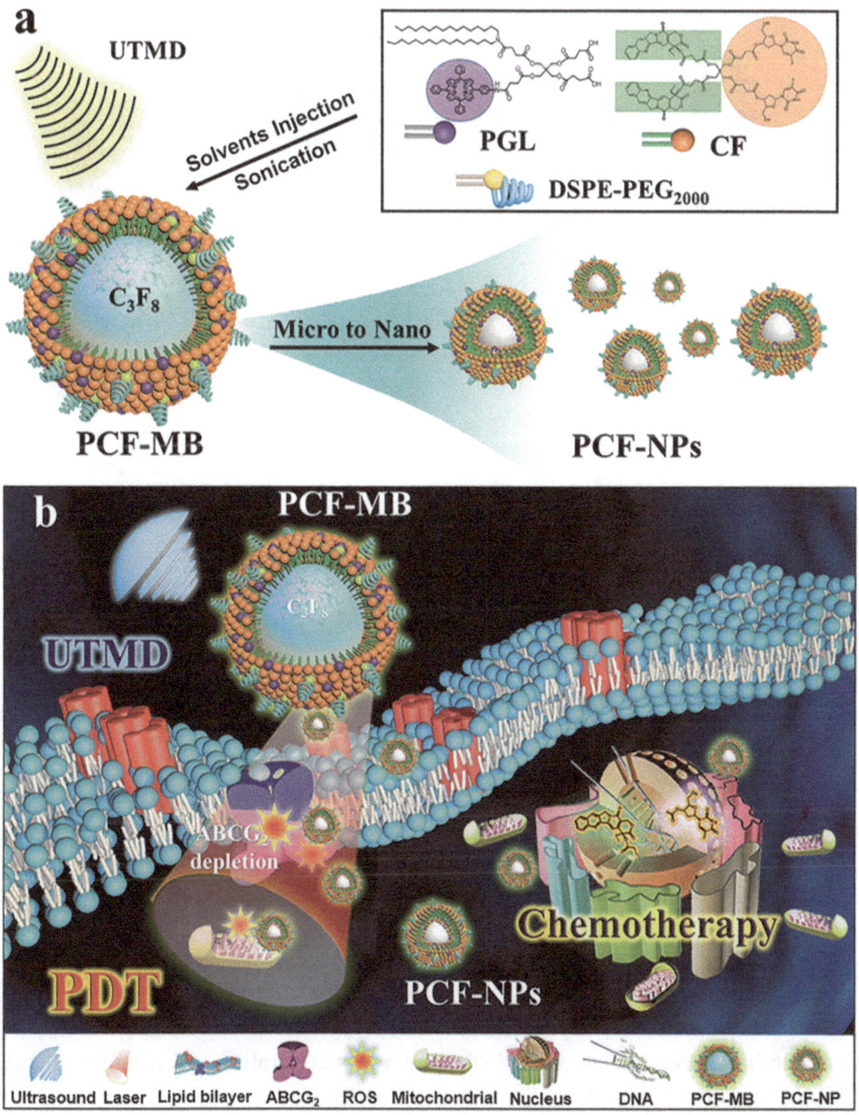

Fig. 6.17 Schematic illustration of the design of PCF-MBs (**a**) and the process of PCF-MBs converted to PCF-NPs and accumulation to tumor in response to US (**b**). (Reused with permission from Chen et al. (2018))

2018). As shown in Fig. 6.17, porphyrin grafted lipid (PGL) and amphiphilic camptothecin-floxuridine conjugate (CF) were self-assembly into PCF-MBs. After intravenous injection, PCF-MBs could be transformed to PCF-NPs when locally exposed and guided by US imaging. Moreover, US could also induce immediate high

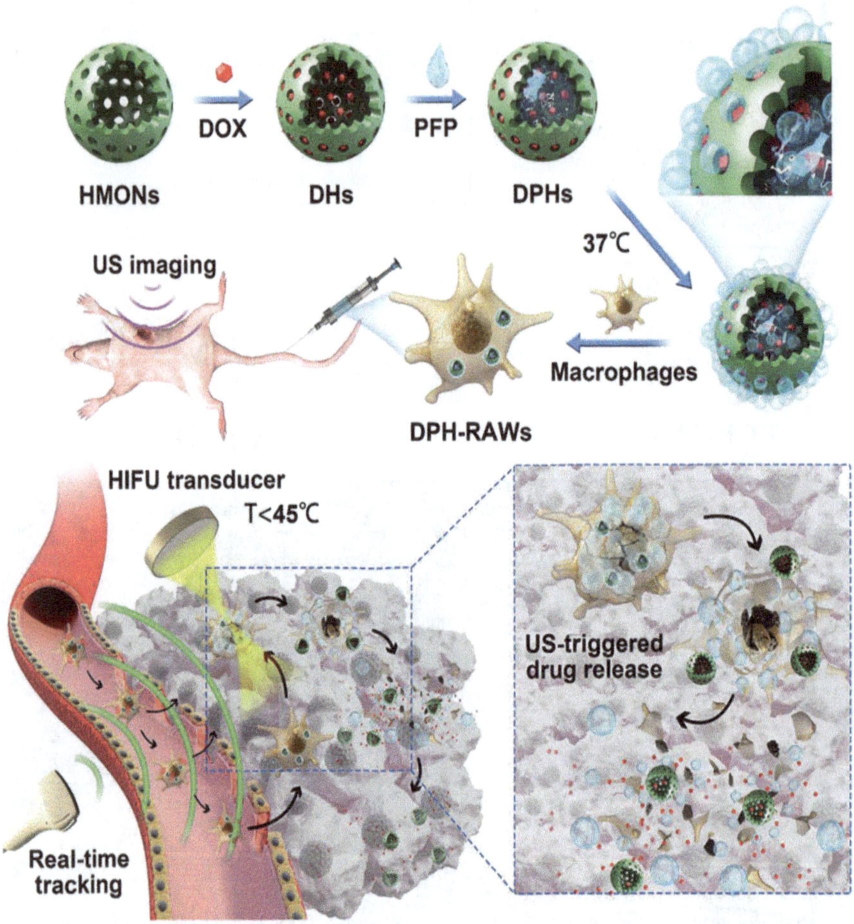

Fig. 6.18 Schematic illustration of the synthesis of DPH-RAWs and drug release from DPH-RAWs in responded to US. (Reused with permission from Xu et al. (2020))

permeability of the capillary wall via the effect of sonoporation which could enhance PCF-NPs accumulation in cancer.

Hyperthermia could be locally generated when US passed through tissues and generated friction (Diederich and Hynynen 1999). The generated hyperthermia could be used to trigger drug release from thermo-sensitive carriers. For example, Yan et al. described a US- triggered macrophage "bombs" (DPF-RAWs) composed of perfluoropentane (PFP) with a phase change temperature of 29 °C (Fig. 6.18) (Xu et al. 2020). After intravenous injection, the DPH-RAWs could be real-time tracked by US contrast imaging because of the part vaporization of PFP in DPHs at 37 °C. Upon the accumulation of DPH-RAWs in the region of interest, the liquid PFP could be further vaporized to microbubbles by high intensity focused ultrasound (HIFU), which could rupture the macrophages to release DOX.

As demonstrated above, enormous stimuli-responsive vehicles with intriguing features have been developed for drug delivery. However, despite these exciting achievements, only limited stimuli-responsive vehicles have successfully practiced in clinical translation. Many stimuli-responsive vehicles still suffered from poor biocompatibility and biodegradability, low sensitivity, small-scale manufacturing, all of which limited their further applications in clinic. Thus, clinical translation-oriented design of stimuli-responsive vehicles should be considered for potential practical applications in the future.

Acknowledgments The authors thank Simiao Wang from School of Life Sciences, Jilin University, for editorial assistance.

References

Bushberg JT, Seibert JA, Leidholdt EM Jr et al (2003) The essential physics of medical imaging. Med Phys 30:1936–1936. https://doi.org/10.1118/1.1585033

Cai Y, Ni D, Cheng W (2020) Enzyme-triggered disassembly of perylene monoimide-based nanoclusters for activatable and deep photodynamic therapy. Angew Chem Int Ed 59:14014–14018. https://doi.org/10.1002/anie.202001107

Cao J, Huang S, Chen Y et al (2013) Near-infrared light-triggered micelles for fast controlled drug release in deep tissue. Biomaterials 34:6272–6283. https://doi.org/10.1016/j.biomaterials.2013.05.008

Chen X-Z, Hoop M, Shamsudhin N et al (2017) Hybrid magnetoelectric nanowires for nanorobotic applications: fabrication, magnetoelectric coupling, and magnetically assisted in vitro targeted drug delivery. Adv Mater 29:1605458. https://doi.org/10.1002/adma.201605458

Chen M, Liang X, Gao C et al (2018) Ultrasound triggered conversion of porphyrin/camptothecin-fluoroxyuridine triad microbubbles into nanoparticles overcomes multidrug resistance in colorectal cancer. ACS Nano 12:7312–7326. https://doi.org/10.1021/acsnano.8b03674

Choi S-W, Zhang Y, Xia Y (2010) A temperature-sensitive drug release system based on phase-change materials. Angew Chem Int Ed 49:7904–7908. https://doi.org/10.1002/anie.201004057

Croissant J, Chaix A, Mongin O et al (2014) Two-photon-triggered drug delivery via fluorescent nanovalves. Small 10:1752–1755. https://doi.org/10.1002/smll.201400042

Diederich CJ, Hynynen K (1999) Ultrasound technology for hyperthermia. Ultrasound Med Biol 25:871–887. https://doi.org/10.1016/s0301-5629(99)00048-4

Gao J, Wang F, Wang S et al (2020) Hyperthermia-triggered on-demand biomimetic nanocarriers for synergetic photothermal and chemotherapy. Adv Sci 7:1903642. https://doi.org/10.1002/advs.201903642

Gordijo CR, Shuhendler AJ, Wu XY (2010) Glucose-responsive bioinorganic nanohybrid membrane for self-regulated insulin release. Adv Funct Mater 20:1404–1412. https://doi.org/10.1002/adfm.200901581

Han H, Valdepérez D, Jin Q et al (2017) Dual enzymatic reaction-assisted gemcitabine delivery systems for programmed pancreatic cancer therapy. ACS Nano 11:1281–1291. https://doi.org/10.1021/acsnano.6b05541

Ihsanullah KM, Kumar BN, Zhao Y et al (2020) Stepwise-activatable hypoxia triggered nanocarrier-based photodynamic therapy for effective synergistic bioreductive chemotherapy. Biomaterials 245:119982. https://doi.org/10.1016/j.biomaterials.2020.119982

Kashyap N, Viswanad B, Sharma G et al (2007) Design and evaluation of biodegradable, biosensitive in situ gelling system for pulsatile delivery of insulin. Biomaterials 28:2051–2060. https://doi.org/10.1016/j.biomaterials.2007.01.007

Li H, Luo R (2009) Modeling and characterization of glucose-sensitive hydrogel: effect of Young's modulus. Biosens Bioelectron 24:3630–3636. https://doi.org/10.1016/j.bios.2009.05.030

Li F, Du Y, Liu J et al (2018) Responsive assembly of upconversion nanoparticles for pH-activated and near-infrared-triggered photodynamic therapy of deep tumors. Adv Mater 30:1802808. https://doi.org/10.1002/adma.201802808

Li Y, Ding J, Xu X et al (2020) Dual hypoxia-targeting RNAi nanomedicine for precision cancer therapy. Nano Lett 20:4857–4863. https://doi.org/10.1021/acs.nanolett.0c00757

Liu G, Zhao X, Zhang Y et al (2019) Engineering biomimetic platesomes for pH-responsive drug delivery and enhanced antitumor activity. Adv Mater 31:1900795. https://doi.org/10.1002/adma.201900795

Liu Z, Liang X, Liu H et al (2020) High-throughput and self-powered electroporation system for drug delivery assisted by microfoam electrode. ACS Nano 14:15458–15467. https://doi.org/10.1021/acsnano.0c06100

Ma N, Xu H, An L et al (2011) Radiation-sensitive diselenide block co-polymer micellar aggregates: toward the combination of radiotherapy and chemotherapy. Langmuir 27:5874–5878. https://doi.org/10.1021/la2009682

Mirvakili SM, Ngo QP, Langer R (2020) Polymer nanocomposite microactuators for on-demand chemical release via high-frequency magnetic field excitation. Nano Lett 20:4816–4822. https://doi.org/10.1021/acs.nanolett.0c00648

Mo R, Jiang T, DiSanto R et al (2014) ATP-triggered anticancer drug delivery. Nat Commun 5:3364. https://doi.org/10.1038/ncomms4364

Qin S, Caskey CF, Ferrara KW (2009) Ultrasound contrast microbubbles in imaging and therapy: physical principles and engineering. Phys Med Biol 54:R27–R57. https://doi.org/10.1088/0031-9155/54/6/r01

Saravanakumar G, Kim J, Kim WJ (2017) Reactive-oxygen-species-responsive drug delivery systems: promises and challenges. Adv Sci 4:1600124. https://doi.org/10.1002/advs.201600124

Servant A, Methven L, Williams RP et al (2013) Electroresponsive polymer–carbon nanotube hydrogel hybrids for pulsatile drug delivery in vivo. Adv Healthc Mater 2:806–811. https://doi.org/10.1002/adhm.201200193

Servant A, Leon V, Jasim D et al (2014) Graphene-based electroresponsive scaffolds as polymeric implants for on-demand drug delivery. Adv Healthc Mater 3:1334–1343. https://doi.org/10.1002/adhm.201400016

Song X-R, Li S-H, Guo H et al (2018) Enhancing antitumor efficacy by simultaneous ATP-responsive chemodrug release and cancer cell sensitization based on a smart nanoagent. Adv Sci 5:1801201. https://doi.org/10.1002/advs.201801201

Tang J, Yao C, Gu Z et al (2020) Super-soft and super-elastic DNA robot with magnetically driven navigational locomotion for cell delivery in confined space. Angew Chem Int Ed 59:2490–2495. https://doi.org/10.1002/anie.201913549

Wan S-S, Zeng J-Y, Cheng H et al (2018) ROS-induced NO generation for gas therapy and sensitizing photodynamic therapy of tumor. Biomaterials 185:51–62. https://doi.org/10.1016/j.biomaterials.2018.09.004

Wang D, Zhou J, Chen R et al (2016) Magnetically guided delivery of DHA and Fe ions for enhanced cancer therapy based on pH-responsive degradation of DHA-loaded Fe3O4@C@MIL-100(Fe) nanoparticles. Biomaterials 107:88–101. https://doi.org/10.1016/j.biomaterials.2016.08.039

Wang M, Zhai Y, Ye H et al (2019) High co-loading capacity and stimuli-responsive release based on cascade reaction of self-destructive polymer for improved chemo-photodynamic therapy. ACS Nano 13:7010–7023. https://doi.org/10.1021/acsnano.9b02096

Wong PT, Chen D, Tang S et al (2015) Modular integration of upconverting nanocrystal–dendrimer composites for folate receptor-specific NIR imaging and light-triggered drug release. Small 11:6078–6090. https://doi.org/10.1002/smll.201501575

Wu Q, Wang L, Yu H et al (2011) Organization of glucose-responsive systems and their properties. Chem Soc Rev 111:7855–7875. https://doi.org/10.1021/cr200027j

Wu Y-n, Zhou M, Li S et al (2014) Magnetic metal–organic frameworks: γ-Fe2O3@MOFs via confined in situ pyrolysis method for drug delivery. Small 10:2927–2936. https://doi.org/10.1002/smll.201400362

Xie H, He M-j, Deng X-Y et al (2016) Design of poly(l-lactide)–poly(ethylene glycol) copolymer with light-induced shape-memory effect triggered by pendant anthracene groups. ACS Appl Mater Interfaces 8:9431–9439. https://doi.org/10.1021/acsami.6b00704

Xiong L, Bi J, Tang Y et al (2016) Magnetic core–shell silica nanoparticles with large radial mesopores for siRNA delivery. Small 12:4735–4742. https://doi.org/10.1002/smll.201600531

Xu Z, Liu H, Tian H et al (2020) Real-time imaging tracking of engineered macrophages as ultrasound-triggered cell bombs for cancer treatment. Adv Funct Mater 30:1910304. https://doi.org/10.1002/adfm.201910304

Yan F, Duan W, Li Y et al (2016) NIR-laser-controlled drug release from DOX/IR-780-loaded temperature-sensitive-liposomes for chemo-photothermal synergistic tumor therapy. Theranostics 6:2337–2351. https://doi.org/10.7150/thno.14937

Yang X, Tang Q, Jiang Y et al (2019) Nanoscale ATP-responsive zeolitic imidazole framework-90 as a general platform for cytosolic protein delivery and genome editing. J Am Chem Soc 141:3782–3786. https://doi.org/10.1021/jacs.8b11996

Yang Y, Zeng W, Huang P et al (2021) Smart materials for drug delivery and cancer therapy. Viewpoints 2:20200042. https://doi.org/10.1002/VIW.20200042

Zhang K, Meng X, Yang Z et al (2020a) Enhanced cancer therapy by hypoxia-responsive copper metal-organic frameworks nanosystem. Biomaterials 258:120278. https://doi.org/10.1016/j.biomaterials.2020.120278

Zhang T-X, Zhang Z-Z, Yue Y-X et al (2020b) A general hypoxia-responsive molecular container for tumor-targeted therapy. Adv Mater 32:1908435. https://doi.org/10.1002/adma.201908435

Chapter 7
Cell Derived/Bionic-Drug Delivery Vehicles

Yuwei Wu, Yanfei Liu, Yixuan Zhang, Minquan Li, Shengnan Qian, Xingjian Gao, Shiyuan Shao, Mengru Zhang, Hongjie Li, and Zhenbao Liu

Abstract Nanomedicines have attracted extensive attention owing to their excellent tissue penetration ability. However, there are still some unsolved problems such as the easy aggregation and rapid clearance by immune system. Cell-derived/bionic vehicles are similar to cell membranes and have low immunogenicity. They are now considered as the promising platforms for evading immune system capture, extending circulation time, and targeting tumor cells through homing tendency proteins in the plasma membrane. In this chapter, the preparation methods, advantages, and the applications of cell-derived/bionic-drug delivery vehicles were summarized. Moreover, based on the recent research progresses, we discussed and proposed the challenges and future directions regarding cell-derived/bionic vehicles.

Yuwei Wu, Yanfei Liu, Yixuan Zhang and Minquan Li contributed equally to this work.

Y. Wu · Y. Zhang · M. Li · M. Zhang
Department of Pharmaceutics, Xiangya School of Pharmaceutical Sciences, Central South University, Changsha, Hunan, P. R. China

Y. Liu
Department of Pharmaceutical Engineering, College of Chemistry and Chemical Engineering, Central South University, Changsha, Hunan, P. R. China

S. Qian · X. Gao · S. Shao · H. Li
Xiangya School of Medicine, Central South University, Changsha, Hunan, P. R. China

Z. Liu (✉)
Department of Pharmaceutics, Xiangya School of Pharmaceutical Sciences, Central South University, Changsha, Hunan, P. R. China

Molecular Imaging Research Center of Central South University,
Changsha, Hunan, P. R. China
e-mail: zhenbaoliu@csu.edu.cn

L. Teng et al. (eds.), *Drug Delivery to Tumors*,
https://doi.org/10.1007/978-981-19-8930-8_7

Keywords Cell membrane · Biomimetic nanoparticles · Drug delivery vehicles · Drug delivery system

7.1 Introduction

The solubility and absorption of insoluble drugs can be improved by nanoparticles (NPs). Moreover, NPs can penetrate the issue through the blood vessels of the lesion site. However, according to the existing literature reports, when NPs are injected into mice, 98% of them will be captured by the liver and spleen and only 2% will actually reach the lesion site, causing systemic toxicity. This is one of the reasons for the poor clinical translation of nanomedicine, and this has become an urgent problem to be solved.

Similar to the cell membrane, cell-derived/bionic vehicles can escape the monitoring of the immune system, thereby reducing the degradation of NPs by the liver and spleen. In cell-derived/bionic vehicles, the stability of NPs is enhanced and the adverse effects caused by NP aggregation are reduced. Moreover, cell-derived/bionic vehicles can enhance their targeting ability. Various cell membranes are used to build drug delivery systems (DDSs). For illustration, cell membranes from red blood cells (RBCs), platelet cells, immune cells, cancer cells, and also stem cells are discussed. Compared with conventional systems, this platform can increase the systematic circulation time and alleviate their removal by the immune system. For example, RBC membrane coating prolongs serum stability, platelet cell or cancer cell membrane coating realizes hemophilic targeting, immune cell membrane coating enables the targeting of metastatic cancer cells, and stem cell membrane coating can effectively prevent NPs from immune system clearance.

In this chapter, we highlight recent research advances in the preparation, modification, drug encapsulation, and applications of the cell-derived/bionic drug delivery vehicles, including exosomes and cell membranes obtained from RBCs, platelet cells, cancer cells, immune cells, stem cells, bacteria, etc. Meanwhile, their advantages and disadvantages are discussed, and suggestions and further directions are presented.

7.2 DDSs Based on Cell Membranes

RBC is the most abundant cell in blood circulation and has been developed as a carrier for various drugs (Xie et al. 2019; Que et al. 2019) due to its excellent physiological characteristics (Gao et al. 2019a; Li et al. 2019). The characteristics of platelet membrane include recognition of blood vessel injury, interaction with circulating tumor cells (CTCs), tumor targeting, and ability to improve systemic circulation. It is known that biomimetic NPs coated with tumor cell membranes can obtain some biological characteristics of the tumor cell membranes and realize tumor targeting (Kroll et al. 2017; Bose et al. 2018). The coating approach with immunocyte membranes and stem cell membranes is feasible enough to overcome the constraints of the introduction of foreign substance from the immune recognition by offering a unique biological interface. The functional molecules on the immune cell membranes can interact with the specific receptors on endothelial cells and some immune cells, and thus, DDSs based on immune cell membrane are promising for targeting the lesion sites (Jin et al. 2018; Chu et al. 2018) and sensitizing the innate and adaptive immune responses (Wang et al. 2018a; Yaman et al. 2020b).

Exosomes are secreted by various cells (Zhan et al. 2021). Now, this name refers in particular to disc-like vesicles which are 40 to 100 nm in diameter. Exosomes contain various kinds of microRNAs (miRNAs), mRNAs, and proteins, with extensive functions, for instance, antigen presentation (Skotland et al. 2017), immune escape (Meldolesi 2018), and tumor progression mediation (Lafitte et al. 2019; Zhang and Yu 2019). Exosomes can be acquired from many cells in normal and pathological states, which means that the parent cells of exosomes can be diverse, such as cancer cells (Ferguson et al. 2018), immune cells (Yang et al. 2020a), stem cells (Riazifar et al. 2019; Maqsood et al. 2020), and so on. The specific formation mechanism of exosomes is still unclear. However, it is generally believed that the process of exosomes generation in different types of cells is similar. Cells produce vesicles through endocytosis, and the vesicles fuse to form early endosomes, which further combine with intracellular substances and become late endosomes. Subsequently, many small vesicles are produced in the late endosomes and gradually evolve into multivescular body, which are finally released into the extracellular to form exosomes. In general, the exosomes can be obtained in two ways: (1) spontaneous excretion and (2) excretion with the biological trigger, chemical trigger, or physical trigger. After that, purification is also a critical procedure. In fact, vesicles in the body have their inherent properties, which can be used to purify them to get the exosomes. According to the properties, scientists proposed different methods, including physical methods, chemical methods, and combination techniques (Ayala-Mar et al. 2019).

Through genetic engineering technology and bioengineering technology, bacterial derivatives with low immunogenicity and low toxicity can be obtained (Zhu et al. 2018a), which include outer membrane vesicle (OMV), bacterial protoplast-derived nanovesicle (PDNV), bacterial minicells, bacterial ghosts, and so on. Bacterial outer membrane vesicle is a specific physiological structure released by gram-negative bacteria, such as salmonella enterica. It is a 20–200 nm spherical double-membrane structure formed by the outer membrane of bacteria with multiplex functions (Zhang and Yu 2019).

7.2.1 DDSs Construction

DDS construction based-on cell membranes can be mostly classified into two steps: extraction of membrane and drug loading.

7.2.1.1 Extraction and Separation of Membranes

Most cell membrane extraction strategies include the following steps: removal of cell contents, membrane vehicle harvesting, and purification. The significant part of membrane extraction is to remove the internal components of cells while retaining the functional components of the membrane. The main cells commonly used for cell membranes are RBCs, immune cells, and cancer cells (Raimondo et al. 2019). The extraction ways for exosomes and OMVs are different. Hypotonic treatment was adopted to collect erythrocyte membranes, cancer cell membranes, and platelet membranes, while density gradient centrifugation was generally used in the extraction and collection of leukocyte membranes and stem cell membranes. Hypotonic or hypertonic treatment (Zhu et al. 2018b; Wang et al. 2018b), homogenization (Zhao et al. 2020c), Percoll gradient method (Zhai et al. 2017), and repeated freeze-thaw method (He et al. 2019a) are the commonly used membrane extraction methods. The steps of membrane extraction can be summarized as lysis of cells, removal of cell contents, and collection and purification of membranes. Collected pellets are usually washed with Tris–HCl, EDTA, and PBS, and the cell membrane was further purified by ultracentrifugation (Yang et al. 2019). The extrusion (Chen et al. 2018a; Han et al. 2018) and ultrasound methods (Chen et al. 2020b) are commonly used to obtain the cell membrane vesicles. It is worth noting that freezing procedure may lead to loss of membrane structure, degradation of protein stability, protein unfolding, and degradation of membrane function (Harris et al. 2019).

7.2.1.2 Isolation of Exosomes

Nowadays, exosomes can be separated by ultracentrifugation, size-exclusion chromatography, filtration, and polymer precipitation. The ultracentrifugation method and density gradient ultracentrifugation method are frequently used where ultracentrifugation is the most common method. In ultracentrifugation method, the cell suspension goes through a series of centrifugations to exclude the dead cells, large extracellular vesicles, large cell debris, and proteins (Purushothaman 2019; Wierz et al. 2019). After these procedures, the exosomes/vesicles can be obtained though their quality cannot be guaranteed because of the similar size between exosomes/vesicles and some proteins (Guan et al. 2020). Moreover, some of the biological functions of exosomes were lost during this process (Diaz et al. 2018). In the density gradient ultracentrifugation method, the particles form zones in different areas with different density gradients. The shortages of ultracentrifugation method are its low yield, time costing, and needing multiple steps. The filtration method is usually combined with ultracentrifugation. A membrane is firstly used to exclude cells and large extracellular vesicle (EVs), and the leftover is processed by ultracentrifugation to harvest the exosome as mentioned above. Size-exclusion chromatography, also called molecular sieve chromatography, has highly varying EVs yields (2–80%)

compared to the ultracentrifugation. This method can almost completely remove unwanted proteins (Si et al. 2016). Polymer precipitation is a technique for the separation and purification of polymers, and it is now also used for exosome isolation with the basis for forming mesh-like net to embed EV (60–180 nm) (Deregibus et al. 2016). Immunomagnetic isolation can capture exosomes through the use of the antibody-labeled magnetic beads (Oksvold et al. 2015). The drawback of this technique is that non-exosomes vesicles carrying the antigens can also bind to the antibody, which reduces the purity of the product (Vlassov et al. 2012). Recently, methods of purification based on acoustic and jet devices have been developed, and they are mainly focused on improving the quality of exosomes. Meanwhile, several applications have been conducted to enhance the production and secretion of exosomes, including hypoxia, electrical stimulation, and chemical treatment (Chinnappan et al. 2020). Currently, the obtained exosomes are characterized by their size and protein content. The most commonly methods used for exosome characterization include nanoparticle tracking analysis, dynamic light scattering, flow cytometry, Western blot, microscopy techniques, and mass spectrometry (Batrakova and Kim 2015).

7.2.1.3 Separation and Extraction Outer-Membrane Vesicles (OMVs)

OMVs are originate released by "pinching-off" segments of the outer membrane (Farjadian et al. 2018). Ultracentrifugation is the most frequently method used for the separation of OMVs. The main principle is to establish different centrifugal forces according to different particles' size and density in the medium and use the principle of equilibrium precipitation to complete the purification (Li et al. 2020).

Genetic engineering was used to reduce the toxicity of OMVs and increase their yield. When OMVs are used as drug delivery vehicles, some of their components, such as LPS, may also influence the safety of the DDS. For example, methods have been established to knock out the msbB gene to reduce their toxicity (Cao and Liu 2020). After purification, some unnecessary or harmful components can be removed, making OMVs suitable for the application. Traditional OMVs purification methods include ultracentrifugation (Wang et al. 2019b), density gradient separation (Gujrati et al. 2014), protein precipitation (Li et al. 2020), and gel filtration (Carvalho et al. 2019). Gel filtration and density gradient centrifugation are the common methods to purify OMVs, while the main methods for obtaining OMV are ultracentrifugation and sedimentation (Cao and Liu 2020).

Since ultrafiltration can remove high molecular weight substances, for instance, viruses, organic and inorganic polymers, the purification by filtration is an optional method. For example, gel filtration isolates molecules with different hydrodynamic radii, such as proteins, polysaccharides, and OMVs (Salih et al. 2014). However, the drawback of this method is that it needs preprocessing, like ultracentrifugation or ultrafiltration.

7.2.2 *Drug Loading*

7.2.2.1 Co-extrusion

The co-extrusion method is widely used for drug-loading into NPs coated with cell membrane. The resulting cell membrane vesicles were mixed with drug-carrying NPs in a proper proportion. The NPs are physically extruded through the polycarbonate membrane to be encapsulated by the cell membrane (He et al. 2019a; Liu et al. 2019). The ratio of membrane surface proteins to NPs can also affect the fusion efficiency (Zhao et al. 2020a). The drawback is that this method can easily cause excessive loss of materials, and thus, multiple co-extrusions are required.

7.2.2.2 Sonication

Sonication is a feasible method for the construction of membrane-coated nanomedicines. The membrane was mixed with the NPs, and the mixture was subjected to probe ultrasonic treatment and further centrifugation to obtain the NPs wrapped in the membrane (Zhao et al. 2020a; Cai et al. 2019). This technique causes less material losses than physical extrusion (Harris et al. 2019). In terms of exosomes, ultrasonic treatment can improve the loading efficiency.

7.2.2.3 Microfluidic Electroporation

In the presence of an intense electric field, the permeability of the cell membrane is instantly improved, and exosomes can load exogenous molecules. A hydrophilic molecule, such as miRNA, is commonly used in electroporation. Diverse voltages and pulse lengths have different impacts on loading efficiency. The investigation showed that when the voltage was 1000 V and the duration was 10 ms, the loading efficiency of 5-Fu and miR-21i into exosomes was the highest. Under this electroporation condition, the loadings of 5-Fu and miR-21i in exosomes were 3.1% and 0.5%, respectively (Fig. 7.1) (Liang et al. 2020).

For in vitro drug loading, electroporation into the OMVs has been demonstrated. Passive diffusion can be used for small molecule drugs with relatively hydrophobic and positively charged, which readily to interact with lipophilic membranes. However, bilayer lipid membranes of hydrophilic molecules may present a barrier; thus, electroporation is the best approach. In this technique, membranes and NPs are rapidly mixed through electroporation with strict requirements for the pulse voltage, duration, and flow rate required during the process (Yang et al. 2019).

Microfluidic electroporation is a perfect combination of biology and physics. To a certain extent, it can maintain the integrity of the membrane and reduce the loss of protein on the cell surface, achieving a better therapeutic effect. Moreover, compared with extrusion, this method has better magnetic resonance imaging (MRI), photothermal therapy (PTT) performance, and colloidal stability.

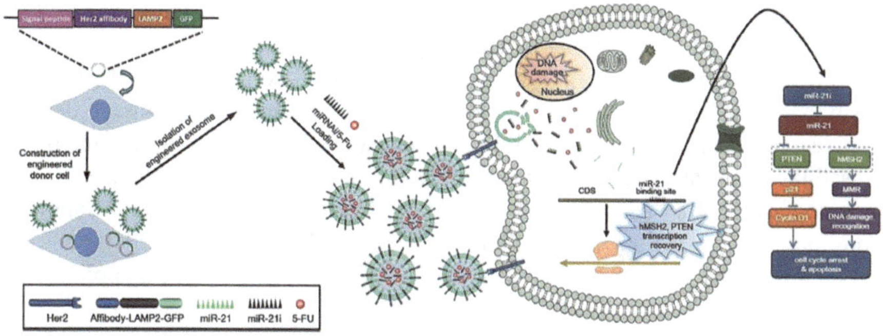

Fig. 7.1 Engineered exosomes-based nanocarriers for 5-FU and miR-21i delivery to HCT-116^{5FR} human colon cancer cells to enhance chemotherapy efficacy simultaneously. (Reprinted with permission from Liang et al. (2020) Copyright, © 2021 BioMed Central Ltd unless otherwise stated. Part of Springer Nature)

7.2.2.4 Spontaneous Formation by Electrostatic Attractions

A method of spontaneous formation by electrostatic attractions has been developed as well. The membrane camouflage NPs (CMCNPs) was spontaneously formed by the electrostatic attraction between the NPs (positive charge) and the membrane (negative charge). Moreover, charge interaction gives this method an advantage: the outer surface of the plasma membrane is located on the outer surface of biomimetic NPs. For instance, Tasciotti's CMCNPs are self-assembled from negatively charged white cell membrane vesicles and positively charged silicon NPs (Chxa et al. 2020).

7.2.2.5 Bath Sonication

In a frequency and power bath sonicator, the fusion membrane and NPs were dispersed in a proper proportion to prepare the membrane-wrapped NPs. Fluorescent probes can interact with membrane-binding proteins and fluoresce, which can be characterized by confocal laser scanning microscopy (Wang et al. 2020a).

7.2.2.6 Other Methods

In the pre-loading methods, the drugs are produced first and loaded into the parental cells, and then, cellular vesicles that containing the drug are produced or isolated from the cells. This method applies to oligonucleotides or proteins loaded in vesicles (Antimisiaris et al. 2018).

In the early stage, the fusion of NPs and membrane vesicles often utilized some characteristics of the membranes themselves. For example, the fusion of RBC membrane and NPs used the surface chemical action of RBCs to functionalize NPs. However, the RBC simulation vector formed by chemical, biological coupling technology often leads to protein denaturation. So the original method has a considerable limitation. Now, cell membrane-templated polymerization is usually used.

Cell membrane-template polymerization includes a complete coated core with easily controlled size and rigidity of fiber-like NPs. It means that membrane-derived vesicles can be used as nanoreactors to encapsulate reaction mixtures, including crosslinkers, initiators, and monomers, and as templates to control the final nanometer size (Zhang et al. 2015). Moreover, it has a broad range of biomedical applications and can overcome the limitations of NPs. Therefore, the cell membrane template method may be applied to a variety of nanostructures with applicability (Xia et al. 2019).

7.3 Advantages of Cell Membrane-Coated NPs

7.3.1 Stability

Stability is one of the key parameters of cell membrane-derived DDSs. The cell membrane encapsulating the drug protects it from being removed and thus reduces drug leakage. In addition, polysaccharides on the surface of erythrocyte membranes are hydrophilic which provide better stability for NPs when covered by such membranes. Hu et al. were the first to obtain cell membranes from erythrocytes by hypotonic treatment. Later, oxygen self-enriched erythrocytes with high oxygen capacity and extended circulation time were evaluated. Transmission electron microscopy showed that proteins contained inside the erythrocyte membrane-encapsulated NPs (65–340 nm in diameter) were protected and stabilized by surface dextran and lipid membranes (Jimenez-Jimenez et al. 2020).

7.3.2 Immune Evasion and Long Circulation

Since the immune system has the function of recognizing and removing foreign bodies and certain abnormal organism components, it prevents the localization of NPs to the site of the lesion. After injection into the bloodstream, NPs are rapidly cleared from the circulation through two main pathways: the mononuclear phagocyte system (MPS) and natural clearance by filtering organs. Some of the cell membrane molecules can contribute to cell membrane coated NPs escaping from uptake by phagocytes (Sushnitha et al. 2020), resulting in a more robust enhanced permeability and retention (EPR) effect. As shown in Fig. 7.2 (Nie et al. 2020), Nie et al. developed a cancer cell membrane (CCM)-encapsulated yolk-shell NP with mesoporous silica NPs (MSN) supported by polyethylene glycolized liposomal yolk, CCM@LM. It was applied in chemotherapy and showed excellent cancer-targeting effects. They further evaluated the protein adsorption capacity of NPs in FBS to

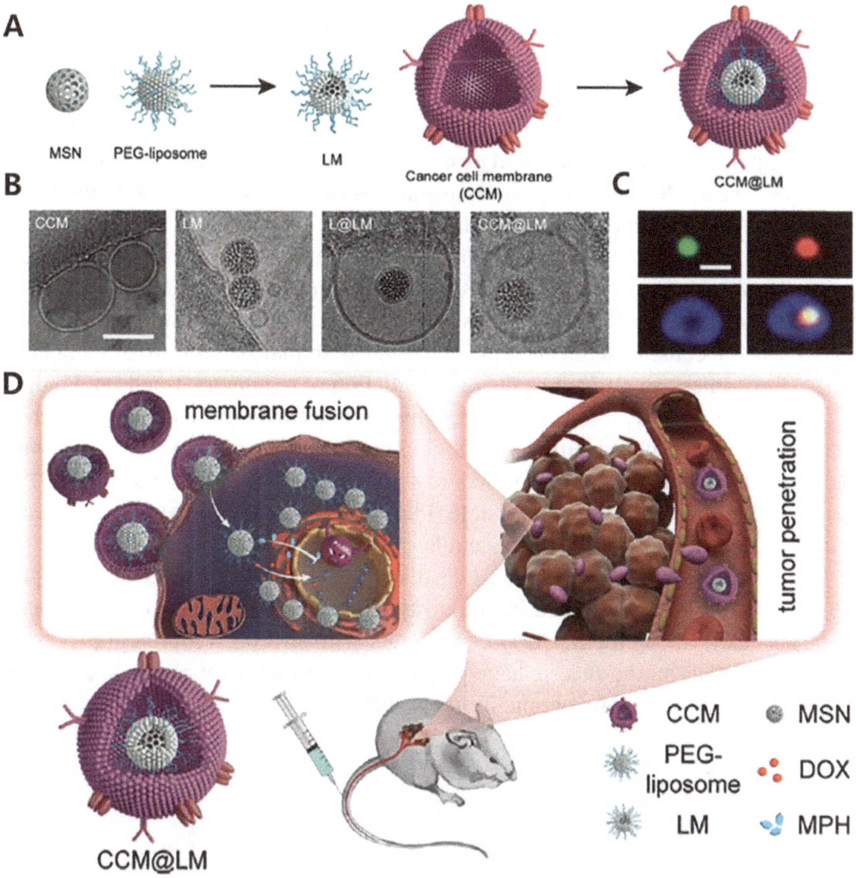

Fig. 7.2 (**a**) Schematic illustration of preparing CCM@LM. (**b**) Cryo-TEM images of NPs. Scale bar: 100 nm. (**c**) Stimulated emission depletion (STED) images of CCM@LM. Scale bar: 100 nm. Green, FITC-labeled MSN; red, DiI-labeled liposome; blue, DiD-labeled CCM. (**d**) CCM@LM evaded attack by the immune system and was directly internalized by membrane fusion and the PEGylated yolk (LM) was subsequently released into the cytoplasm. (Reprinted with permission from Nie et al. (2020) Copyright, © 2020 Elsevier B.V. All rights reserved)

check if CCM@LM was able to escape the immune system. Results showed that CCM@LM protein uptake was minimal and that effects in particle size and zeta potential were not significant, indicating that CCM@LM would produce sufficient immune evasion capacity in circulation. The cellular uptake of NPs in a mouse macrophage-like cell line (J774A.1) was also researched. The fluorescence intensity in LM group was strong in J774A.1 and it is higher than that in CCM@LM, revealing an effective escape of phagocytosis by CCM@LM. Another case used the stem cell membrane to encapsulate Au-Ag NPs which were pretreated by the polydopamine, and the NPs exhibited immune evasion and long circulation, less cytotoxicity, and better efficiency in photothermal therapy (Meng et al. 2020).

Exosomes are produced by autologous cells, and this indicates that exosomes and maternal cell membrane structure are similar. Therefore, it is difficult for exosomes to induce the immune system to produce an immune response (Pang et al. 2019; Theodoraki et al. 2018). Kamerkar et al. found that the expression of CD47 protein on exosomes can protect the exosomes from phagocytosis, thus increasing circulating half-life of exosome and promoting tumor targeting in pancreatic cancer (Kamerkar et al. 2017). Moreover, exosomes from some sources can regulate the immune system to reduce its toxicity to the organism (Kamerkar et al. 2017; Poggio et al. 2019). For instance, immature dendritic cell-derived exosomes (imDECs) contain miR-682 which can target ROCK2 (the target of miR-682) on the surface of cells. Consequently, it can suppress the IL-17^+CD4^+T cells and promote the Foxp3^+CD4^+T cells, which results in the extended animal survival time (Pang et al. 2019). Exosomes can also adjust the differentiation of regulatory T cells and induce T cell dysfunction. Through genetic engineering technology and bioengineering technology, bacterial derivatives with low immunogenicity and low toxicity, which include outer membrane vesicle (OMV) (Kim et al. 2017a; Toyofuku et al. 2019), bacterial protoplast-derived nanovesicle (PDNV) (Kim et al. 2017b), bacterial minicells (Frazer and Curtiss 3rd 1975), and bacterial ghosts (Gong et al. 2020b).

Moreover, the circulation time of NPs disguised as red blood cells has been prolonged. This effect is mediated by the array of membrane proteins on their membrane surface. Among them, CD47 can interact with siRPα glycoprotein expressed by phagocytes to inhibit macrophage phagocytosis (Koichiro et al. 2018; Xia et al. 2019) (Fig. 7.3). Other membrane proteins such as homologous restriction protein

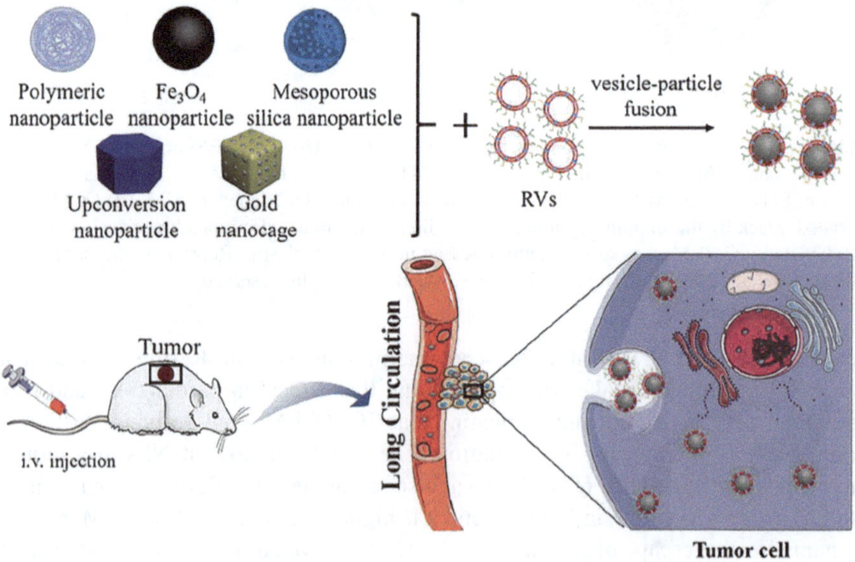

Fig. 7.3 NPs encapsulated in RBC membranes masquerading as RBCs realize long-term circulation in vivo by interacting with the surrounding environment. (Reprinted with permission from Xia et al. (2019) Copyright, © 2019 Chinese Pharmaceutical Association and Institute of Materia Medica, Chinese Academy of Medical Sciences. Production and hosting by Elsevier B.V.)

(HRP), C8 binding protein (C8bp), membrane cofactor protein (MCP), decay accelerating factor (DAF), complement receptor 1 (CR1), and CD59 on the RBC plasma membranes also matter in the defense of complement system attacks (Xia et al. 2019). In addition, since sialic acid resides on the erythrocyte membrane, the surface of the RBC carries a negative charge, thus creating electrostatic repulsion between the RBCs which prevents the RBCs from adhering to each other (Koichiro et al. 2018). These properties provide a longer circulation time to NPs carried by RBCs. Su and colleagues reported the macrophage membrane coating biomimetic NPs loaded with Saikosaponin D (SCMNPs) for breast cancer treatment. SCMNPs has the character of macrophage-homing because there are receptors and ligands on the macrophage membrane, and the NPs can target cancer cells that overexpress transferrin receptors by recognizing T7 peptides. Thus, SCMNPs evade phagocytosis by the reticuloendothelial system (RES) are concentrated in tumor sites. However, leukocytes can express a particular membrane compound called major histocompatibility complex (MHC) which makes them less biocompatible than RBCs. Therefore, the immunogenicity of the NPs should be taken into consideration. Meanwhile, the infectious disease screening and blood compatibility testing should be conducted when allogeneic blood is applied as a membrane source (He et al. 2020; Liang et al. 2020).

7.3.3 Targeting Ability

According to the research conducted by Nie et al., a CCM with surface expression of endogenous molecules not only generated longer cycling and immune evasion capabilities but also achieved the ability to homo-target CCM-encapsulated NPs. To confirm this, Nie et al. examined the NP uptake capacity of six cell lines and found that MCF-7 cells internalized CCM@LM and CCM more efficiently than other cells, suggesting that the NPs could selectively target homotypic tumor cells (Nie et al. 2020). Targeted tumor therapy can take advantage of exosome homing in which exosomes from cancer cells preferentially fuse to their parent cells. Qiao et al. further validated the excellent tumor homing of exosomes by injecting HT1080 or Hela exosomes carrying Doxil into mice (Qu et al. 2018).

Some other molecules on exosome membranes also contribute to the targeted delivery, such as the cell adhesion factor integrin. There is an integrin combination in tumor exosomes that facilitates the formation of a metastable microenvironment in different organs, especially in the lung and liver. $\alpha6\beta4$ plays an essential role in pulmonary metastasis and $\alpha v\beta5$ is critical in hepatic metastasis (Mu et al. 2019). In contrast, bacterial-derived minicells are non-living and non-nucleated cells with nanometer size (~400 nm in diameter). They are produced by inhibiting the polar sites of cell division, which is achieved by controlling the mutations in genes that normally divide bacterial cells. As compared with donor bacterial cells, minicells contain ribosomes, membranes, proteins and nucleic acids, and commonly plasmids, with the exception of the genome (Liang et al. 2021). After intravenous injection, minicells are targeted to the tumor microenvironment through the EPR effect and achieved high accumulation in tumor tissue when bispecific antibodies were modified on the minicell surface. There are a variety of immune cell surface markers involved in tumor immunization, including antigens, adhesion molecule receptors, etc. They are associated with the accurate recognition of tumor cells and immune response. Through the guidance of various receptors on macrophage cell surface, such as C–C chemokine receptor 2 (CCR2) and intercellular adhesion molecule-1 (ICAM-1, CD54), NPs encapsulated by minicells accumulate in tumor tissues and inflammation sites. The combination of Mac-1 and activated lymphocyte function-associated antigen-1 (ILA-1) on the neutrophil surface with ICAM-1 on the vascular endothelial cell surface promotes the migration of neutrophils across vascular endothelial cells.

The mechanism mentioned above indicates that the immunocyte-mimicking delivery system can specifically target to inflammation site in vivo. Although there are fewer antigen-specific receptors on the membrane of NK cells, they can be substituted by other receptors which are able to identify cancer cells, such as NKp44, NKG2D, etc. Wherein, NKG2D is a homodimer expressed on tumor cells that recognizes a variety of ligands, such as UL16-binding protein (ULBP) and MHC class I chain-associated gene A (Chen et al. 2018b). With the help of MHC molecules, T cell receptors (TCRs) recognize mutant antigens that are specifically expressed in tumor tissues. Thus, T cell membrane-encapsulated NPs showed high tumor tissue specificity. Macrophages accumulate in the tumor microenvironment and it is directly related to tumorigenesis and metastasis. A membrane molecule CD49d, known as α4 integrins, is expressed on the surface of macrophage membranes, and it can specifically recognize the vascular cell adhesion molecule-1 (VCAM-1), which is present only in cancer cells (Gong et al. 2020a).

The blood-brain barrier (BBB) is a layer composed of endothelial cells connected through tight junctions that can constrain the transport of molecules and is selective and semi-permeable, thus affecting the drug delivery efficiency into the central nervous system (CNS). Generally, exosomes have strong penetrability to pass the BBB (Xiaowei 2018; Qu et al. 2018). Chen et al. found that glioblastoma cells-derived exosomes are rich in VEGF-A, which can interdict the expression of claudin-5 and occluding, enhancing the permeability of the BBB (Chen et al. 2018b). Lai et al. (2019) applied macrophage plasma membrane modified IR-792 NPs (MDINPs) for the treatment of orthotopic glioblastoma. The photothermal therapy (PTT) was guided by NIR-Ib fluorescence imaging. As shown in Fig. 7.4, the binding of macrophage membrane protein outside MDINPs to the receptors VCAM-1 and ICAM-1 expressed by cerebrovascular endothelial cells could downregulate the junction-related protein (occlusion band 1) and activate the signaling pathway, thus allowing MDINPs to easily cross the BBB. NPs encapsulated by macrophage membranes can be guided by VCAM-1 and ICAM-1 on the membrane to reach inflammatory and tumor sites. ICAM-1 can antagonize tight junctions, thus promoting the passage of coated NPs through the BBB (Nordal and Wong 2004) (Fig. 7.4).

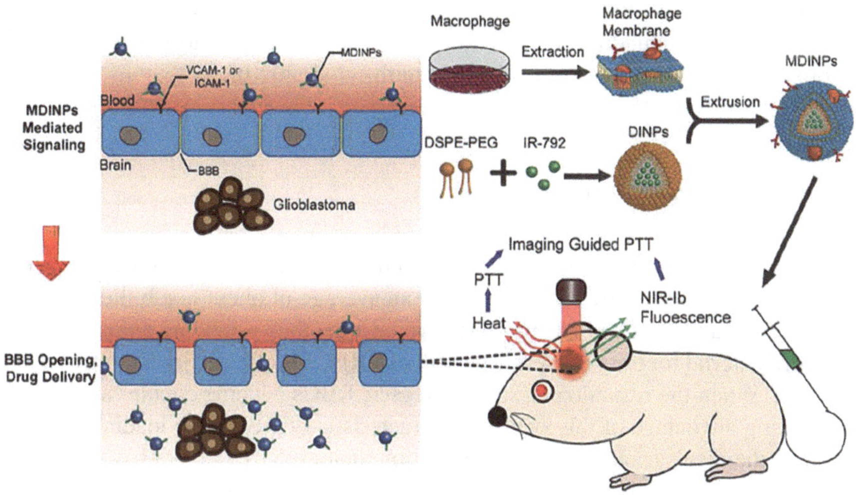

Fig. 7.4 Schematic representation of the preparation of MDINPs and their crossing the BBB into glioblastoma to exert the PTT effect activated by NIR-Ib fluorescence. (Reprinted with permission from Lai et al. (2019) Copyright, © 2019 Published by Elsevier Ltd.)

P-selection is a platelet surface receptor that selectively binds to CD44 which is overexpressed on cancer cells (Fox et al. 2019). Platelet cell membrane-encapsulated NPs can treat bacterial infections through the natural properties of platelets and can be targeted to areas of vascular damage in the body. They can be used to bind injured arteries and release drug-laden NPs to prevent malignant tumor proliferation. Wang et al. prepared porous NPs with the chitosan oligosaccharide (CS)-poly(lactic-co-glycolic acid) (PLGA) copolymer and loaded them with the anticancer drug bufalin (Bu). The synthesized NPs were subsequently wrapped through platelet membranes (PLTM) to synthesize PLTM-CS-pPLGA/BuNPs. Their characterization showed that the formulation exhibited targeting and higher cellular uptake efficiency due to the ability of p-selectin on the surface of PLTM to specifically bind to the CD44 receptor of H22 hepatocellular carcinoma cells (He et al. 2018). Dehaene et al. fabricated a kind of biomembrane with RBCs and platelet membranes and prepared RBC-platelet hybrid membrane-encapsulated NPs ([RBC-P]NPs) which owned the properties of both source cells after being thoroughly characterized (Dehaini et al. 2017). Cancer-related deaths usually occur because circulating tumor cells (CTCs) originated from the primary tumor spread through the circulation to other tissues or organs causing metastasis, and metastatic CTCs also promote platelet activation and fibrin deposition to form thrombi locally to protect themselves against immune clearance (Placke et al. 2012; Palumbo et al. 2000). Silica (Si) particles functionalized by encapsulating platelet-derived vesicles are biocompatible and can target CTCs by adhering to CTC-associated microthrombi and activated platelets. Moreover, platelet membrane-encapsulated Si particles could bind to each other with tumor necrosis factor-related apoptosis-inducing ligand (TRAIL) which is highly expressed on the membrane of NK cells, cytotoxic T cells, and activated neutrophils. Upon binding, they could activate immune cells to kill tumor tissue through the targeting effect of CTCs (Sushnitha et al. 2020).

7.3.4 *Easy to Obtain*

The resource of the cell membrane can be obtained easily. The most abundant cells in the circulation are RBCs which contain an average of about five billion per milliliter in human blood and have a huge surface area, thus providing an ideal encapsulation material for drug carriers. The shape of RBCs is affected by the surrounding medium. When the osmotic pressure decreases, RBCs become cupular and spherical. Taking advantage of this swelling characteristic of RBCs, the loading of drugs or other chemicals can be realized. In addition, mature RBCs do not have a nucleus and organelles, which makes the subsequent extraction and purification process easier (Xia et al. 2019). The tumor cell membranes are also not hard to extract and isolate, and the yield of membranes can be guaranteed due to the rapid proliferation of tumor cells (Yang et al. 2020b). Moreover, tumor cells are easy to culture in vitro (Han et al. 2020; Nie et al. 2020).

7.3.5 *Other Advantages*

Some of the other superiorities of the cell membrane-derived DDS cannot be ignored. Immunostimulatory molecules of bacterial origin, such as pathogen-associated molecular patterns (PAMPs), containing lipopolysaccharide (LPS) and flagellin, have been demonstrated to have antitumor effects (Liang et al. 2021). OMVs can directly interact with various kinds of immunocyte populations, including neutrophils, dendritic cells (DCs), and macrophages (Lapinet et al. 2000). They prime mover pro-inflammatory cell factor production by macrophages and sensitize macrophages to lead to adaptive immune responses. OMVs from various pathogens induce the production of cytokines and the maturation of DCs (Kaparakis-Liaskos and Ferrero 2015).

Functional molecules on the cell membrane surface also contribute to interactions with immune cells, and it may influence disease progression through immune regulation. The macrophage surface markers include VCAM-1, CCR2, and ICAM-1. Macrophage-encapsulated NPs can evade clearance by the RES and direct the accumulation of NPs to inflammation sites and also suppress local inflammation via pro-inflammatory cytokines (Gao et al. 2020; Gong et al. 2020a). Since mature DC membranes have multiple peptide/MHC complexes on the surface, they also have antigen-presenting functions and can accurately activate T cells, which is a significant merit of this class of cell membranes as a carrier encapsulated membrane (Wang et al. 2020c). Meanwhile, DCs have adhesion and co-stimulatory molecules, including CD40, CD44, ICAM-3/CD50, and members of the integrin family, which can not only mediate cell adhesion but also promote T-cell-DC interactions by reducing the negative charge on their surface (Apostolopoulos et al. 2013; Wang et al. 2020c).

NK cells are unique white blood cells that can recognize and eliminate abnormal cells in the absence of the MHC or antibodies (Vivier et al. 2011). Gong et al. (2020a) reported an NK cell membrane-encapsulated NPs (NK-NPs) loaded with the photosensitizer TCPP. This preparation enhances M1 macrophage polarization and kills cancer cells by generating reactive oxygen species (ROS) and nitrogen radicals due to the ability of NK cell membrane proteins to specifically bind to macrophage membrane proteins such as TNF or TLR 4 (Fig. 7.5).

PD-L1 is a molecule that is aberrantly expressed on the tumor cell surface and inhibits PD-1 activity which is a specific receptor expressed on effector T cells. By removing PD-L1 from the surface of exosomes, Mauro et al. found that they were effect in tumor inhibition in models that were previously resistant to anti-PD-L1 antibodies. They also found that tumor cells in an exosomal PD-L1-deficient environment also inhibited the proliferation of distant wild-type tumor cells either simultaneously or after several months (Fig. 7.6) (Poggio et al. 2019).

It is an attractive strategy to engineer cellular membranes with functional molecules to control cellular functions and processes. Among them, synthetic deoxyribonucleic acids (DNAs) are often used in biomedical applications to design cell membranes. The DNAs can respond to stimuli like light and metal ions, and they can be used to activate immune responses, mimic membrane structures, biosense cell surfaces, and build cell clusters (He et al. 2021) (Fig. 7.7).

7.4 Applications

Due to the advantages of its low immunogenicity, good mobility, and high stability, the cell-derived DDSs have proven to be a useful route for drug administration. Such delivery systems typically use the cell membranes of living cells as carrier to exert drug therapeutic effects by transporting drugs or NPs to the target site (Wu et al. 2022). This review summarized the mechanisms of action and characteristics of cell-derived DDSs and also reviewed their applications in disease treatment (Fig. 7.8).

7.4.1 RBC Membrane

Since the natural erythrocyte membrane surface contains a series of immune regulatory proteins, the erythrocyte membrane made from the RBCs owns the ability of long circulation in the body. Moreover, the erythrocyte membrane can form a shell to protect the synthesized NPs from direct contact with the immune system (Koichiro

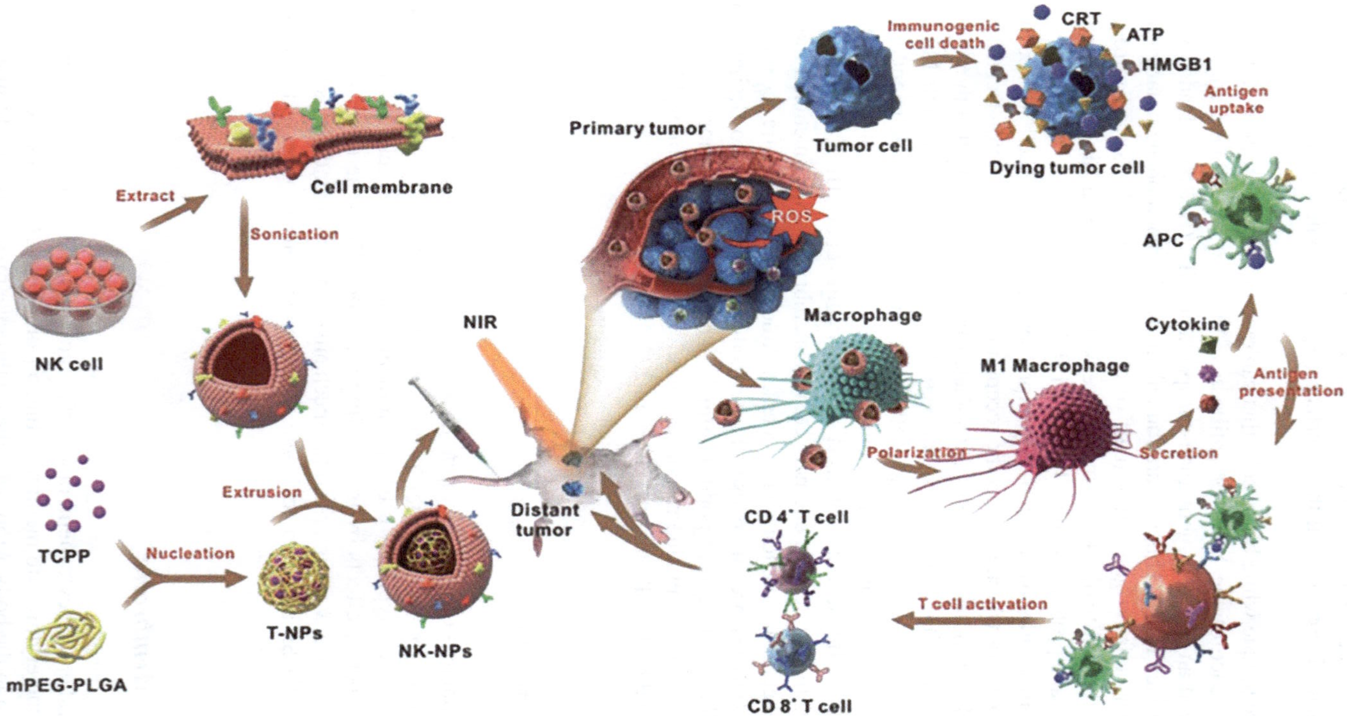

Fig. 7.5 Schematic representation of the preparation of NK cell membrane-encapsulated NPs and their antitumor treatment by combined photodynamic therapy and immunotherapy. (Reprinted with permission from Gong et al. (2020a) Copyright, © 1996–2021 MDPI (Basel, Switzerland) unless otherwise stated)

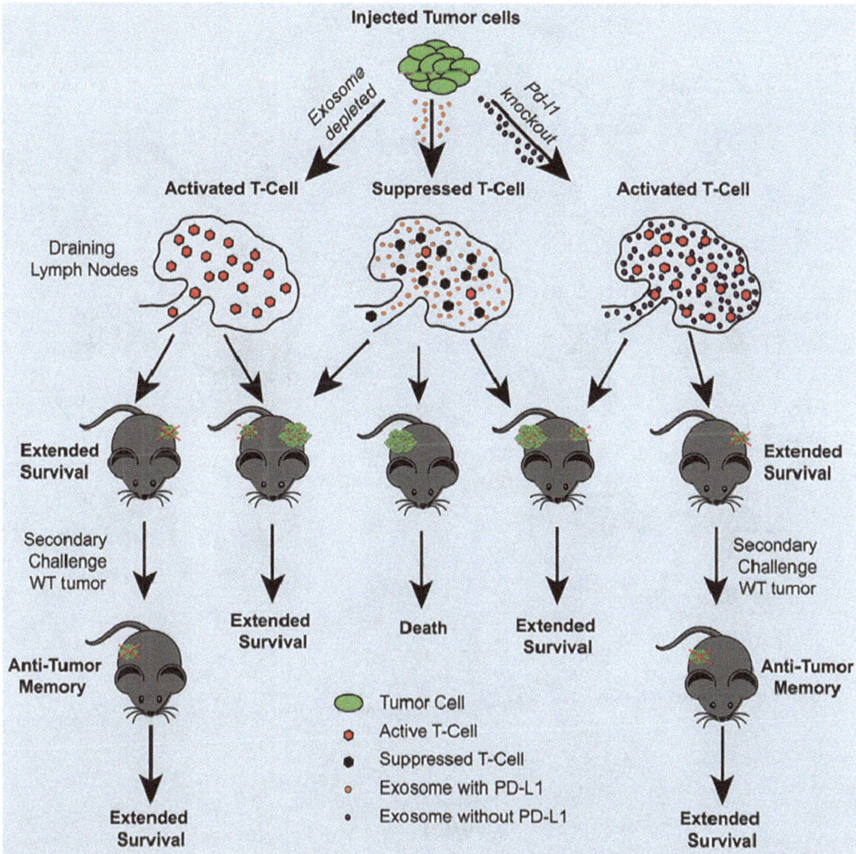

Fig. 7.6 Exosomal PD-L1 can promote T-cell activity in draining lymph nodes through genetic blockade and induce a systemic anti-tumor immune response to inhibit tumor growth. (Reprinted with permission from Poggio et al. (2019) Copyright, © 2019 The Authors. Published by Elsevier Inc.)

et al. 2018; Xia et al. 2019). Therefore, it is important to apply erythrocyte membrane-derived DDSs to a wide range of tumor chemotherapy and PDT where different drugs can be delivered to achieve different effects.

Xiao et al. coated silica NPs (RS) on erythrocyte membrane (RBM) to jointly deliver doxorubicin (DOX) and indocyanine green (ICG) to treat lung cancer. ICG expedited drug release from DDS through its inherent photothermal effect. More importantly, compared with the free form of drugs, the RBM-modified DDS could enhance the apoptosis and blockade of anti-DOX A549 cells (A549/DOX) by enhancing the cellular internalization of drugs (Xiao et al. 2020). Wang et al. prepared a pH-responsive erythrocyte-cancer cell hybrid membrane-encapsulated copolymer micelle (dextran poly(histidine) copolymer) loaded with the selective CSF-1R inhibitor BLZ-945 (DH@ECm). Because of its suitable particle size

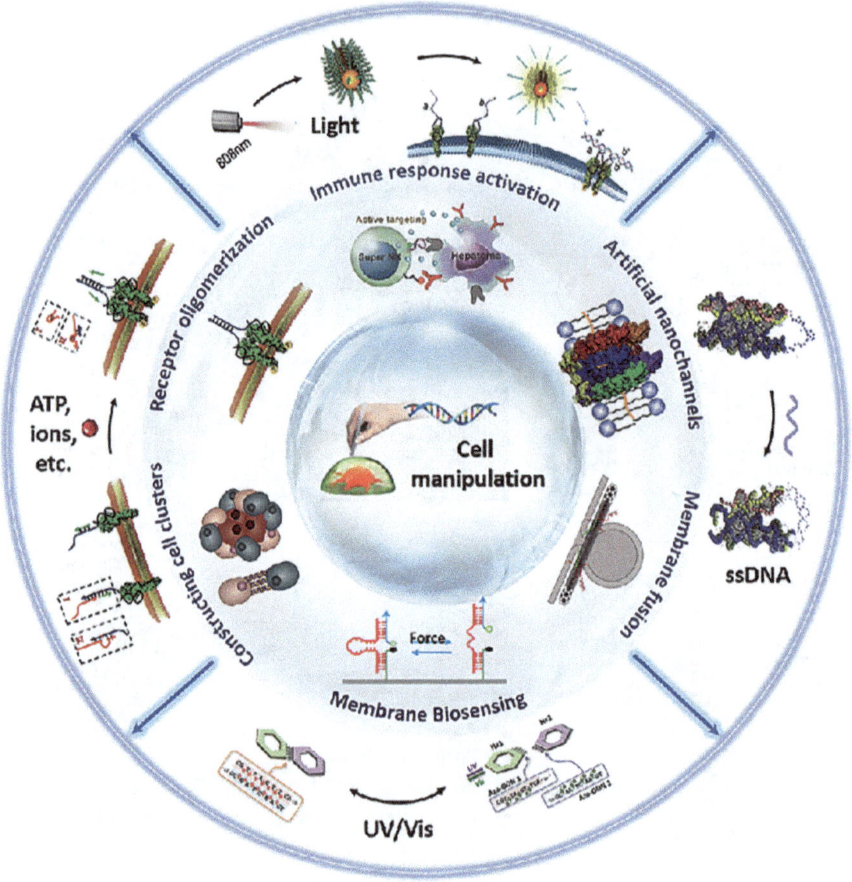

Fig. 7.7 Schematic diagram of a method to manipulate cellular functions and processes through DNA-functionalized cell membranes. (Reprinted with permission from He et al. (2021) Copyright, © The Royal Society of Chemistry 2021)

(around 190 nm), specific targeting, and good immune camouflage, the formulation exhibited unique membrane escape effects and pH-responsive properties when it was in the tumor acidic microenvironment. In addition, the formulation reversed the tumor immune microenvironment in vivo by elevating $CD8^+$ T cells and showed good immunotherapeutic effect and tumor immunosuppression (Wang et al. 2020b). Hao et al. prepared a bionic gene delivery system with the nano-erythrocyte membrane encapsulated in the outer layer of the gene complex. It could improve its biocompatibility and prolong its circulation cycle in vivo through electrostatic interactions, thus opening a new way for the study of the camouflage gene delivery system of the erythrocyte membrane (Hao et al. 2018). In addition, Wan et al. used nanovesicles prepared from CD47 KO mouse erythrocytes (KO-RDNVs) to encapsulate sodium clodronate and found that it had higher biocompatibility and lower

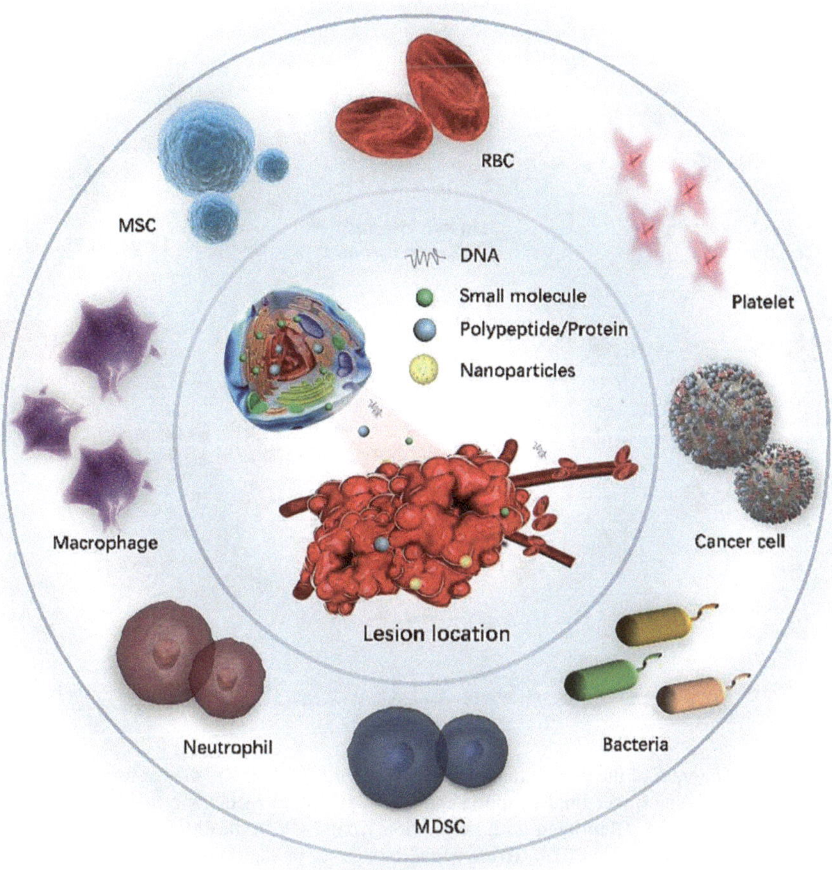

Fig. 7.8 Summary diagram of the different types of cell membrane carriers and their loaded cargoes for cell-derived DDSs. (Reprinted with permission from Wu et al. (2022) Copyright, © 2022 The Authors)

in vivo toxicity compared to sodium clodronate-encapsulated liposomes (Wan et al. 2018). The engineered erythrocyte membrane processed not only the advantages of the erythrocyte membrane DDS but also higher targeting ability. The polymeric NPs are coated by natural RBC membranes through a "top-down approach." Functionalization of erythrocyte membranes by physical insertion of ligand-linker-lipid couplings in the erythrocyte membrane does not damage existing surface proteins (Fang et al. 2013) (Fig. 7.9a). Table 7.1 includes some applications of RBCs membrane-coated DDSs.

Liu's group reported photolytic pyropheophorbide a (PA)-inserted RBC membrane-coated curcumin dimeric prodrug (CUR2-TK)-PLGA NPs [(CUR2-TK)-PLGA@RBC-PA] for improved anti-cancer therapy. With the application of light irradiation, PA incorporated in the NPs generated ROS. The ROS further

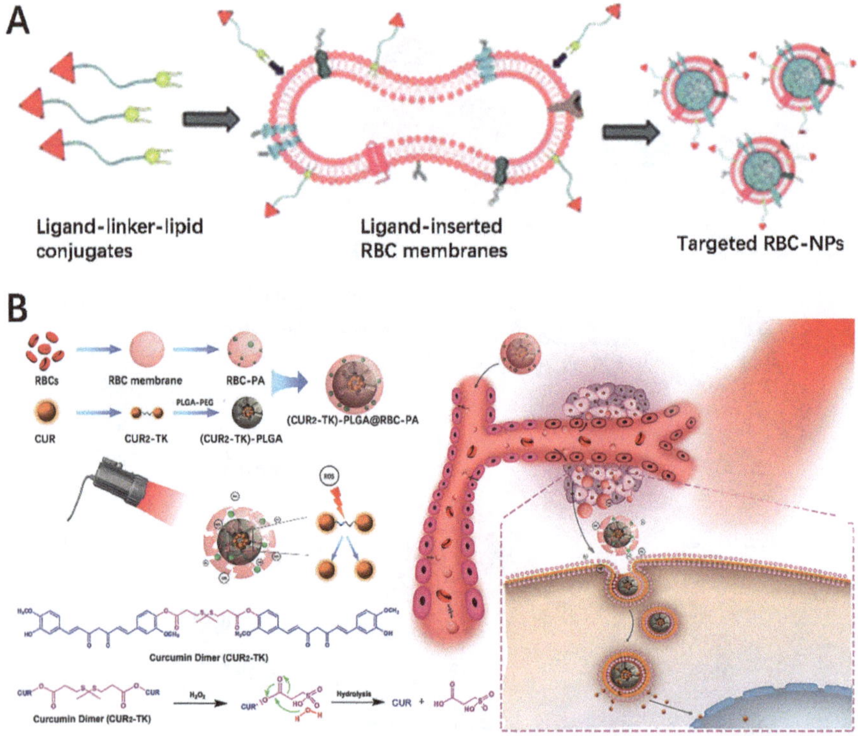

Fig. 7.9 (**a**) Diagram of the preparation of RBC-NPs. Targeted RBC-NPs were formed by physically inserting ligand-linker-lipid coupling into the erythrocyte membrane to functionalize the erythrocyte membrane. (Reprinted with permission from Fang et al. (2013) Copyright, © The Royal Society of Chemistry 2013). (**b**) Schematic diagram of the preparation of [(CUR2-TK)-PLGA@RBC-PA], which achieved combined chemotherapy-PDT by light-activated erythrocyte membrane cleavage and controlled drug release. (Reprinted with permission from Liu et al. (2022) Copyright, © 2022 American Chemical Society)

induced the degradation of the cell membrane, which enhanced uptake of NPs in tumor cells. Meanwhile, the ROS produced can be exploited for photodynamic therapy (PDT). This combined effect on the chemotherapy and PDT could enhance anticancer efficiency and efficacy (Liu et al. 2022) (Fig. 7.9b).

7.4.2 Platelet Membrane

Platelets contribute to many physiological functions, such as hemostasis, immune escape, subendothelial adhesion, and pathogen interaction (Miron and Zhang 2018). For instance, core-shell nano vehicle (designated PM-NV) coated with platelets membrane are used for targeting, sequential and site-specific delivery of active

Table 7.1 Applications of RBCs membrane-coated DDSs

NPs/drugs	Preparation methods	Applications	Properties and effects	Refs
The red blood cell membrane (RBM)-coated silica NPs (RS), doxorubicin (Dox), and indocyanine green (ICG)	Chemical precipitation method in the water-in-oil microemulsion	Tumor photothermal-assisted chemotherapy	The photothermal effect; accelerates the drug release from the DDS; enhances cell internalization and cell arrest; enhances anticancer efficacy	Xiao et al. (2020)
Erythrocyte-cancer cell hybrid membrane camouflaged pH-responsive copolymer micelle (dextran-grafted-poly (histidine) copolymer)	Self-assembly, rotary evaporation	Tumor chemo-photothermal therapy	Preferable immune camouflage and tumor homologies targeting characteristics	Wang et al. (2020b)
Encapsulating photocatalyst titanium dioxide colloid and photothermal agent gold nanorods (Au NRs) in erythrocyte membrane vesicles	Seed growth method proposed, extruded, repeatedly extruded NPs and membrane	Tumor photothermal therapy	Au NRs controlled release of the drug; kill tumor cells by photocatalysis producing the reactive oxygen species (ROS)	Li and Zhang (2020)
An erythrocyte and tumor cell membrane camouflage nanocarrier coloaded with a photosensitizer (indocyanine green) and a hypoxic prodrug (tirapazamine)	The MSN particles were mixed with the membranes and ultrasonically coated with the membrane	Tumor photodynamic therapy with chemotherapy	Less macrophage clearance, increased tumor environment hypoxia by NIR irradiation of photosensitizers, and the hypoxic environment activated TPZ to kill tumor cells	Wang et al. (2020a)
RBC-coated bovine serum albumin NPs	Dispersing and fusing RBCm vesicles with GIB NPs to obtain RmGIB NPs	Tumor synergistic chemo-photothermal therapy	Pronounced long-term circulation; avoiding premature leakage of drugs	Wang et al. (2020a)

proteins or some small molecular drugs. It consists of the following parts: (1) an inner core component based on nano-gel (2) an outer shell part based on the platelet membranes. Then, two anticancer agents-TRAIL and Dox are loaded into PM-NV (named as TRAIL-Dox-PM-NV) (Pei et al. 2020). Pei et al. engineered platelet membrane-encapsulated PLGA NPs IR780@PLGA/Dox (PM-NPs) loaded with drugs and photothermal agents for combination chemotherapy with PTT, and this nano-DDS not only allowed drugs to circulate longer in vivo but also enhanced their

Table 7.2 Applications of platelets membrane-coated DDSs

NPs/drugs	Preparation methods	Applications	Properties and effects	Refs
Platelet membrane (PM)-wrapped core-shell nano vehicle loaded with TRAIL and Dox	Single emulsion method and use of acid-degradable crosslinker	Tumor targeting chemotherapy	PM-NV could actively target tumor sites and deliver anticancer therapeutics in turn to their most active destinations	Pei et al. (2020)
Platelet membrane-coated PLGA NPs, near-infrared (NIR) fluorescent dye IR780 and DOX	Single emulsification method	Tumor chemotherapy combined with PTT	Circulates in vivo longer, tumor-targeting, high tumor inhibition efficacy	Pei et al. (2020)
Platelet membrane envelope loaded with l-arginine and γ-Fe_2O_3 magnetic NPs	Natural platelets membrane envelope using extrusion method loaded with l-arginine and γ-Fe_2O_3 magnetic NPs (PAMNs)	Ischemic stroke therapy	Rapid localization of ischemic stroke lesions depending on the guidance of an external magnetic field	Zhai et al. (2017)
Platelet-shaped Janus nanocages with a mesoporous silica shell, loaded with DOX	Platelets mix under stirring. A sol-gel process is conducted at 40 °C for 12 h and collected by centrifugation and washed with ethanol	Tumor chemotherapy and photothermal therapy	Hydrophobic performance to selectively capture hydrophobic species, and photothermal effects of near-infrared irradiation	Tang et al. (2016)
Biocompatible silica (Si) particles were functionalized with membrane-derived vesicles from activated platelets	Stirred overnight, seven freeze-thaw cycles and sonication	Tumor targeting chemotherapy	Platelets and the tumoricidal activity of immune cells produce a targeted therapeutic effect and deliver cancer-killing drugs at a high local concentration within the CTC thrombus	Chang et al. (2020)

tumor-targeting ability and anti-tumor efficacy (Pei et al. 2020) Table 7.2 includes some applications of platelets membrane-coated DDSs.

7.4.3 *Cancer Cell Membrane*

Zhao et al. constructed an effective treatment strategy (CCM/CS/R-D), which coated silica (CS) NPs (CCM/CS) with the CCM and co-delivered Ca^{2+} channel RNAi with DOX, for cervical cancer therapy. The formulation could promote cellular uptake and lysosomal escape, and the transfection efficiency of siRNA was excellent (comparable to that of PEI 25K). The results showed that CCM/CS/R-D had significantly improved cancer targeting ability and anticancer efficacy compared with CS/R-D or single DDS (Zhao et al. 2020c). Liu et al. applied the MSN core as a container to encapsulate drugs efficiently (doxorubicin, DOX) and used the $CaCO_3$ intermediate layer as a pH-sensitive gatekeeper to control drug release. Data showed that the DOX/MSN@CaCO3@CM DDS can be effectively delivered to LNCaPAI prostate cancer cells, and the tumor inhibition effect is more significant than that of free Dox. In addition, MSN/DOX@CaCO3@CM could effectively induce apoptosis of prostate cancer cells (Liu et al. 2019). Chen et al. reported a zeolite-imidazolium framework hybrid NP (Chang et al. 2020) that could deliver cisplatin (DDP) and oleanolic acid (Mengesha et al. 2007) in combination. This platform (HP/DDP/OLA) could effectively treat bladder cancer (SW780), and it was superior to free drugs alone or mono delivery systems in inducing sw780 cell apoptosis and reversing multidrug resistance (Chen et al. 2020a). Nie et al. evaluated a DDS of MSN loaded with polyethylene glycol liposomes and CCM coated NPs for chemotherapy. CCM@LM encapsulated in low doses of DOX and the PARP inhibitor mefuparib hydrochloride (MPH) showed a stronger tumor inhibitory effect than the Doxil. Their research emphasized that NPs have broad prospects in cancer chemotherapy due to their undergoing facilitating of tumor penetration and powerful intracellular transport (Nie et al. 2020).

Some of the applications of cancer cell membrane-coated DDSs are listed in Table 7.3.

Mem@GOx@ZIF-8@BDOX was developed on zeolitic imidazolate framework (ZIF-8). It was assembled with cancer cell membrane and glucose oxidase (GOX) and further loaded with prodrug of hydrogen peroxide (H2O2)-sensitive BDOX and used in cancer treatment (Cheng et al. 2019) (Fig. 7.10).

7.4.4 *Immune Cell Membrane*

Immunocytes are these cells involved in or related to immune response and can be categorized into many groups. Different kinds of immune cells are summarized in Fig. 7.11 (Wu et al. 2022). By imitating the immune evasion and targeting properties of immune cells, these NPs can more effectively target tumors and then deliver the encapsulated payloads. There are molecules on the immune cell membrane that can provide recognition, called cell surface markers, including antigens, adhesion molecules, receptors, and other cell surface molecules. By the hybrid membrane technology, the fusion membrane of immunocytes with another type of cells showed an excellent capability to improve the efficacy of drug-loaded NPs in disease treatment.

Cao et al. developed macrophage plasma membranes coated albumin NPs (RANPs) containing paclitaxel (PTX) to target therapy against malignant melanoma. In cell line B16F10, RANPs loaded with PTX had higher cytotoxicity and apoptosis rate than albumin NPs without membrane. The internalization efficiency of RANPs in B16F10 cells was also higher. In addition, the anti-tumor effect of

Table 7.3 Applications of cancer cell membrane-coated DDSs

Cells	NPs/drugs	Methods of preparation	Therapeutic applications	Main features	Refs
Cervical cancer cell	Silica NPs, siRNA, DOX	Sonication	Chemotherapy	High stability, low hemolysis, acid-responsive drug release, high accumulation, overcome the multidrug resistance	Zhao et al. (2020c)
Hela human cervix carcinoma cell	PLGA NPs doxorubicin, PD-L1 siRNA	Extrusion	Personalized cancer treatment	Excellent self-recognition internalization, precise tumor therapy	He et al. (2019a)
LNCaP-AI cells	Mesoporous silica nanoparticle core, DOX	Extraction, and ultrasonication	Acid-responsive drug delivery	Biocompatible, simple structured, tumor acidic environment responsive	Liu et al. (2019)
Gastric cancer	Silica NPs, Chlorins e6	Sonication	Specific photodynamic therapy	Good dispersion and stability, photodynamic targeted drug delivery, superior tumor-targeting ability	Yang et al. (2019)
MCF-7 breast cancer cell membrane	Carbon@silica NPS loaded with doxorubicin (DOX)	Extrusion	Synergistic photothermal and chemotherapy	Excellent synergistic photothermal and chemotherapy	Mengyun Zhou et al. (2020)
MCF-7 breast cancer cell membrane	Glucose oxidase (GOx) onto zeolitic imidazolate framework (ZIF-8)	Extraction	Chemotherapy	Cutting off glucose supply and metabolism pathways	Cheng et al. (2019)
SGC7901 cells	Silica NPs loaded with Ce6	Extrusion	Chemotherapy and PDT	Effective tumor-targeted PDT of gastric cancer	Yang et al. (2019)
A549/DOX cells	miR495 with doxorubicin (DOX)	Ultracentrifugation, sonication	Chemotherapy	Drug accumulation increased in A549/DOX cells	He et al. (2019b)

Pancreatic adenocarcinoma cell (BxPC-3) and pancreatic stellate cell (HPSC)	DOX-loaded silica nanospheres and silica nanorods (NRs) NPs	Ultrasonication, ultracentrifugation	Chemotherapy	Superior endocytosis efficiency, effectively accumulate in the endoplasmic reticulum, enhanced tumor accumulation	Zhang et al. (2019b)
4T1	Paclitaxel (PTX)-loaded poly (lactic acid) (Placke et al.) NPs (PPNs)	Centrifugation, extraction	Chemotherapy	Accumulate and long-term exist, inhibition of tumor progression by increasing the area of apoptosis and necrosis in tumor tissue	Han et al. (2020)
MDA-MB-831	DOX loaded MPEG-PLGA NPs	Hypotonic solution, centrifugation, extraction	Chemotherapy	Extended circulation and retention	Kumar et al. (2019)
B16F10 tumor cell membrane	CpG-loaded aluminum phosphate NPs	Ultrasonication, ultracentrifugation	Chemotherapy	Improvement of a colloidal dispersion of aluminum phosphate NPs	Gan et al. (2020)
Bladder cancer	miR-451 and adriamycin-loaded calcium carbonate NPs	Extraction	Chemotherapy	Increase the DDS target retention in tumor tissue	Wei et al. (2020)
143B cell line	Silica NPs loaded with indocyanine green (ICG)	Ultracentrifugation, ultrasonication	Photothermal therapy (PTT)	Advanced stability in a physiological environment; superior anticancer efficacy	Zhang et al. (2019a)

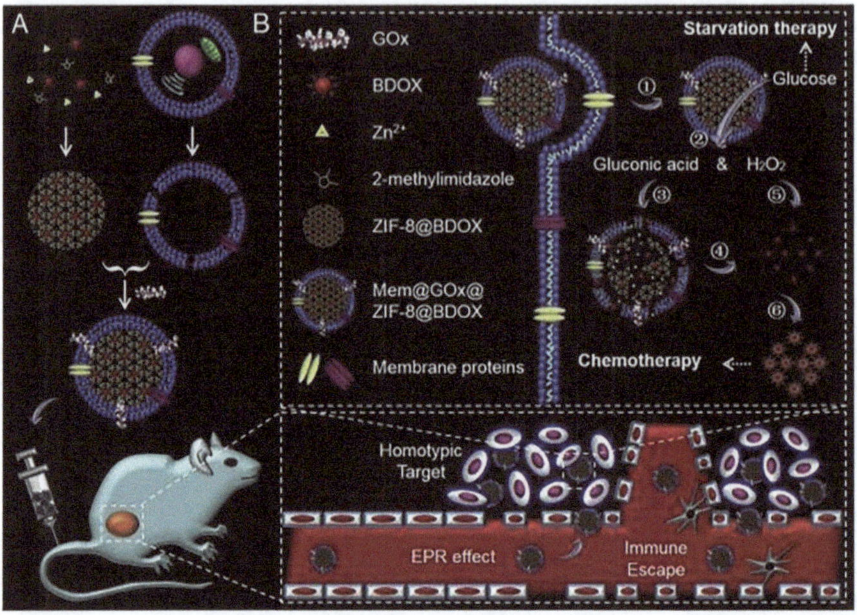

Fig. 7.10 Schematic illustration of (**a**) the fabrication procedure of Mem@GOx@ZIF-8@BDOX nanomotors and the (**b**) the mechanisms in cancer treatment. (Reprinted with permission from Cheng et al. (2019) Copyright, © 2019 Elsevier Ltd. All rights reserved)

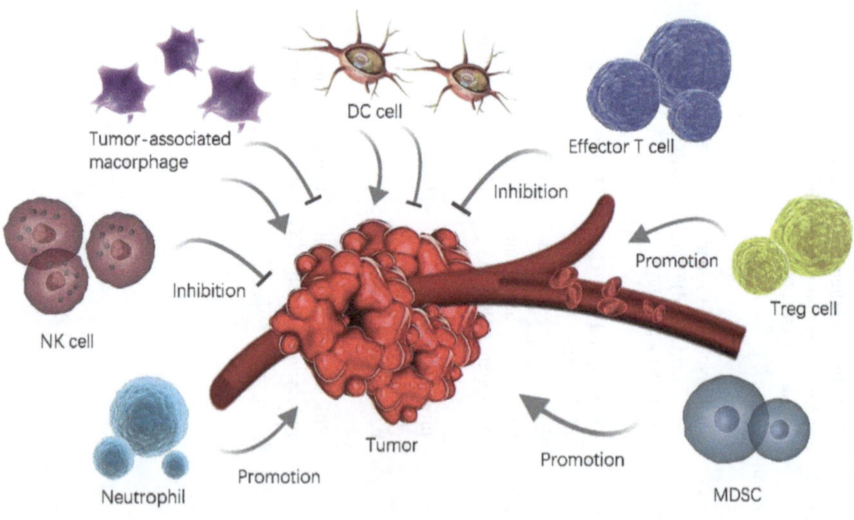

Fig. 7.11 Different kinds of immune cells and their anti-tumor effects. (Reprinted with permission from Wu et al. (2022) Copyright, © 2022 The Authors)

RANPs loaded with PTX was also robust in animal experiments (Cao et al. 2020). Wang et al. reported gallium nano-swimmers coated with leukocyte membrane (LMGNS), which were capable of ultrasound-driven motion, anti-biofouling, and the recognition and targeting of the cancer cell. LMGNSs coated with leukocyte membrane avoided biofouling in the motion of blood and could recognize cancer cells. In addition, these LMGNSs had the ability to actively search for cancer cells and achieved enhanced anticancer efficiency when combined with other therapies (Meng et al. 2020). Zhou and colleagues developed a neutrophil membrane-encapsulated PEG-PLGA NP loaded with celastrol (NNPs/CLT) for the treatment of acute pancreatitis. The results showed that the formulation could cross the blood-pancreatic barrier for distribution into the pancreas and achieved higher tissue accumulation as a result of neutrophil membrane encapsulation. In addition, the systemic toxicity of CLT was significantly reduced due to the encapsulation of NNPs carriers (Zhou et al. 2019).

Ma et al. coated CAR-T cell membrane on the basis of mesoporous silica and loaded IR780 NPs to construct a novel DDS that can specifically recognize GPC3+ liver cancer cells. The experimental results showed that the targeting of obtained NPs was better than that of uncoated MSNs. Therefore, this new DDS achieved better photothermal anti-tumor ability and tumor targeting ability in the treatment of HCC (Gong et al. 2020a). Zhao et al. constructed a sOKGM-PS-miR-31I/cur DDS for targeted treatment of colorectal cancer using oxidized Konjac glucomannan (SOKGM) microspheres encapsulated with α-lactalbumin peptosomes (PSs) and microRNA (MiR)-31 inhibitor (miR-31I). The results revealed that the DDS had the advantages of mucus-penetrating and mucoadhesive and could achieve effective rectal delivery, thus providing a feasible and effective treatment strategy for CRC (Zhao et al. 2020a). Hu et al. established a DDS containing CpG oligodeoxynucleotides and cationic agarose (C-agarose). CpG gel and C-agarose treatment suppressed the tumor growth and inhibited the metastasis size in the metastatic lung nodules and tumor-draining lymph node (TDLN) (Hu et al. 2019). Besides, the current generation of the immunocyte biomimetic nano-DDS could also be used to treat many other diseases. Some of the applications and immune cells involved are listed in Table 7.4.

Due to the wrapping of M1m, (C/I)BP@B-A(D)&M1m could avoid the clearance by MPS, thus achieved successful accumulation in the tumor tissue. The above process is known as membrane escape and often occurs in the microacidic environment inside and outside the cell. The Ce6, IND, and A-B(D) in this formulation are released by laser irradiation and form "caterpillar-like" nanostructures that are highly toxic to deep tumor cells (Hu et al. 2020) (Fig. 7.12).

In NK cells, the NK cell markers can help DOX-loaded NK cell membrane camouflaged liposomes to recognize cancer cells, fuse with cancer cells, and further release DOX (Pitchaimani et al. 2018) (Fig. 7.13).

Table 7.4 Applications of immunocyte membrane coated NPs in disease treatments

Cells	NPs/drugs	Preparation methods	Applications	Properties and effects	Refs
Macrophage	Mesoporous silica nano-capsules loaded with Dox	Extrusion	Tumor; chemotherapy	Improved tumor tropic accumulation, prolonged circulation time, and enhanced primary tumor resistance	Xuan et al. (2015)
	PTX loaded albumin NPs	Co-extruded	B16F10 tumor; chemotherapy	Improved tumor targetability, enhancing the cellular uptake efficiency in B16F10 cells	Cao et al. (2020)
	Persistent luminescence nanoparticle (PLNP)@metal-organic framework (MOF)-derived mesoporous carbon (MC) core-shell nanocomposites (PLMC)	Extrusion	Tumor; autofluorescence-free imaging-guided; chemotherapy	Evading the mononuclear phagocyte system, auto fluorescence-free persistent luminescence imaging-guided drug delivery	Ma et al. (2020)
	Near-infrared Ib (NIR-Ib) fluorescence dye IR-792 NPs (MDINPs)	Extrusion	Glioblastoma; targeted tumor imaging; photothermal therapy	Effectively penetrating BBB, obviously observing orthotopic GBM and suppressing the growth of GBM	Lai et al. (2019)
	DOX loaded (C/I)BP@B-A(D) NPs	Sonication and extrusion	4T1 and B16F10 tumor models; photochemotherapy	Laser-responsive, size-changeable, on-demand drug release and prolonged circulation retention	Hu et al. (2020)
	Amphotericin B (AmB)	Incubation	Visceral leishmaniasis; chemotherapy	Increasing the drug concentration specifically at the infected site, lower collateral toxicity toward healthy cells	Kumar and Bose (2019)

Monocyte	PLGA nano-particles loaded with Dox	Extrusion	Tumor; chemotherapy	Increasing cellular uptake of doxorubicin	Krishnamurthy et al. (2016)
	Gallium nano swimmer loaded with Dox	Sonication	Tumor; photochemotherapy	Capacities of active motion, cancer cell recognition, imaging, anti-biofouling, drug delivery	Meng et al. (2020)
T cell	IR780 loaded mesoporous silica NPs	Genetic engineering	Tumor; photothermal therapy	Combining excellent tumor targeting ability with photothermal effect	Gong et al. (2020a)
	Trametinib loaded PLGA	Surface modification	Melanoma; chemotherapy	Cyto-compatibility and natural targeting were recruited to the tumor regions and displayed a distinctive accumulation signal	Yaman et al. (2020a)
	Poly-lactic-co-glycolic acid nano-particles	Sonication	Photochemotherapy	Increasing the expression of adhesion molecules in the tumor vessels	Zhang et al. (2017)
NK cell	Photosensitizer loaded NPs	Extrusion	Tumor; photodynamic therapy	Selectively accumulation in the tumor, capability to eliminate primary tumors as well as inhibit distant tumors due to an abscopal effect	Wang et al. (2020b)
	DOX loaded cationic liposomes	Extrusion	MCF-7 tumor; chemotherapy	Non-immunogenic prolongs circulation half-life and tumor homing	Pitchaimani et al. (2018)
	Sparfloxacin (SPX) loaded PCL-PEG NPs	Sonication	Lung inflammation; chemotherapy	Prolong the half-life of SPX and reduce its cytotoxicity against normal tissue	Cao et al. (2019)

(continued)

Table 7.4 (continued)

Cells	NPs/drugs	Preparation methods	Applications	Properties and effects	Refs
Neutrophils	CTL-loaded PEG-PLGA NPs	Extrusion	Pancreatic carcinoma; chemotherapy	Overcoming the blood-pancreas barrier; site-specific release of CLT	Cao et al. (2019)
	Poly(latic-co-glycolic acid) nano-particles	Extrusion	4T1 tumor; chemotherapy	The ability to selectively remove CTCs from the blood	Kang et al. (2017)
Dendritic cell	DCM/histidine modified stearic acid-grafted chitosan (HCtSA)/OVA micelles	Extrusion from the mixed solution	Tumor; biomimetic vaccines	pH-responsive biomimetic vaccines with small size, targeting lymph nodes and inducing potent cellular immunity	Liao et al. (2020)
Proinflammatory leukocytes	Dox/siLPCAT1 loaded EYLNs	Incubation and bath-sonication	Esophageal cancer; chemo and gene therapy	Strong synergistic effect of chemo and gene therapy, less toxicity, and the stronger EPR effect	Jun et al. (2020)
Leukocyte-tumor cell	PTX	Co-extrusion	Tumor; chemotherapy	Prominent accumulation within the tumor; Inhibiting tumor growth while not causing systemic adverse effects	He et al. (2018)

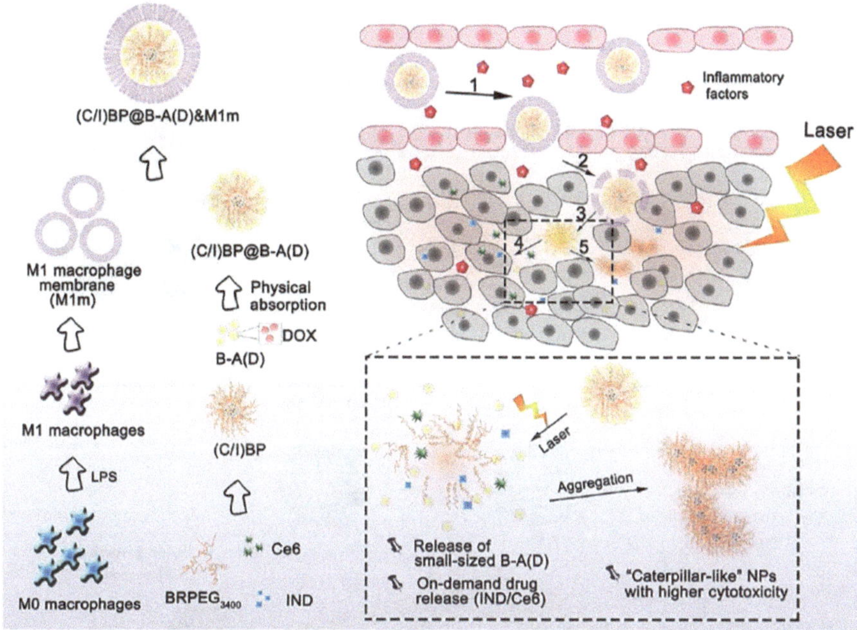

Fig. 7.12 Schematic illustration of tumor-targeted combination therapy. (Reprinted with permission from Hu et al. (2020) Copyright, © 2020 Elsevier Ltd. All rights reserved)

7.4.5 Mesenchymal Stem Cell Membrane (MSCs)

Mesenchymal stem cells (MSCs) are multipotent progenitor cells that are capable of self-renewal and pluripotent differentiation (Wu et al. 2019). Their membrane has a promising application in gene transfection and drug delivery due to its properties of homing and tumor tropism. MSCs may directly produce paracrine influences on the tumor cells, have an important role in promoting tumor vascularization, and respond to the chemotactic signals delivered by tumor cells (Gao et al. 2016). Some of the MSC membrane applications are listed in Table 7.5.

7.4.6 Bacterial Membrane (Outer Membrane Vesicle (OMV))

OMVs have been found to have great potential in transporting small molecule drugs. They have the advantages of good constancy, small size, strong survivability, and high specificity. Under certain physiological conditions, OMVs can slowly release drugs and reduce the toxicity, resulting from drug penetration into the circulatory system. OMV does not decompose in suspension due to its thermodynamic

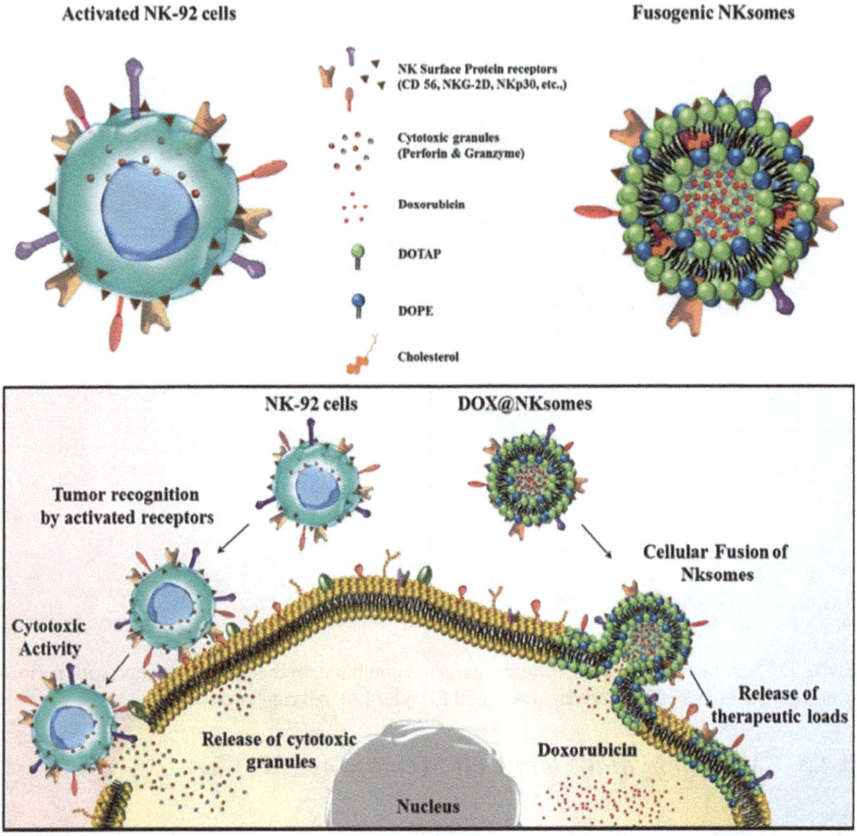

Fig. 7.13 Schematic illustration of activated NK cells and NK-92 cell membrane-derived fusogenic liposomes. (Reprinted with permission from Pitchaimani et al. (2018) Copyright, © 2018 Elsevier Ltd. All rights reserved)

stability, while general liposomes have hidden trouble with drug leakage and are easy to degrade in circulation (Cao and Liu 2020).

Kurban et al. found that OMVs-encapsulated DOX could be effectively transported to NSCLC A549 cells with significant tumor growth inhibition, good tolerability, and pharmacokinetic characteristics and had great potential in tumor chemotherapy and immunotherapy (Kuerban et al. 2020). Gu et al. proposed the engineering of transforming the OMVs derived from *E. coli* and used lysozyme detoxification (named TEVS) as the carrier for transdermal delivery. Using the integrin α(v)β (Frazer and Curtiss 3rd 1975) (αVβ3) targeted peptide, TEVS was modified with indocyanine green (ICG) (p-TEVS-G). The obtained formulation could penetrate the intact cuticle easily, suggesting the great potential of OMVs in the treatment of superficial tumors (Gu et al. 2020) Table 7.6 shows the applications of Neisseria meningitis type B OMV vaccines.

Table 7.5 Applications of MSC membrane coated NPs in disease treatment

NPs/drugs	Preparation methods	Applications	Properties and effects	Refs
Gelatin nanogels	Extrusion	Tumor targeting chemotherapy	A natural and high tumor affinity, increasing its accumulation at tumor sites in the body, decreasing the clearance	Gao et al. (2016)
Upconversion NPs	Extrusion	Photodynamic therapy	Circulates in the body for a longer time, tumor-targeting, high tumor inhibition efficacy	Hao et al. (2019)
Fe_3O_4 NPs, siRNA	Extrusion	Photothermal therapy, chemotherapy	Photothermal functionality, the capability of magnetic resonance	Mu et al. (2018)
Polydopamine NPs	Extrusion	Tumor chemotherapy	Uptaken nonspecifically by macrophage lesser, longer blood retention, and more efficient tumor site aggregation	Zhang et al. (2020)

7.4.7 Exosomes

Currently, exosomes are mostly used for loading miRNAs, nanoscale drugs, and metal oxides, but rarely for loading macromolecular compounds such as proteins. Drug-loaded exosomes are promising DDSs for the treatment of cancer and are regarded as good natural drug carriers. To enhance immunotherapy of PDAC, Zhou et al. developed a dual delivery biosystem based on exosomes to reverse tumor immunosuppression of M2-like tumor-associated macrophages (M2-TAMs) by destroying the Galectin-9/Dectin1 axis. The biosystem utilized exosomes secreted from bone marrow MSCs (BM-MSC) as a vector by modifying oxaliplatin (OXA) on its surface. Galectin-9 siRNA was electroporated into exosome to constitute an immunogenic cell death (ICD) trigger, and results showed that these exosomes could enhance tumor targeting and drug accumulation (Zhou et al. 2021). Liang et al. took advantage of engineered exosomes to synchronously deliver the miR-21i and anticancer drug 5-FU to HER2-expressing tumor cells and effectually reversed drug resistance and observably improved the outcome of 5-FU resistive colon cancer cells (Liang et al. 2020). Wang et al. showed that after co-incubation of M1-Exos-containing macrophages with breast cancer cells in vitro, the pro-inflammatory environment provided by M1-Exos increased caspase-3 expression in breast cancer cells, resulting in the production of a large quantity of pro-inflammatory cytokines and ultimately enhanced tumor inhibition. In tumor-bearing mouse models, M1-Exos loaded with PTX successfully transported the chemotherapeutic agent PTX and enhanced its antitumor effects. In addition to treating cancer, Tian et al. modified curcumin-loaded C(RGdyK) peptides on the surface of exosomes via a simple, rapid, bioorthogonal chemical approach, and after intravenous injection, the

Table 7.6 Summary of *Neisseria meningitis* type B OMV vaccines in applications

Vaccine	Type OMV	Purification methods	Extraction methods	Main antigens	Targeted bacteria	OMV origin	Other contents	Refs
MenBvac®	dOMV	Ultracentrifugation	Deoxycholate extraction method	*Serosubtype* outer membrane protein Porin A (Por A), Porin B (Por B), reduction modifiable protein (RMP), and Opc A, a protein involved in bacterial invasion	Group B *meningococcal* disease	Wild-type strain 44/76-SL	Aluminum hydroxide, OM protein, thiomersal and sucrose	Kulp and Kuehn (2010); Bjune et al. (1991)
VA-Mengoc-BC®	dOMV			OMPs from serogroup B *meningococci*, a polysaccharide from *serogroup C meningococci*	Group B *Neisseria meningitidis*	Wild-type strain CU-385	Aluminum hydroxide, and sodium thiomersalate	Sierra et al. (1991); Pérez Rodríguez and Meneses (1999)
MeNZB®	dOMV	Ultracentrifugation	Detergent deoxycholate		Group B *meningococci*, subtype P1.7b,4	NZ98/254 strain		Oster et al. (2005); Petousis-Harris et al. (2017)

system was able to target transport to areas of ischemic brain injury and trigger an inflammatory response, thereby significantly inhibiting apoptosis in the diseased areas (Tian et al. 2018).

In addition, exosomes can also play a therapeutic role in other diseases. For instance, the construction of MSC-exosome can accurately explain the process in which exosome is involved in wound healing (Wu et al. 2018; Wang et al. 2019a). In recent years, it has been a hot spot to research engineering exosomes. Besides the original advantages of exosomes, remarkable targeting is also exhibited by the engineering of exosomes for DDSs. Through proper modification of exosomes, its effectiveness and stability in imaging probe and therapeutics can be improved, and cell uptake can be enhanced at the same time (Luan et al. 2017). Tian et al. conjugated the functional ligands onto the surface of exosomes. The cyclo(Arg-Gly-Asp-D-Tyr-Lys) peptide [c(RGDyK)] and integrin $\alpha v\beta 3$ could be linked on the surface of MSC-derived exosomes. After administration, cRGD-exo-cur efficiently inhibited apoptosis and inflammation in the lesion area, and the effect was better than that of exosome as carrier or curcumin alone. In addition, cRGD-Exo was also developed as a targeted DDS for cerebral ischemia (Tian et al. 2018). Table 7.7 includes a summary of the use of exosomes as delivery systems.

Table 7.7 Applications of exosomes for drug delivery

EXO origin	Drugs	Methods	Size	Applications	Refs
5-FU-resistant cell line	Exosomes derived from 5-FU resistant cells	Centrifugation	–	Colon cancer treatment	Petousis-Harris et al. (2017)
Bone marrow mesenchymal stem cell	Galectin-9 siRNA, and superficially modified with oxaliplatin (OXA) prodrug	Electroporation	Before: 122 ± 35.3 nm After: 140 ± 37.6 nm	Pancreatic ductal adenocarcinoma	Zhou et al. (2021)
Human adipose mesenchymal stem cell	MiR-375 which is overexpressed after lentiviral transfection	Transfection by lentiviral	–	Bone regeneration	Chen et al. (2019)
M1-macrophages	Paclitaxel	Sonication	Before: 75.3 nm After: 172.8 nm	Anti-tumor	Wang et al. (2019a)
Brain endothelial cell	Paclitaxel	Protein isolation kit and total exosome RNA	Before: 74.5 ± 1.5 nm After: less than 150 nm	Brain cancer treatment	Yang et al. (2015)
Plasma	MiR 140	Frozen at 80 °C for 2 h	Before: 32 ± 25 nm After: 37.1 nm	GD and cartilage repair	Won Lee et al. (2020)
Genetically engineered mammalian cell	Gold NPs	Fabricated by a serial extrusion method	Before: 225 ± 33.4 nm After: 105 ± 10.1 nm	Highly effective brain targeting	Khongkow et al. (2019)

7.4.8 Antibacterial Applications of Biofilms

This part adds a little about its anti-inflammatory application. This table lists the application of DDSs constructed from various biofilms and their derivatives in antimicrobial therapy in recent years. The main idea is to use the biofilm characteristics to deliver drugs targeted to the location of the infection, such as the use of macrophages and neutrophils cell chemotaxis, which can also be achieved by modifying biofilm drugs in the position of the fixed-point release of infection, for example, combined with bacteria and responsive gelatin NPs (GNPs), gelatinases can result in pathogen infection in situ degradation of GNPs.

Wu et al. proposed a novel PLGA NPs (RBCM-PLGA-TR-701NPS, RPTR-701ns) DDS coated with RBC membrane (RBCM), which could achieve continuous drug release and continuous absorption of bacterial toxins (Wu et al. 2021). In Table 7.8, some of the antimicrobial applications of cell membrane-coated DDSs are listed.

7.5 Challenges

Though of great use in DDSs, cellular bionics/derivatives also have some defects. Firstly, some cellular bionics/derivatives are difficult to extract, and the acquisition methods are too complicated, which represents a major expenditure of time and the cryogenic method will cause the loss of cell structure (Harris et al. 2019). Local high temperatures and metal exfoliation from ultrasonic dispersion technology can deactivate proteins on cell membranes. When a DDS is prepared by extrusion, the composition of the membrane is redistributed, and this will change the character of the membrane. All of these will change the integrity of the cell membrane. For RBCM, sialic acid (SA) is the main source of negative charge. After the loss of SA, the repulsive force between the membranes decreases, and this will affect the microscopic rheological properties of RBCM and increase the shear elastic modulus and surface viscosity of RBCM. The redistribution of membrane structure will lead to a burst of phosphatidylserine inside the membrane, which will affect the efficiency of drug delivery (Baier et al. 2021). Moreover, the amplification of the biological process of therapeutic cells requires more efforts and attention due to the large-scale expansion of cells from in vitro culture to bioreactors (Thanuja et al. 2018).

Meanwhile, the drug-loading of drugs needs extensive exploration. How to further reduce the dosage of drugs and how to improve efficiency is of great importance. It is still difficult to determine the appropriate dose and realize individualized administration. Applications of cellular bionics/derivatives require a comprehensive understanding of the distinctions of various membranes, the effects of membrane-nanoparticle interactions, and the ways of drug loading. Moreover, the cell membrane-derived/bionic vehicle from different sources has special pertinence as the carrier, so the choice of cell source is crucial. The tumor cell membrane and

Table 7.8 Antimicrobial applications of cell membrane-coated DDSs

Cells	NPs/drugs	Preparation methods	Therapeutic applications	Properties and effects	Refs
Macrophage	Gold-silver nanocage (GSNC)	Surface coating	Photothermal therapy	Superior bacterial-targeting ability, excellent biocompatibility	Jun et al. (2020)
Neutrophil	NPs loaded with sparfloxacin (SPX)	Surface coating	Targeted drug delivery	Precise targeting ability to inflammatory sites, controlled-release	Wang et al. (2020c)
Red blood cell	Ru-complex-modified selenium NPs (Ru-Se NPs)	Combined with bacterial responsive gelatin NPs (GNPs)	Targeted drug release	Releasing in the pathogen-infected area in situ, intense fluorescence imaging capability, accurately monitor the infection treatment process	Lin et al. (2019)
Extracellular vesicle secreted by S. aureus	Nanoparticle	Surface coating	Targeted drug delivery	Internalized at higher efficiency by *S. aureus*-infected macrophage	Gao et al. (2019b)
Human U937 monocytes	Dexamethasone	The alkaline solution, sonication, and buoyant density gradient ultracentrifugation	Anti-inflammatory therapy	High production yields, mitigate the symptoms of OMV-induced SIRS in vivo	Go et al. (2019)
Neutrophil-like differentiated human promyelocytic leukemia cells (dHL-60)	Antibiotics (penicillin)	Surface coating	Antibiotic therapy	Efficient incorporation, bactericidal activity, monocyte chemotaxis, and macrophage polarization	Kim et al. (2021)

exosomes secreted by tumor cells have a better homing ability when used as drug carriers. Macrophage cell membranes and exosomes secreted by tumor cells have cell adhesion molecules and chemokines, which can reach inflammation and tumor sites. Besides, making cellular bionics/derivatives release the drug within the expected time is also very important.

Moreover, the issue of safety requires more attention because donor cells may pose a threat to the organism. Considering the problems in drug delivery with cells, the manufacturing and regulatory standards should be enacted (Yurkin and Wang 2017). Meanwhile, adverse effects and immunoreaction should also be evaluated when bacteria and viruses were used as drug delivery vehicles (Mengesha et al. 2007). When using the erythrocyte membrane as a DDS, strict disinfection and blood-type compatibility testing are needed to avoid harmful immune reactions.

7.6 Conclusion

Based on the researches on cell bionics/derivatives drug carriers, this chapter mainly introduces the process of constructing DDSs, their advantages, and their applications. Because of their stability, low immunogenicity, biocompatibility, and ability to target tissues, they present a reliable platform for disease treatment. It is promising that the cell membrane-derived/bionic vehicles will be used for clinical applications in the future.

Acknowledgments This work was supported by the National Natural Science Foundation of China (32271464), the Hunan Provincial Natural Science Foundation for Distinguished Young Scholars (2022JJ10086), Hunan Provincial High-Level Health Talents (20240304088), the Joint Fund of the Hunan Provincial Natural Science Foundation and the Hunan Medical Products Administration (2023JJ60501), the Huxiang Young Talent Program of Hunan Province (2018RS3005), the Innovation-Driven Project of Central South University (2020CX048), the Changsha Science and Technology Plan Project (kq2005001, kq2004086), the National Natural Science Foundation of China (81301258), the Natural Science Foundation of Hunan Province (2019JJ60071, 2020JJ4680), the Natural Science Foundation of Changsha (kq2202131), the Shenghua Yuying Project of Central South University, the Hunan Provincial Graduate Research and Innovation Project (2020zzts819), the Undergraduate Innovation Project of Hunan Province (S2020105330414), the Undergraduate Innovation Project of China (202010533015X), and the Open-End Fund for the Valuable and Precision Instruments of Central South University. The authors thank Yaxin Cui, MS, from School of Life Sciences, Jilin University for editorial assistance.

References

Antimisiaris SG, Mourtas S, Marazioti A (2018) Exosomes and exosome-inspired vesicles for targeted drug delivery. Pharmaceutics 10:218. https://doi.org/10.3390/pharmaceutics10040218

Apostolopoulos V, Thalhammer T, Tzakos AG et al (2013) Targeting antigens to dendritic cell receptors for vaccine development. J Drug Deliv 2013:869718. https://doi.org/10.1155/2013/869718

Ayala-Mar S, Donoso-Quezada J, Gallo-Villanueva RC et al (2019) Recent advances and challenges in the recovery and purification of cellular exosomes. Electrophoresis 40:3036–3049. https://doi.org/10.1002/elps.201800526

Baier D, Muller T, Mohr T et al (2021) Red blood cell stiffness and adhesion are species-specific properties strongly affected by temperature and medium changes in single cell force spectroscopy. Molecules 26:2771. https://doi.org/10.3390/molecules26092771

Batrakova EV, Kim MS (2015) Using exosomes, naturally-equipped nanocarriers, for drug delivery. J Control Release 219:396–405. https://doi.org/10.1016/j.jconrel.2015.07.030

Bjune G, Hoiby EA, Gronnesby JK et al (1991) Effect of outer membrane vesicle vaccine against group B meningococcal disease in Norway. Lancet 338:1093–1096. https://doi.org/10.1016/0140-6736(91)91961-s

Bose RJ, Paulmurugan R, Moon J et al (2018) Cell membrane-coated nanocarriers: the emerging targeted delivery system for cancer theranostics. Drug Discov Today 23:891–899. https://doi.org/10.1016/j.drudis.2018.02.001

Cai D, Liu L, Han C et al (2019) Cancer cell membrane-coated mesoporous silica loaded with superparamagnetic ferroferric oxide and Paclitaxel for the combination of Chemo/Magnetocaloric therapy on MDA-MB-231 cells. Sci Rep 9:14475. https://doi.org/10.1038/s41598-019-51029-8

Cao ZP, Liu JY (2020) Bacteria and bacterial derivatives as drug carriers for cancer therapy. J Control Release 326:396–407. https://doi.org/10.1016/j.jconrel.2020.07.009

Cao X, Hu Y, Luo S et al (2019) Neutrophil-mimicking therapeutic NPs for targeted chemotherapy of pancreatic carcinoma. Acta Pharm Sin B 9:575–589. https://doi.org/10.1016/j.apsb.2018.12.009

Cao X, Tan T, Zhu D et al (2020) Paclitaxel-loaded macrophage membrane camouflaged albumin NPs for targeted cancer therapy. Int J Nanomedicine 15:1915–1928. https://doi.org/10.2147/IJN.S244849

Carvalho AL, Fonseca S, Miquel-Clopes A et al (2019) Bioengineering commensal bacteria-derived outer membrane vesicles for delivery of biologics to the gastrointestinal and respiratory tract. J Extracell Vesicles 8:1632100. https://doi.org/10.1080/20013078.2019.1632100

Chang X, Xing L, Wang Y et al (2020) Monocyte-derived multipotent cell delivered programmed therapeutics to reverse idiopathic pulmonary fibrosis. Sci Adv 6:eaba3167. https://doi.org/10.1126/sciadv.aba3167

Chen H, Sha H, Zhang L et al (2018a) Lipid insertion enables targeted functionalization of paclitaxel-loaded erythrocyte membrane nanosystem by tumor-penetrating bispecific recombinant protein. Int J Nanomedicine 13:5347–5359. https://doi.org/10.2147/IJN.S165109

Chen Z, Wang H, Xiong C et al (2018b) Hypoxic glioblastoma release exosomal VEGF-A induce the permeability of blood-brain barrier. Biochem Biophys Res Commun 502(3):324–331. https://doi.org/10.1016/j.bbrc.2018.05.140

Chen S, Tang Y, Liu Y et al (2019) Exosomes derived from miR-375-overexpressing human adipose mesenchymal stem cells promote bone regeneration. Cell Prolif 52:e12669. https://doi.org/10.1111/cpr.12669

Chen D, Cai L, Guo Y et al (2020a) Cancer cell membrane-decorated zeolitic-imidazolate frameworks codelivering cisplatin and oleanolic acid induce apoptosis and reversed multidrug resistance on bladder carcinoma cells. ACS Omega 5:995–1002. https://doi.org/10.1021/acsomega.9b02261

Chen HY, Deng J, Wang Y et al (2020b) Hybrid cell membrane-coated NPs: a multifunctional biomimetic platform for cancer diagnosis and therapy. Acta Biomater 112:1–13. https://doi.org/10.1016/j.actbio.2020.05.028

Cheng H, Jiang XY, Zheng RR et al (2019) A biomimetic cascade nanoreactor for tumor targeted starvation therapy-amplified chemotherapy. Biomaterials 195:75–85. https://doi.org/10.1016/j.biomaterials.2019.01.003

Chinnappan M, Srivastava A, Amreddy N et al (2020) Exosomes as drug delivery vehicle and contributor of resistance to anticancer drugs. Cancer Lett 486:18–28. https://doi.org/10.1016/j.canlet.2020.05.004

Chu D, Dong X, Shi X et al (2018) Neutrophil-based DDSs. Adv Mater 30:e1706245. https://doi.org/10.1002/adma.201706245

Chxa B, Pjya B, Ycza B et al (2020) Cell membrane-camouflaged NPs as drug carriers for cancer therapy. Acta Biomater 105:1–14. https://doi.org/10.1016/j.actbio.2020.01.036

Dehaini D, Wei X, Fang RH et al (2017) Erythrocyte-platelet hybrid membrane coating for enhanced nanoparticle functionalization. Adv Mater 29:1606209. https://doi.org/10.1002/adma.201606209

Deregibus MC, Figliolini F, D'Antico S et al (2016) Charge-based precipitation of extracellular vesicles. Int J Mol Med 38:1359–1366. https://doi.org/10.3892/ijmm.2016.2759

Diaz GBC, Lucas M, Cheng Y et al (2018) Protein digestion, ultrafiltration, and size exclusion chromatography to optimize the isolation of exosomes from human blood plasma and serum. J Vis Exp (134):57467. https://doi.org/10.3791/57467

Fang RH, Hu CM, Chen KN et al (2013) Lipid-insertion enables targeting functionalization of erythrocyte membrane-cloaked NPs. Nanoscale 5:8884–8888. https://doi.org/10.1039/c3nr03064d

Farjadian F, Moghoofei M, Mirkiani S et al (2018) Bacterial components as naturally inspired nano-carriers for drug/gene delivery and immunization: set the bugs to work? Biotechnol Adv 36(4):968–985. https://doi.org/10.1016/j.biotechadv.2018.02.016

Ferguson SW, Megna JS, Nguyen J (2018) Composition, physicochemical and biological properties of exosomes secreted from cancer cells. In: Diagnostic and therapeutic applications of exosomes in cancer. Academic Press, pp 27–57. https://doi.org/10.1016/B978-0-12-812774-2.00003-1

Fox SC, May JA, Dovlatova N et al (2019) How does measurement of platelet P-selectin compare with other methods of measuring platelet function as a means of determining the effectiveness of antiplatelet therapy? Platelets 30(3):290–295. https://doi.org/10.1080/09537104.2018.1434311

Frazer AC, Curtiss R 3rd (1975) Production, properties and utility of bacterial minicells. Curr Top Microbiol Immunol 1975(69):1–84

Gan J, Du G, He C et al (2020) Tumor cell membrane enveloped aluminum phosphate NPs for enhanced cancer vaccination. J Control Release 326:297–309. https://doi.org/10.1016/j.jconrel.2020.07.008

Gao C, Lin Z, Jurado-Sanchez B et al (2016) Stem cell membrane-coated nanogels for highly efficient in vivo tumor targeted drug delivery. Small 12:4056–4062. https://doi.org/10.1002/smll.201600624

Gao C, Lin Z, Wang D et al (2019a) Red blood cell-mimicking micromotor for active photodynamic cancer therapy. ACS Appl Mater Interfaces 11:23392–23400. https://doi.org/10.1021/acsami.9b07979

Gao F, Xu LL, Yang BQ et al (2019b) Kill the real with the fake: eliminate intracellular Staphylococcus aureus using nanoparticle coated with its extracellular vesicle membrane as active-targeting drug carrier. ACS Infect Dis 5:218–227. https://doi.org/10.1021/acsinfecdis.8b00212

Gao C, Huang Q, Liu C et al (2020) Treatment of atherosclerosis by macrophage-biomimetic NPs via targeted pharmacotherapy and sequestration of proinflammatory cytokines. Nat Commun 11:2622. https://doi.org/10.1038/s41467-020-16439-7

Go G, Lee J, Choi DS et al (2019) Extracellular vesicle-mimetic ghost nanovesicles for delivering anti-inflammatory drugs to mitigate gram-negative bacterial outer membrane vesicle-induced systemic inflammatory response syndrome. Adv Healthc Mater 8:e1801082. https://doi.org/10.1002/adhm.201801082

Gong P, Wang Y, Zhang P et al (2020a) Immunocyte membrane-coated NPs for cancer immunotherapy. Cancers 13:77. https://doi.org/10.3390/cancers13010077

Gong SS, Nan N, Sun YK et al (2020b) Protective immunity elicited by VP1 chimeric antigens of bacterial ghosts against hand-foot-and-mouth disease virus. Vaccine 8:61. https://doi.org/10.3390/Vaccines8010061

Gu TW, Wang MZ, Niu J et al (2020) Outer membrane vesicles derived from E. coli as novel vehicles for transdermal and tumor targeting delivery. Nanoscale 12(36):18965–18977. https://doi.org/10.1039/d0nr03698f

Guan S, Yu H, Yan G et al (2020) Characterization of urinary exosomes purified with size exclusion chromatography and ultracentrifugation. J Proteome Res 19:2217–2225. https://doi.org/10.1021/acs.jproteome.9b00693

Gujrati V, Kim S, Kim SH et al (2014) Bioengineered bacterial outer membrane vesicles as cell-specific drug-delivery vehicles for cancer therapy. ACS Nano 8:1525–1537. https://doi.org/10.1021/nn405724x

Han X, Wang C, Liu Z (2018) Red blood cells as smart delivery systems. Bioconjug Chem 29:852–860. https://doi.org/10.1021/acs.bioconjchem.7b00758

Han L, Xu Y, Guo X et al (2020) Cancer cell membrane-coated biomimetic platform for targeted therapy of breast cancer in an orthotopic mouse model. J Biomater Sci Polym Ed 31:1538–1551. https://doi.org/10.1080/09205063.2020.1764163

Hao X, Li Q, Wang H et al (2018) Red-blood-cell-mimetic gene delivery systems for long circulation and high transfection efficiency in ECs. J Mater Chem B 6:5975–5985. https://doi.org/10.1039/c8tb01789a

Hao M, Kong C, Jiang C et al (2019) Polydopamine-coated Au-Ag nanoparticle-guided photothermal colorectal cancer therapy through multiple cell death pathways. Acta Biomater 83:414–424. https://doi.org/10.1016/j.actbio.2018.10.032

Harris JC, Scully MA, Day ES (2019) Cancer cell membrane-coated NPs for cancer management. Cancers 11:1836. https://doi.org/10.3390/cancers11121836

He H, Guo C, Wang J et al (2018) Leutusome: a biomimetic nanoplatform integrating plasma membrane components of leukocytes and tumor cells for remarkably enhanced solid tumor homing. Nano Lett 18:6164–6174. https://doi.org/10.1021/acs.nanolett.8b01892

He C, Zhang X, Yan R et al (2019a) Enhancement of cisplatin efficacy by lipid-CaO2 nanocarrier-mediated comprehensive modulation of the tumor microenvironment. Biomater Sci 7:4260–4272. https://doi.org/10.1039/c9bm00797k

He J, Gong C, Qin J et al (2019b) Cancer cell membrane decorated silica nanoparticle loaded with miR495 and doxorubicin to overcome drug resistance for effective lung cancer therapy. Nanoscale Res Lett 14:339. https://doi.org/10.1186/s11671-019-3143-3

He ZH, Zhang YT, Feng NP (2020) Cell membrane-coated nanosized active targeted DDSs homing to tumor cells: a review. Mater Sci Eng C Mater Biol Appl 106:110298. https://doi.org/10.1016/J.Msec.2019.110298

He Q, Liu Y, Li K et al (2021) Deoxyribonucleic acid anchored on cell membranes for biomedical application. Biomater Sci 9:6691–6717. https://doi.org/10.1039/d1bm01057c

Hu J, Xu J, Li M et al (2019) Targeting lymph node sinus macrophages to inhibit lymph node metastasis. Mol Ther Nucleic Acids 16:650–662. https://doi.org/10.1016/j.omtn.2019.04.016

Hu C, Lei T, Wang Y et al (2020) Phagocyte-membrane-coated and laser-responsive NPs control primary and metastatic cancer by inducing anti-tumor immunity. Biomaterials 255:120159. https://doi.org/10.1016/j.biomaterials.2020.120159

Jimenez-Jimenez C, Manzano M, Vallet-Regi M (2020) NPs coated with cell membranes for biomedical applications. Biology (Basel) 9:406. https://doi.org/10.3390/Biology9110406

Jin K, Luo Z, Zhang B et al (2018) Biomimetic NPs for inflammation targeting. Acta Pharm Sin B 8:23–33. https://doi.org/10.1016/j.apsb.2017.12.002

Jun Y, Tang Z, Luo C et al (2020) Leukocyte-mediated combined targeted chemo and gene therapy for esophageal cancer. ACS Appl Mater Interfaces 12:47330–47341. https://doi.org/10.1021/acsami.0c15419

Kamerkar S, LeBleu VS, Sugimoto H et al (2017) Exosomes facilitate therapeutic targeting of oncogenic KRAS in pancreatic cancer. Nature 546:498–503. https://doi.org/10.1038/nature22341

Kang T, Zhu QQ, Wei D et al (2017) NPs coated with neutrophil membranes can effectively treat cancer metastasis. ACS Nano 11:1397–1411. https://doi.org/10.1021/acsnano.6b06477

Kaparakis-Liaskos M, Ferrero RL (2015) Immune modulation by bacterial outer membrane vesicles. Nat Rev Immunol 15:375–387. https://doi.org/10.1038/nri3837

Khongkow M, Yata T, Boonrungsiman S et al (2019) Surface modification of gold NPs with neuron-targeted exosome for enhanced blood-brain barrier penetration. Sci Rep 9:8278. https://doi.org/10.1038/S41598-019-44569-6

Kim OA-O, Park HT, Dinh NTH et al (2017a) Bacterial outer membrane vesicles suppress tumor by interferon-γ-mediated antitumor response. Nat Commun 8(1):626. https://doi.org/10.1038/s41467-017-00729-8

Kim OY, Dinh NTH, Park HT et al (2017b) Bacterial protoplast-derived nanovesicles for tumor targeted delivery of chemotherapeutics. Biomaterials 113:68–79. https://doi.org/10.1016/j.biomaterials.2016.10.037

Kim JK, Youn YJ, Lee YB et al (2021) Extracellular vesicles from dHL-60 cells as delivery vehicles for diverse therapeutics. Sci Rep 11:8289. https://doi.org/10.1038/s41598-021-87891-8

Koichiro H, Shota Y, Wataru S et al (2018) Red blood cell-shaped microparticles with a red blood cell membrane demonstrate prolonged circulation time in blood. ACS Biomater Sci Eng 4(8):2729–2732. https://doi.org/10.1021/acsbiomaterials.8b00197

Krishnamurthy S, Gnanasammandhan MK, Xie C et al (2016) Monocyte cell membrane-derived nanoghosts for targeted cancer therapy. Nanoscale 8:6981–6985. https://doi.org/10.1039/c5nr07588b

Kroll AV, Fang RH, Jiang Y et al (2017) Nanoparticulate delivery of cancer cell membrane elicits multiantigenic antitumor immunity. Adv Mater 29:1703969. https://doi.org/10.1002/adma.201703969

Kuerban K, Gao X, Zhang H et al (2020) Doxorubicin-loaded bacterial outer-membrane vesicles exert enhanced anti-tumor efficacy in non-small-cell lung cancer. Acta Pharm Sin B 10:1534–1548. https://doi.org/10.1016/j.apsb.2020.02.002

Kulp A, Kuehn MJ (2010) Biological functions and biogenesis of secreted bacterial outer membrane vesicles. Ann Rev Microbiol 64:163–184. https://doi.org/10.1146/annurev.micro.091208.073413

Kumar P, Bose PP (2019) Macrophage ghost entrapped amphotericin B: a novel delivery strategy towards experimental visceral leishmaniasis. Drug Deliv Transl Res 9:249–259. https://doi.org/10.1007/s13346-018-00602-1

Kumar P, Van Treuren T, Ranjan AP et al (2019) In vivo imaging and biodistribution of near infrared dye loaded brain-metastatic-breast-cancer-cell-membrane coated polymeric NPs. Nanotechnology 30:265101. https://doi.org/10.1088/1361-6528/Ab0f46

Lafitte M, Lecointre C, Roche S (2019) Roles of exosomes in metastatic colorectal cancer. Am J Physiol Cell Physiol 317:C869–C880. https://doi.org/10.1152/ajpcell.00218.2019

Lai J, Deng G, Sun Z et al (2019) Scaffolds biomimicking macrophages for a glioblastoma NIR-Ib imaging guided photothermal therapeutic strategy by crossing Blood-Brain Barrier. Biomaterials 211:48–56. https://doi.org/10.1016/j.biomaterials.2019.04.026

Lapinet JA, Scapini P, Calzetti F et al (2000) Gene expression and production of tumor necrosis factor alpha, interleukin-1beta (IL-1beta), IL-8, macrophage inflammatory protein 1alpha (MIP-1alpha), MIP-1beta, and gamma interferon-inducible protein 10 by human neutrophils stimulated with group B meningococcal outer membrane vesicles. Infect Immun 68:6917–6923. https://doi.org/10.1128/IAI.68.12.6917-6923.2000

Li S, Zhang L (2020) Erythrocyte membrane nano-capsules: biomimetic delivery and controlled release of photothermal-photochemical coupling agents for cancer cell therapy. Dalton Trans 49:2645–2651. https://doi.org/10.1039/c9dt04335g

Li T, Li N, Ma Y et al (2019) A blood cell repelling and tumor cell capturing surface for high-purity enrichment of circulating tumor cells. J Mater Chem B 7:6087–6098. https://doi.org/10.1039/c9tb01649j

Li M, Zhou H, Yang C et al (2020) Bacterial outer membrane vesicles as a platform for biomedical applications: an update. J Control Release 323:253–268. https://doi.org/10.1016/j.jconrel.2020.04.031

Liang G, Zhu Y, Ali DJ et al (2020) Engineered exosomes for targeted co-delivery of miR-21 inhibitor and chemotherapeutics to reverse drug resistance in colon cancer. J Nanobiotechnology 18:10. https://doi.org/10.1186/s12951-019-0563-2
Liang K, Liu Q, Kong Q (2021) New technologies in developing recombinant-attenuated bacteria for cancer therapy. Biotechnol Bioeng 118:513–530. https://doi.org/10.1002/bit.27596
Liao Y, Zhang Y, Blum NT et al (2020) Biomimetic hybrid membrane-based nanoplatforms: synthesis, properties and biomedical applications. Nanoscale Horiz 5:1293–1302. https://doi.org/10.1039/d0nh00267d
Lin A, Liu Y, Zhu X et al (2019) Bacteria-responsive biomimetic selenium nanosystem for multidrug-resistant bacterial infection detection and inhibition. ACS Nano 13:13965–13984. https://doi.org/10.1021/acsnano.9b05766
Liu CM, Chen GB, Chen HH et al (2019) Cancer cell membrane-cloaked mesoporous silica NPs with a pH-sensitive gatekeeper for cancer treatment. Colloids Surf B Biointerfaces 175:477–486. https://doi.org/10.1016/j.colsurfb.2018.12.038
Liu Y, Wen N, Li K et al (2022) Photolytic removal of red blood cell membranes camouflaged on NPs for enhanced cellular uptake and combined chemo-photodynamic inhibition of cancer cells. Mol Pharm 19:805–818. https://doi.org/10.1021/acs.molpharmaceut.1c00720
Luan X, Sansanaphongpricha K, Myers I et al (2017) Engineering exosomes as refined biological nanoplatforms for drug delivery. Acta Pharmacol Sin 38:754–763. https://doi.org/10.1038/aps.2017.12
Ma W, Zhu D, Li J et al (2020) Coating biomimetic NPs with chimeric antigen receptor T cell-membrane provides high specificity for hepatocellular carcinoma photothermal therapy treatment. Theranostics 10:1281–1295. https://doi.org/10.7150/thno.40291
Maqsood M, Kang MZ, Wu XT et al (2020) Adult mesenchymal stem cells and their exosomes: sources, characteristics, and application in regenerative medicine. Life Sci 256:118002. https://doi.org/10.1016/J.Lfs.2020.118002
Meldolesi J (2018) Exosomes and ectosomes in intercellular communication. Curr Biol 28:R435–R444. https://doi.org/10.1016/j.cub.2018.01.059
Meng T, Jiang R, Wang S et al (2020) Stem cell membrane-coated Au-Ag-PDA nanoparticle-guided photothermal acne therapy. Colloids Surf B Biointerfaces 192:111145. https://doi.org/10.1016/j.colsurfb.2020.111145
Mengesha A, Dubois L, Chiu RK et al (2007) Potential and limitations of bacterial-mediated cancer therapy. Front Biosci 12:3880–3891. https://doi.org/10.2741/2357
Mengyun Zhou YX, Li X, Xin D et al (2020) Cancer cell membrane camouflaged semi-yolk@spiky-shell nanomotor for enhanced cell adhesion and synergistic therapy. Small 16(39):e2003834. https://doi.org/10.1002/smll.202003834
Miron RJ, Zhang Y (2018) Autologous liquid platelet rich fibrin: a novel DDS. Acta Biomater 75:35–51. https://doi.org/10.1016/j.actbio.2018.05.021
Mu X, Li J, Yan S et al (2018) siRNA delivery with stem cell membrane-coated magnetic NPs for imaging-guided photothermal therapy and gene therapy. ACS Biomater Sci Eng 4:3895–3905. https://doi.org/10.1021/acsbiomaterials.8b00858
Mu W, Wang Z, Zoller M (2019) Ping-pong-tumor and host in pancreatic cancer progression. Front Oncol 9:1359. https://doi.org/10.3389/fonc.2019.01359
Nie D, Dai Z, Li J et al (2020) Cancer-cell-membrane-coated NPs with a yolk-shell structure augment cancer chemotherapy. Nano Lett 20:936–946. https://doi.org/10.1021/acs.nanolett.9b03817
Nordal RA, Wong CS (2004) Intercellular adhesion molecule-1 and blood-spinal cord barrier disruption in central nervous system radiation injury. J Neuropathol Exp Neurol 63:474–483. https://doi.org/10.1093/jnen/63.5.474
Oksvold MP, Neurauter A, Fau-Pedersen KW, Pedersen KW (2015) Magnetic bead-based isolation of exosomes. Methods Mol Biol 1218:465–481. https://doi.org/10.1007/978-1-4939-1538-5_27
Oster P, Lennon D, O'Hallahan J et al (2005) MeNZB: a safe and highly immunogenic tailor-made vaccine against the New Zealand Neisseria meningitidis serogroup B disease epidemic strain. Vaccine 23:2191–2196. https://doi.org/10.1016/j.vaccine.2005.01.063

Palumbo JS, Kombrinck KW, Drew AF et al (2000) Fibrinogen is an important determinant of the metastatic potential of circulating tumor cells. Blood 96:3302–3309. https://doi.org/10.1182/blood.V96.10.3302

Pang XL, Wang ZG, Liu L et al (2019) Immature dendritic cells derived exosomes promotes immune tolerance by regulating T cell differentiation in renal transplantation. Aging 11:8911–8924. https://doi.org/10.18632/aging.102346

Pei W, Huang B, Chen S et al (2020) Platelet-mimicking drug delivery NPs for enhanced chemo-photothermal therapy of breast cancer. Int J Nanomedicine 15:10151–10167. https://doi.org/10.2147/IJN.S285952

Pérez Rodríguez A, Meneses FO (1999) VA-MENGOC-BC vaccine: its impact on meningococcal disease in children 1-4 years old. Rev Cubana Med Trop 51(3):189–193

Petousis-Harris H, Paynter J, Morgan J et al (2017) Effectiveness of a group B outer membrane vesicle meningococcal vaccine against gonorrhoea in New Zealand: a retrospective case-control study. Lancet 390:1603–1610. https://doi.org/10.1016/S0140-6736(17)31449-6

Pitchaimani A, Nguyen TDT, Aryal S (2018) Natural killer cell membrane infused biomimetic liposomes for targeted tumor therapy. Biomaterials 160:124–137. https://doi.org/10.1016/j.biomaterials.2018.01.018

Placke T, Örgel M, Fau-Schaller M, Schaller M, Fau-Jung G et al (2012) Platelet-derived MHC class I confers a pseudonormal phenotype to cancer cells that subverts the antitumor reactivity of natural killer immune cells. Cancer Res 72:440–448. https://doi.org/10.1158/0008-5472.CAN-11-1872

Poggio M, Hu T, Pai CC et al (2019) Suppression of exosomal PD-L1 induces systemic anti-tumor immunity and memory. Cell 177:414–427.e13. https://doi.org/10.1016/j.cell.2019.02.016

Purushothaman A (2019) Exosomes from cell culture-conditioned medium: isolation by ultracentrifugation and characterization: methods and protocols. Methods Mol Biol 1952:233–244. https://doi.org/10.1007/978-1-4939-9133-4_19

Qu MLQ, Huang L, Fu Y et al (2018) Dopamine-loaded blood exosomes targeted to brain for better treatment of Parkinson's disease. J Control Release 287:156–166. https://doi.org/10.1016/j.jconrel.2018.08.035

Que X, Su J, Guo P et al (2019) Study on preparation, characterization and multidrug resistance reversal of red blood cell membrane-camouflaged tetrandrine-loaded PLGA NPs. Drug Deliv 26:199–207. https://doi.org/10.1080/10717544.2019.1573861

Raimondo S, Giavaresi G, Lorico A et al (2019) Extracellular vesicles as biological shuttles for targeted therapies. Int J Mol Sci 20(8):1848. https://doi.org/10.3390/ijms20081848

Riazifar M, Mohammadi MR, Pone EJ et al (2019) Stem cell-derived exosomes as nanotherapeutics for autoimmune and neurodegenerative disorders. ACS Nano 13:6670–6688. https://doi.org/10.1021/acsnano.9b01004

Salih M, Zietse R, Hoorn EJ (2014) Urinary extracellular vesicles and the kidney: biomarkers and beyond. Am J Physiol Renal Physiol 306:F1251–F1259. https://doi.org/10.1152/ajprenal.00128.2014

Si JX, Shao SQ, Shen YQ et al (2016) Macrophages as active nanocarriers for targeted early and adjuvant cancer chemotherapy. Small 12:5108–5119. https://doi.org/10.1002/smll.201601282

Sierra GV, Campa HC, Varcacel NM et al (1991) Vaccine against group B Neisseria meningitidis: protection trial and mass vaccination results in Cuba. NIPH Ann 14:195–207; discussion 208–10

Skotland T, Sandvig K, Llorente A (2017) Lipids in exosomes: current knowledge and the way forward. Prog Lipid Res 66:30–41. https://doi.org/10.1016/j.plipres.2017.03.001

Sushnitha M, Evangelopoulos M, Tasciotti E et al (2020) Cell membrane-based biomimetic NPs and the immune system: immunomodulatory interactions to therapeutic applications. Front Bioeng Biotechnol 8:627. https://doi.org/10.3389/fbioe.2020.00627

Tang L, Yang S, Liang F et al (2016) Janus nanocage toward platelet delivery. ACS Appl Mater Interfaces 8:12056–12062. https://doi.org/10.1021/acsami.6b03208

Thanuja MY, Anupama C, Ranganath SH (2018) Bioengineered cellular and cell membrane-derived vehicles for actively targeted drug delivery: so near and yet so far. Adv Drug Deliv Rev 132:57–80. https://doi.org/10.1016/j.addr.2018.06.012
Theodoraki MN, Yerneni SS, Hoffmann TK et al (2018) Clinical significance of PD-L1(+) exosomes in plasma of head and neck cancer patients. Clin Cancer Res 24:896–905. https://doi.org/10.1158/1078-0432.CCR-17-2664
Tian T, Zhang HX, He CP et al (2018) Surface functionalized exosomes as targeted drug delivery vehicles for cerebral ischemia therapy. Biomaterials 150:137–149. https://doi.org/10.1016/j.biomaterials.2017.10.012
Toyofuku M, Nomura N, Eberl L (2019) Types and origins of bacterial membrane vesicles. Nat Rev Microbiol 17:13–24. https://doi.org/10.1038/s41579-018-0112-2
Vivier E, Raulet DH, Moretta A et al (2011) Innate or adaptive immunity? The example of natural killer cells. Science 331:44–49. https://doi.org/10.1126/science.1198687
Vlassov AV, Magdaleno S, Fau-Setterquist R, Setterquist R, Fau-Conrad R et al (2012) Exosomes: current knowledge of their composition, biological functions, and diagnostic and therapeutic potentials. Biochim Biophys Acta Rev Cancer 1820:940–948. https://doi.org/10.1016/j.bbagen.2012.03.017
Wan X, Zhang S, Wang F et al (2018) Red blood cell-derived nanovesicles for safe and efficient macrophage-targeted drug delivery in vivo. Biomater Sci 7:187–195. https://doi.org/10.1039/c8bm01258j
Wang C, Wen D, Gu Z (2018a) Cellular bioparticulates with therapeutics for cancer immunotherapy. Bioconjug Chem 29:702–708. https://doi.org/10.1021/acs.bioconjchem.7b00619
Wang Y, Chen X, He D et al (2018b) Surface-modified nanoerythrocyte loading DOX for targeted liver cancer chemotherapy. Mol Pharm 15:5728–5740. https://doi.org/10.1021/acs.molpharmaceut.8b00881
Wang P, Wang H, Huang Q et al (2019a) Exosomes from M1-polarized macrophages enhance paclitaxel antitumor activity by activating macrophages-mediated inflammation. Theranostics 9:1714–1727. https://doi.org/10.7150/thno.30716
Wang S, Gao J, Wang Z (2019b) Outer membrane vesicles for vaccination and targeted drug delivery. Wiley Interdiscip Rev Nanomed Nanobiotechnol 11:e1523. https://doi.org/10.1002/wnan.1523
Wang H, Zhang C, Zhang Y et al (2020a) An efficient delivery of photosensitizers and hypoxic prodrugs for a tumor combination therapy by membrane camouflage NPs. J Mater Chem B 8:2876–2886. https://doi.org/10.1039/d0tb00235f
Wang H, Liu Y, He R et al (2020b) Cell membrane biomimetic NPs for inflammation and cancer targeting in drug delivery. Biomater Sci 8:552–568. https://doi.org/10.1039/c9bm01392j
Wang K, Lei Y, Xia D et al (2020c) Neutrophil membranes coated, antibiotic agent loaded NPs targeting to the lung inflammation. Colloids Surf B Biointerfaces 188:110755. https://doi.org/10.1016/j.colsurfb.2019.110755
Wang P, Jiang F, Chen B et al (2020d) Bioinspired red blood cell membrane-encapsulated biomimetic nanoconstructs for synergistic and efficacious chemo-photothermal therapy. Colloids Surf B Biointerfaces 189:110842. https://doi.org/10.1016/j.colsurfb.2020.110842
Wang Y, Luan Z, Zhao C et al (2020e) Target delivery selective CSF-1R inhibitor to tumor-associated macrophages via erythrocyte-cancer cell hybrid membrane camouflaged pH-responsive copolymer micelle for cancer immunotherapy. Eur J Pharm Sci 142:105136. https://doi.org/10.1016/j.ejps.2019.105136
Wang Y, Xiang Y, Xin VW et al (2020f) Dendritic cell biology and its role in tumor immunotherapy. J Hematol Oncol 13:107. https://doi.org/10.1186/s13045-020-00939-6
Wei S, Gao J, Zhang M et al (2020) Dual delivery nanoscale device for miR-451 and adriamycin co-delivery to combat multidrug resistant in bladder cancer. Biomed Pharmacother 122:109473. https://doi.org/10.1016/j.biopha.2019.109473
Wierz M, Pierson S, Gargiulo E et al (2019) Purification of leukemia-derived exosomes to study microenvironment modulation. Methods Mol Biol 1884:231–245. https://doi.org/10.1007/978-1-4939-8885-3_16

Won Lee G, Thangavelu M, Joung Choi M et al (2020) Exosome mediated transfer of miRNA-140 promotes enhanced chondrogenic differentiation of bone marrow stem cells for enhanced cartilage repair and regeneration. J Cell Biochem 121:3642–3652. https://doi.org/10.1002/jcb.29657

Wu P, Zhang B, Shi H et al (2018) MSC-exosome: a novel cell-free therapy for cutaneous regeneration. Cytotherapy 20:291–301. https://doi.org/10.1016/j.jcyt.2017.11.002

Wu M, Le W, Mei T et al (2019) Cell membrane camouflaged NPs: a new biomimetic platform for cancer photothermal therapy. Int J Nanomedicine 14:4431–4448. https://doi.org/10.2147/IJN.S200284

Wu X, Li Y, Raza F et al (2021) Red blood cell membrane-camouflaged tedizolid phosphate-loaded PLGA NPs for bacterial-infection therapy. Pharmaceutics 13:99. https://doi.org/10.3390/pharmaceutics13010099

Wu Y, Liu Y, Wang T et al (2022) Living cell for drug delivery. Eng Regen 3:131–148. https://doi.org/10.1016/j.engreg.2022.03.001

Xia Q, Zhang Y, Li Z et al (2019) Red blood cell membrane-camouflaged NPs: a novel DDS for antitumor application. Acta Pharm Sin B 9:675–689. https://doi.org/10.1016/j.apsb.2019.01.011

Xiao J, Weng J, Wen F et al (2020) Red blood cell membrane-coated silica NPs codelivering DOX and ICG for effective lung cancer therapy. ACS Omega 5:32861–32867. https://doi.org/10.1021/acsomega.0c01541

Xiaowei D (2018) Current strategies for brain drug delivery. Theranostics 8(6):1481–1493. https://doi.org/10.7150/thno.21254

Xie XT, Wang HJ, Williams GR et al (2019) Erythrocyte membrane cloaked curcumin-loaded NPs for enhanced chemotherapy. Pharmaceutics 11:429. https://doi.org/10.3390/pharmaceutics11090429

Xuan MJ, Shao JX, Dai LR et al (2015) Macrophage cell membrane camouflaged mesoporous silica nanocapsules for in vivo cancer therapy. Adv Healthc Mater 4:1645–1652. https://doi.org/10.1002/adhm.201500129

Yaman S, Ramachandramoorthy H, Oter G et al (2020a) Melanoma peptide MHC specific TCR expressing T-cell membrane camouflaged PLGA NPs for treatment of melanoma skin cancer. Front Bioeng Biotechnol 8:943. https://doi.org/10.3389/fbioe.2020.00943

Yaman S, Chintapula U, Rodriguez E et al (2020b) Cell-mediated and cell membrane-coated NPs for drug delivery and cancer therapy. Cancer Drug Resist 3:879–911. https://doi.org/10.20517/cdr.2020.55

Yang T, Martin P, Fogarty B et al (2015) Exosome delivered anticancer drugs across the blood-brain barrier for brain cancer therapy in Danio rerio. Pharm Res 32:2003–2014. https://doi.org/10.1007/s11095-014-1593-y

Yang J, Teng Y, Fu Y et al (2019) Chlorins e6 loaded silica NPs coated with gastric cancer cell membrane for tumor specific photodynamic therapy of gastric cancer. Int J Nanomedicine 14:5061–5071. https://doi.org/10.2147/IJN.S202910

Yang B, Wei L, Wang Y et al (2020a) Oxidation-strengthened disulfide-bridged prodrug nanoplatforms with cascade facilitated drug release for synergetic photochemotherapy. Asian J Pharm Sci 15:637–645. https://doi.org/10.1016/j.ajps.2019.09.001

Yang Y, Sun B, Zuo S et al (2020b) Trisulfide bond-mediated doxorubicin dimeric prodrug nanoassemblies with high drug loading, high self-assembly stability, and high tumor selectivity. Sci Adv 6:eabc1725. https://doi.org/10.1126/sciadv.abc1725

Yurkin ST, Wang Z (2017) Cell membrane-derived NPs: emerging clinical opportunities for targeted drug delivery. Nanomedicine (Lond) 12:2007–2019. https://doi.org/10.2217/nnm-2017-0100

Zhai Y, Su J, Ran W et al (2017) Preparation and application of cell membrane-camouflaged NPs for cancer therapy. Theranostics 7:2575–2592. https://doi.org/10.7150/thno.20118

Zhan Q, Yi K, Li X et al (2021) Phosphatidylcholine-engineered exosomes for enhanced tumor cell uptake and intracellular antitumor drug delivery. Macromol Biosci 21:e2100042. https://doi.org/10.1002/mabi.202100042

Zhang L, Yu D (2019) Exosomes in cancer development, metastasis, and immunity. Biochim Biophys Acta Rev Cancer 1871:455–468. https://doi.org/10.1016/j.bbcan.2019.04.004

Zhang J, Gao W, Fang RH et al (2015) Synthesis of nanogels via cell membrane-templated polymerization. Small 11:4309–4313. https://doi.org/10.1002/smll.201500987

Zhang L, Li R, Chen H et al (2017) Human cytotoxic T-lymphocyte membrane-camouflaged NPs combined with low-dose irradiation: a new approach to enhance drug targeting in gastric cancer. Int J Nanomedicine 12:2129–2142. https://doi.org/10.2147/IJN.S126016

Zhang J, Miao Y, Ni W et al (2019a) Cancer cell membrane coated silica NPs loaded with ICG for tumour specific photothermal therapy of osteosarcoma. Artif Cells Nanomed Biotechnol 47:2298–2305. https://doi.org/10.1080/21691401.2019.1622554

Zhang W, Yu M, Xi Z et al (2019b) Cancer cell membrane-camouflaged nanorods with endoplasmic reticulum targeting for improved antitumor therapy. ACS Appl Mater Interfaces 11:46614–46625. https://doi.org/10.1021/acsami.9b18388

Zhang M, Zhang F, Liu T et al (2020) Polydopamine NPs camouflaged by stem cell membranes for synergistic chemo-photothermal therapy of malignant bone tumors. Int J Nanomedicine 15:10183–10197. https://doi.org/10.2147/IJN.S282931

Zhao Q, Sun X, Wu B et al (2020a) Construction of homologous cancer cell membrane camouflage in a nano-DDS for the treatment of lymphoma. J Nanobiotechnol 19(1):8. https://doi.org/10.1186/s12951-020-00738-8

Zhao R, Du S, Liu Y et al (2020b) Mucoadhesive-to-penetrating controllable peptosomes-in-microspheres co-loaded with anti-miR-31 oligonucleotide and Curcumin for targeted colorectal cancer therapy. Theranostics 10:3594–3611. https://doi.org/10.7150/thno.40318

Zhao ZA, Mei JA, Qw B et al (2020c) Ca2+ signaling modulation using cancer cell membrane coated chitosan NPs to combat multidrug resistance of cancer. Carbohydr Polym 238:116073. https://doi.org/10.1016/j.carbpol.2020.116073

Zhou X, Cao X, Tu H et al (2019) Inflammation-targeted delivery of celastrol via neutrophil membrane-coated NPs in the management of acute pancreatitis. Mol Pharm 16:1397–1405. https://doi.org/10.1021/acs.molpharmaceut.8b01342

Zhou W, Zhou Y, Chen X et al (2021) Pancreatic cancer-targeting exosomes for enhancing immunotherapy and reprogramming tumor microenvironment. Biomaterials 268:120546. https://doi.org/10.1016/j.biomaterials.2020.120546

Zhu CQC, Liu XU, Chen R et al (2018a) Research progress of bacteria-derived products vectors in tumor therapy. Chin J Bioprogress Eng 16(5):26–32

Zhu DM, Xie W, Xiao YS et al (2018b) Erythrocyte membrane-coated gold nanocages for targeted photothermal and chemical cancer therapy. Nanotechnology 29:084002. https://doi.org/10.1088/1361-6528/aa9ca1

Chapter 8
Cellular Drug Delivery Vehicle

Juan Li, Yunyi Liu, and Xiaoxiao Hu

Abstract The cellular drug delivery vehicle has great prospects in biomedical applications. The cellular drug delivery system uses specific cells as drug carriers and utilizes the congenital disease sensing and homing characteristics of these cells to deliver therapeutic and diagnostic agents to the diseased site to achieve high targeting efficiency. In this review, the research progress of stem cells, monocytes/macrophage cells, T cells, red blood cells (RBC), bacteria, and dendritic cells as drug carriers is summarized in recent years, and future challenges are discussed.

Keywords Cellular drug delivery · Cell carrier · Cell therapy · Drug delivery systems

8.1 Introduction

In the past few decades, drug delivery systems (DDS) have been extensively studied because they can significantly improve the pharmacokinetics and pharmacodynamics to enhance the therapeutic and diagnostic effects of cancer (Su et al. 2015). DDS usually uses carriers to carry therapeutic drugs to improve the drug bioavailability, reduce toxicity, increase the circulating half-life of drugs in vivo, limit biodistribution, achieve specific targeting, control drug release, and reduce immunogenicity.

J. Li · Y. Liu
State Key Laboratory of Chemo/Biosensing and Chemometrics, College of Biology, Molecular Science and Biomedicine Laboratory and Aptamer Engineering Center of Hunan Province, Hunan University, Changsha, Hunan, China

X. Hu (✉)
Innovation Institute of Industrial Design and Machine intelligence, Quanzhou-Hunan University, Quanzhou, Fujian, China

Shenzhen Research Institute, Hunan University, Shenzhen, Guangdong, China

Research Institute of Hunan University in Chongqing, Chongqing, China
e-mail: xxhu@hnu.edu.cn

L. Teng et al. (eds.), *Drug Delivery to Tumors*,
https://doi.org/10.1007/978-981-19-8930-8_8

The ideal delivery carriers should be biocompatible, degradable, easily modified, disease-specific, withstanding complex environment in vivo and so on (Yong et al. 2019). However, traditional DDS includes nanoparticles (NPs), liposomes, micelles, hydrogels, and targeted ligands or molecularly modified nanomaterials, such as peptides, antibodies, and aptamers, suffer from difficulties in fulfilling all the basic requirements of ideal delivery carriers, which ultimately fails to meet the ever-increasing modern medical needs (Sengupta and Balla 2018). Therefore, the development of a new type of DDS with truly specific targeting is a huge challenge for modern medicine.

Recently, cellular drug delivery vehicle, also called cell-mediated DDS, has become a promising strategy to solve the above-mentioned challenges (Yu et al. 2020). Different cells have different physiological characteristics, such as a long-term circulation effect of red blood cells, the function of targeting and enriching inflammation and tumor sites of immune cells, and the homing characteristics of stem cells, etc., which can be used as a variety of therapeutic drug delivery vehicle. In addition, compared with traditional drug delivery systems, cellular drug delivery vehicle has outstanding advantages such as low immunogenicity, small toxic and side effects, high biocompatibility, high drug-carrying capacity, and natural crossing of physiological barriers. Therefore, cellular drug delivery vehicle has become a new frontier in medicine (Xue et al. 2017). This chapter summarizes the research progress and future development prospects of stem cells, monocytes/macrophage cells, T cells, red blood cells (RBC), bacteria, and dendritic cells as drug carriers.

8.2 Stem Cells

Stem cells are cells with proliferation and differentiation potential. According to different stages of development, stem cells can be divided into embryonic stem cells and adult stem cells. Compared with embryonic stem cells, adult stem cells possess outstanding advantages such as uncomplicated obtainment and culture in vitro, immunosuppressive, "homing" characteristics, etc, and therefore have been successfully used in targeted therapy research for multiple types of cancer treatments and are less likely to cause tumors formation after transplantation. Homing is one of the key characteristics of effective clinical application of stem cells. Homing refers to the characteristics of endogenous or exogenous stem cells migrating and returning to the injured site which could be caused by ischemia, hypoxia, or injury (Liesveld et al. 2020). Tumor is similarly to a non-healing wound that can mobilize stem cells to homing in the tumor microenvironment (TME). However, the mechanism of "homing" has not been fully elucidated. It is generally believed that chemokines, cytokines, adhesion factors, and vascular endothelial growth factor (VEGF) in the tumor microenvironment may interact with the surface of stem cells. The SDF-1/CXCR4 axis formed by stromal cell-derived factor-1 (SDF-1) and its specific receptor CXC chemokine receptor 4 (CXCR4) is considered to play a key role in the homing process of MSCs (Wynn et al. 2004). There are two main design hypotheses of stem cells as vectors, namely, genetic modification and non-gene modification.

Gene modification is mainly the use of genetic engineering technology to deliver prodrug converting enzyme and anti-cancer proteins. The prodrug converting enzyme delivery system targets tumor sites specifically through stem cells which are transduced with exogenous enzyme genes (suicide genes). Then the gene product enzyme of the suicide gene converts the low-toxic prodrug into cytotoxic substance through its enzymatic activity thereby playing the role of targeted killing of tumor cells (Cihova et al. 2011). The currently used prodrug converting enzyme systems mainly include Cytosine Deaminase/5-Fluorocytosine (CD/5FC), Herpes Simplex Virus Thymidine Kinase (HSV-TK)/Ganciclovir, tomato thymidine kinase 1(toTK1)/zidovudine (AZT), rabbit carboxylesterase (rCE)/CPT-11 (Altaner et al. 2014; Gutova et al. 2017; Mooney et al. 2017), etc. Anti-cancer proteins are another way of genetic modification of stem cells. The anti-cancer proteins currently used in stem cell carriers mainly include cytotoxic factors such as tumor necrosis factor-related apoptosis-inducing ligand (TRAIL), anti-angiogenic factors, interleukins, and interferons. Tumor necrosis factor-related apoptosis-inducing ligand (TRAIL) is a type II transmembrane death ligand that can selectively induce tumor cell apoptosis without affecting normal cells. Stem cells engineered to express TRAIL have achieved certain therapeutic effects in melanoma lung metastases, pancreatic cancer, cervical cancer, Ewing sarcoma, esophageal cancer, and other cancers (Wu et al. 2019; Rossignoli et al. 2019; Grisendi et al. 2010; Guiho et al. 2016; Li et al. 2014). In addition, interleukins are a type of immune regulation-related cytokines, which mainly inhibit tumor growth by stimulating the immune system and activating immune cells. Interferon also has similar anti-tumor effects. Gene recombination to make stem cells express interleukin and interferon has also been widely used in anti-tumor research (Nakamura et al. 2004; Ren et al. 2008). In recent years, it is hoped that inducing RNA from stem cells to tumor cells can inhibit the development of tumors cells by downregulating the expression of related genes. Such hypothesis has gradually become a focus of attention amongst researchers.

Non-genetic modifications include oncolytic virus therapy and loading of nanoparticles or drugs. Oncolytic virus therapy induces the viral genome selectively into tumor cells and will therefore replicate in cancer cells and dissolve cancer cells without affecting normal cells (Goradel et al. 2021). However, this method will reduce the efficiency of oncolysis due to the production of antiviral antibodies. Moreover, the rapid isolation of the virus by the reticuloendothelial system will inhibit the circulation of the virus to the tumor (Zendedel et al. 2019). As a viral vector, one of its main advantages (Mader et al. 2009) is that stem cells can protect oncolytic viruses from the invasion of antiviral antibodies. In addition, using the homing properties of stem cells, loading oncolytic viruses can improve the targeting of oncolytic viruses (Everson et al. 2016). In addition to loading oncolytic viruses, due to the tolerance of certain stem cells to chemotherapeutic drugs and their unique endocytosis, these stem cells can be used for the targeted delivery of chemotherapeutic drugs and solve the problems of the short half-life and the poor tumor targeting of these chemotherapeutic drugs (Ho et al. 2018; Ouyang et al. 2020).

Even though stem cells have been widely used as vectors, there are still many problems (Golinelli et al. 2020), such as (a) how to quantify the stem cells that reached the tumor site. (b) Due to the multidirectional differentiation function of stem cells, the tumorigenicity of stem cells is also a very controversial issue (Zhang et al. 2018b; Lye et al. 2016). (c) Although stem cells have low immunogenicity, some stem cells will still be eliminated in the body. The dense tumor stroma can also hinder the distribution of stem cells in tumor tissues. (d) When stem cells load drugs into the body, the way of drug release and the controllability of the release site are not completely clear. (e) In terms of stem cells carrying therapeutic genes, viral vectors have problems such as cytotoxicity and immunogenicity, while non-viral vectors still have problems such as low loading rates. Therefore, how to achieve safe and efficient stem cells to carry therapeutic genes is still a problem that needs to be solved urgently.

8.3 Monocytes/Macrophages

Monocytes/macrophages are important immune cells in the body. Monocytes are polarized into classically activated tumor-associated macrophages under the stimulation of Toll-like receptor-related ligands and gamma-interferon (interferon-γ, IFN-γ) in the tumor microenvironment (TME) (M1 type) (Lee et al. 2016). These are polarized into selective activation tumor-associated macrophages (M2 type) under the stimulation of cytokines such as IL-4, IL-6, IL-13 (An and Yang 2020; Chen et al. 2021) (see Fig. 8.1 in Hao et al. 2012). M2 type macrophages produce

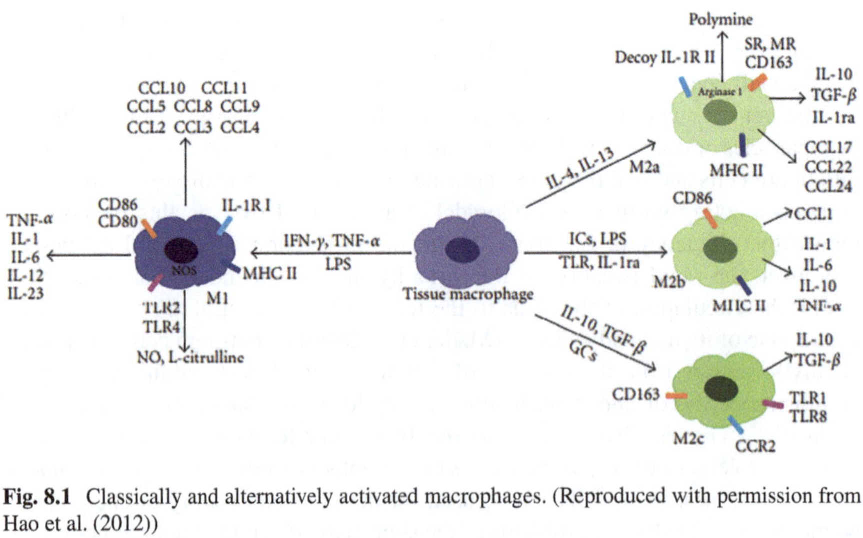

Fig. 8.1 Classically and alternatively activated macrophages. (Reproduced with permission from Hao et al. (2012))

cancer-promoting effects and are mainly used in tumor immunotherapy (Hao et al. 2012). In comparison, M1 type macrophages exert a tumor suppressor effect and can be recruited by factors released from the tumor microenvironment (TME) such as monocyte chemoattractant protein-1 (CCL-2, also known as MCP-1) (Qiu et al. 2021; Guo et al. 2020). Therefore, researchers mainly use monocytes and M1 macrophages as cell carriers to deliver drugs (Xia et al. 2020). There are three main mechanisms for mononuclear/macrophage to deliver drugs, which are directly loading anticancer drugs, loading anticancer drug nanocarriers, and using engineered macrophages to load drug carriers.

Monocytes/macrophages can swallow drugs into the tumor microenvironment (Xia et al. 2020; Su et al. 2015). But free drugs are easily degraded in phagolysosomes, and anti-cancer drugs are toxic to monocytes/macrophages as carriers (Pang et al. 2018). Therefore, most drug delivery systems based on monocytes/macrophages use monocytes/macrophages as indirect anticancer drug carriers, that is, load NPs containing drugs in order to improve the efficiency of drug delivery. In addition to improving the efficiency of drug delivery, macrophages can also carry drugs across the blood-brain barrier (BBB) (Liang Pang et al. 2016). In addition, macrophage vectors can infiltrate tumors and target drugs to the anaerobic areas inside tumors (Wang et al. 2020a). Macrophage drug-loaded chemotherapy can also be combined with other therapies for synergistic treatment of tumors (Nguyen et al. 2020). Nevertheless, nano-carrier drugs still have the phenomenon that the drug release efficiency and the activity of phagocytes are reduced due to the phagocytosis of phagocytes (Zhang et al. 2018a; Xu et al. 2020). The plastic "backpack (BP)" design can avoid the unfavorable phenomena due to phagocytosis, because the drugs or NPs is connected to the cell surface (Klyachko et al. 2017). It is worth mentioning that Ling Guo et al. (2019) discovered a drug delivery mechanism in engineered M1 macrophages (M1-dox) loaded with doxorubicin, which consists of M1-Dox transferring the drug (doxorubicin) into tumor cells through the tunnel nanotube pathway. This also provides new ideas for the delivery of drugs to monocytes/macrophages.

In addition to using monocytes/macrophages to load drugs or drug carriers, the use of genetic engineering to modify monocytes/macrophages to express special proteins has also been extensively studied (Luo et al. 2018; Zhang et al. 2019).

Although monocytes/macrophage vectors have been successfully used in preclinical research, their use has some limitations, for example, how to ensure drug loading, control drug release efficiency and rate, and avoid immunogenicity and tumorigenicity (Zheng et al. 2019) without disrupting cell targeting. Monocytes/macrophages have strong plasticity, and their phenotypes can be transformed into each other in different microenvironments. In the tumor microenvironment, monocytes/macrophages may transform into M2 type macrophages that promote tumor growth (An and Yang 2020). Therefore, although monocytes/macrophages are perfect as cell carriers, there are still many potential problems and risk factors that need to be explored for clinical treatment (Krishnamurthy et al. 2016).

8.4 T Cells

T cells, also known as T lymphocytes, are differentiated from lymphatic stem cells in the thymus. They are the most numerous and complex immune cells among lymphocytes, and have a variety of biological functions, such as direct killing of target cells. They help to inhibit the production of antibodies by B cells, the response to specific antigens and mitogens, and the production of cytokines. There are many classification standards for T cells. According to their functions, they can be divided into helper T cells (Th) and cytotoxic T cells (Tc). According to different differentiation antigens (CD molecules), they can be divided into CD4+ and CD8+ subgroups. Among them, Th cells are also called CD4+ cells because they express cluster of differentiation 4 (CD4) on the surface. It is activated by reacting with the peptide antigen presented by MHC II (major histocompatibility complex). MHC II is expressed on the surface of antigen presenting cells (APCs). Once activated, it can secrete cytokines to regulate or assist the immune response. Tc cells are also known as CD8+ cells. The cells expressing cluster of differentiation 8 (CD8) on their surface can directly bind to antigens through MHC I. T lymphocytes will target secondary lymphoid organs and inflammatory tissues in the body, and they have been widely studied for cell therapy in cancer treatment (Su et al. 2015). At present, there are two main research ideas for T cells as cell carriers, namely, adoptive cell transfer therapy (ACT) and T cells loaded with nano-drugs.

Adoptive immune cell therapy is a type of immunotherapy base on T cells. It consists of the body's immune cells undergoing a series of operations (induction, modification, amplification, functional identification) in vitro to screen out immune cells that can efficiently and specifically attack tumors. Then it is injected back into the patient's body to inhibit or even kill tumor cells. Chimeric antigen receptor-modified T-cell (CAR-T) immunotherapy and T cell receptor (TCR) genetically engineered T-cell (TCR-T) are the two latest immune cell technologies for adoptive immunotherapy. The principle is to isolate T cells from the peripheral blood of the body for in vitro amplification, genetically engineer them to have specific antigens that target tumor cells, activate them, and finally inject into the body for anti-tumor therapy (Liu et al. 2017). The main difference between the two is that TCR-T still requires major histocompatibility complex (MHC) molecules to present tumor cell-specific antigens. In essence, TCR-T is a T cell with a clear antigen specificity. CAR-T directly gives T cells an antibody that can bind to the specific antigen of tumor cells, so it overcomes the MHC limitation. Here, the components of the chimeric antigen receptor (CAR) consist of the hinge region (responsible for forming the correct conformation and forming dimers), the intracellular signal transduction region (responsible for transducing the activation signal of T cells), the extracellular antigen recognition region, and transmembrane structure region (Tahmasebi et al. 2020). The extracellular tumor antigen binding region is usually determined by single chain variable fragment (scFv) of the antigen recognition region. The scFv determines the specificity of CAR. When T cells encounter a specific tumor antigen, the scFv segment will bind to the tumor antigen and then directly activate a large number of T cells and stimulate the secretion of a large number of cytokines, so that the tumor cells are directly attacked and killed. The same principle can also be used to induce the expression of specific genes after the antigen is recognized, which also makes CAR-T cells and TCR-T cells become the drug delivery vehicles for therapeutic genes (such as cytokines IL-2, IL-7, IL-12, IL-15, IL-18, IL-21, etc.) (Combes et al. 2020; Chmielewski et al. 2011; Chinnasamy et al. 2012; Pegram et al. 2015; Hu et al. 2017; Shum et al. 2017; Hoyos et al. 2010; Markley and Sadelain 2010).

CAR-T cell therapy is a promising new treatment method for hematological malignancies (Wang et al. 2017). However, CAR-T cells are rarely effective in the treatment of solid tomor therapy due to the fact that injected CAR-T cells do not easily reach the tumor site and has limited infiltration and survival in the immunosuppressive and hypoxic tumor microenvironment (TME) (Wang et al. 2020b). In addition, although TCR-T cells are widely used (Tran et al. 2014, 2016; Yong-Chen et al. 2017), the tumor microenvironment inhibiting the anti-tumor activity and cell proliferation of T cells, targeted/untargeted toxicity due to antigen loss or recurrence and recognition of healthy tissues still remain the biggest problem of T cell immunotherapy (Han et al. 2020; Wang et al. 2017; Bielamowicz et al. 2018).

T cells already have a complete in vitro expansion system. In addition to using the tumor cell killing effect of T cells, T cells can also be used as carriers to deliver nano-drugs to tumor sites. Compared with the direct application of nanocarriers for drug delivery, this method has the advantages of reducing cytotoxicity, increasing the accumulation of drugs at tumor sites, and enhancing tumor targeting (Bonnie Huang et al. 2015; Jones et al. 2017). In addition to coupling with nanoparticles, TCR-T can also carry oncolytic virus, which can effectively protect oncolytic virus from antibodies neutralization and transfer them to tumor cells (Melzer et al. 2019). In addition, researchers have coupled T cells with aptamers that specifically target tumors to improve their targeting and enhance their anti-tumor effects (Liu et al. 2020).

8.5 Red Blood Cell (RBC) Carrier

In 1953, Gardos attempted to load ATP into the "erythrocyte ghosts"; this attempt became a beginning for coating of the erythrocyte membrane, opening up a new strategy of drug delivery (Villa et al. 2016). Subsequently, the term "carrier RBCs" was introduced in 1979 (Xia et al. 2019). In 1994, Lejeune et al. (1994) reported an RBC membrane-derived liposome, termed "nanoerythrosomes," which was obtained by physically squeezing RBC ghosts through membranes with defined pore sizes.

RBCs have been considered as an excellent carrier for drug delivery because of lacking nucleus and organelles. Compared with synthetic drug carriers, RBCs have the characteristics of long circulation in blood, easy to obtain, larger specific surface area and volume, excellent biocompatibility, and easy to clearance (Hayashi et al. 2018; Biagiotti et al. 2011). The RBC carrier allows for the sustained release of loaded drugs through the bloodstream to deliver the therapeutics drugs to location of lesions. Moreover, the double concave shape of RBCs makes drug vehicles to have a bigger specific surface area to deliver various active substances including small molecule drugs, enzymes, peptides, etc. (Glassman et al. 2020). Because of its nature of not having nuclear and mitochondrial DNA, it keeps RBCs undisturbed by other cellular compositions.

Due to a certain degree of artificial operation, RBCs can change its structure to target the reticuloendothelial system (RES), especially in liver and spleen, which is a passive targeting drug deliver strategy. This peculiarity has been used to treat a number of RES diseases, including lysosomal storage disease, liver metastases, RES parasitic disease, poisoning caused by hepatic porphyria, etc. (Yan et al. 2017).

The preparation of RBC carrier is based on its osmotic properties. In brief, the RBCs are placed in gradient hypoosmotic solution to reach the critical point of hemolysis. Under such condition, pores start to appear on the cell membrane, through which the drug entered the RBC. Once equilibrium is reached, RBCs are transferred to an isotonic solution. The RBCs will be returned to their normal form, and the membrane surface will close to retain the drug in the citosol. Wang et al. (Zhang et al. 2020) developed a drug delivery vector (RV). Subsequently, the first-line chemotherapy drug doxorubicin (DOX) and the common photothermal agent IR 780 were loaded into the erythrocyte membrane enrichment vector for combination therapy (RV/I-D). DOX and IR 780 were dissolved in dimethyl sulfoxide (DMSO), and RV/I-D was obtained by taking advantage of the low osmotic properties of RBCs. RV can help RV/I-D avoid premature excretion due to increased tumor accumulation. Then, IR780's intracellular photothermal properties make it synergistic with DOX to fight cancer, which can effectively treat DOX-resistant prostate cancer (PC-3/DOX).

Electroporation is widely used at present. The basic principle of electroporation is to induce the permeability change of erythrocyte membrane in isotonic solution by electric shock. Transmembrane potential differences cause the formation of pores in the RBCs membrane, through which drug molecules enter the RBC.

Shi et al. (Usman et al. 2018) developed a novel strategy using extracellular vesicles produced and isolated from a large number of RBCs for the delivery of RNA drugs, including antisense oligonucleotides, Cas9 mRNA, and guide RNA. Extracellular vesicle inclusion of therapeutic RNA drugs and Cas9 mRNAs was induced by electroshock, and RBCEV RNA drug delivery demonstrated highly robust microRNA inhibition and CRISPR-Cas9 genome editing in human cells and xenograft mouse models without observed cytotoxicity.

In addition to electroshock and low osmosis methods, there are also methods of drug encapsulation using cell penetrating membrane peptides (see Fig. 8.2 in Usman (2018)).

Due to the characteristic of large specific surface area of RBC carrier, drug delivery using RBC surface is also a method of drug delivery. The most common is direct chemical coupling, such as the use of the nonspecific chemical crosslinker glutaraldehyde. Glutaraldehyde binds the drug molecule to the free amino group on the RBC membrane and releases the drug when the RBC enters the bloodstream under the

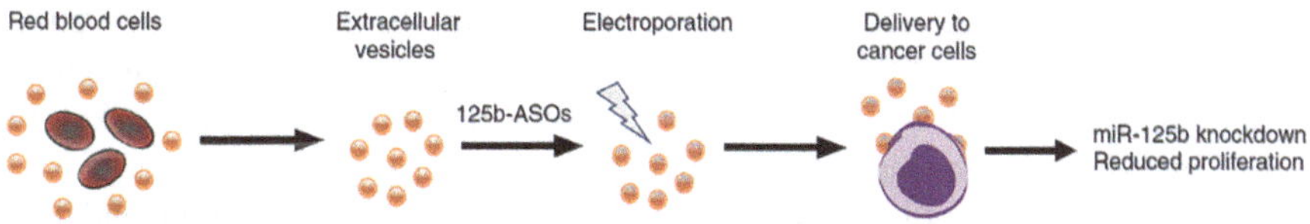

Fig. 8.2 RBC Extracellular vesicles (RBCEVs) deliver ASOs to leukemia and breast cancer cells for miR-125b inhibition. Experimental scheme of ASOs delivery to cancer cells using RBCEVs. (Reproduced with permission from Usman et al. (2018))

action of hydrolase (Rossi et al. 2019). In addition to chemical coupling, there are methods for RBC cell surface adsorption. For example, Brenner et al. (Ferguson et al. 2020) developed dual targeted RBC hitchhiking (DTRH), which adsorbs drug nanoparticles to RBCs and then delivers them to patients through intravenous injection or arterial intubation. After entering the bloodstream, the nanoparticles are desorbed due to shear stress or direct contact with endothelial cells. This method can effectively improve the serious uptake of nanocarrier drugs to the liver and spleen, poor targeting in vivo, and other problems. Combined with the clinical use of vascular intubation therapy, the organ targeting ability of nanocarrier drugs can be greatly improved. However, the drug delivery system also has some problems, such as weak connections and desorption. At present, the targeting of heart, brain, and other organs by using this system requires the aid of blood transfusion and intubation, so it is suitable for acute and serious diseases, but not for most chronic diseases.

In addition to methods of encapsulation and surface ligation, drug and antibody/peptide complexes are currently prepared in vitro and then injected into the human body (Li et al. 2019). Targeting therapeutic drugs to circulating RBCs by means of antibodies can avoid problems caused by in vitro manipulation and reinfusion to a certain extent. The method targets multiple antigens on the surface of RBCs, such as the immune adhesion receptor CRL, the Kell protein, the Band anion channel protein, blood group glycoproteins, the sugar transporter family, and other highly expressed antigens (Rossi et al. 2019).

At present, there are still some problems that need to be overcome. First, it is difficult to control the quality of drug-carrying RBCs, which is also not suitable for long-time storage in batch. Second, drug delivery may impair the structural and functional integrity of RBCs, leading to some local and systemic side effects. Third, the range of drugs available for RBC delivery remains limited. However, due to the characteristics of biocompatibility and personalized carrier, erythrocyte carrier drugs are still a choice with great application potential.

8.6 Bacteria Carrier

With the development of bacterial biology and molecular biology, people have realized that bacteria can be used as anti-tumor weapons (Cao and Liu 2020). At present, there are four classes of bacteria-related oncology drug carriers: living bacteria or bacterial cells, bacterial derivatives, and bacterial shadow (Suh et al. 2019; Mostaghaci et al. 2017; Toyofuku et al. 2019). Due to the hypoxic environment of tumor, the use of anaerobic bacteria can achieve the specific targeting effect on tumor, improve the efficiency of treatment, and reduce the damage to normal tissue. Because bacteria can be dispersed in the whole tumor tissue within the cell and have a strong ability of retrograde tumor, the use of bacteria for tumor tracer or targeted delivery of anti-tumor drugs to tumor cells can greatly improve the integration of chemotherapy drug efficacy and diagnosis and treatment (Zhou 2016).

At present, it has been found that a variety of bacteria can be used in anti-tumor therapy, such as bifidobacterium, clostridium, and salmonella. *Escherichia coli* are more likely to colonize in hypoxic and necrotic areas than peripheral neoplastic proliferative tissues, and it can accurately target the hypoxic areas of solid tumors. Inefficient delivery of oxygen and nutrients caused by the disturbed vascular system in tumors provides favorable conditions for the colonization of anaerobes or facultative anaerobes. Under hypoxic conditions, tumor growth factor-β secreted by tumor sites can inhibit the activation and infiltration of neutrophils and thus impair the bactericidal activity of macrophages and neutrophils (Chen et al. 2018).

Bacteria play a crucial role in treating tumors by targeting the necrotic regions within the tumor. The bacteria accumulate in these areas and attract neutrophils, forming a protective barrier that separates the necrotic tissue from the living tumor cells, which helps prevent the necrotic tissue from bacterial infection and promote tumor necrosis. The greater the neutrophils the stronger infiltration. When the neutrophils of the mice were consumed by the antibody, the area of bacteria induced tumor necrosis was significantly increased, and the number of bacteria in the tumor was increased accordingly. However, living bacteria, as an intrusion, can trigger human immune response when used directly as a carrier, and the presence of endotoxin may also cause toxic reaction of host cells. Compared with live bacteria, bacterial derivatives have the characteristics of low immunogenicity and low toxicity, and their preparation process is simple and can be produced on a large scale. Therefore, bacterial derivatives have a good prospect for application in drug delivery system as carriers (Kang et al. 2022; Cao and Liu 2020) (Table 8.1).

Bacteria will naturally secrete the nano-lipoproteins on the cell membrane outside the membrane at each stage of the growth process, which is a common phenomenon in the growth process of bacteria, especially when the bacteria are stimulated by the outside world, they are more likely to secrete various vesicles. These vesicles, ranging from 20 to 250 nm in diameter, are usually spherical and are primarily released by the outer membrane of bacteria in Gram-negative bacteria, hence the name OMVS (Jan 2017). The main component of OM in most Gram-negative bacteria is lipopolysaccharide, so their OMVs are a very potent activator for immune cells. OMVs can induce adaptive immune responses in normal host cells and can be used in the development of vaccines and adjuvants. On the other hand, in the process of drug delivery, the ideal carrier should be non-immunogenic or non-inflammatory to avoid causing inflammatory response in the patient. Kadurugamuwa et al. (Metruccio et al. 2016) found that Pseudomonas aeruginosa vesicles contain autolysin—cytoplasmic hydrolase, which can lyse other

Table 8.1 Delivery of anticancer drugs mediated by bacteria

Bacterium	Therapeutic agent	Modification strategy	Type of treatment	References
Salmonella typhimurium	Paclitaxel-loaded liposomes	Biotin-streptavidin	Chemotherapy	Nguyen et al. (2016)
Salmonella typhimurium YS1646	Doxorubicin loaded-low-temperature sensitive liposome	Biotin-streptavidin	Chemo-immunotherapy	Ektate et al. (2018)
Escherichia coli MG1655	Doxorubicin and Fe_3O_4 nanoparticles	Surface charge and noncovalent interactions	Chemotherapy	Park et al. (2017)
Escherichia coli MG1655	Doxorubicin and SPIONs loaded soft RBCs	Biotin-avidin-biotin	Chemotherapy	Alapan et al. (2018)
Listeria monocytogenes	188-rhenium	Listeria-binding antibodies	Radiotherapy	Quispe-Tintaya et al. (2013)
Listeria monocytogenes	32-phosphorus	Metabolic labeling	Radiotherapy	Chandra et al. (2017)
Salmonella typhimurium VNP20009	Polydopamine	Coating	Photothermal-therapy	Chen et al. (2018)
Salmonella typhimurium YB1	Indocyanine green	Biotic/abiotic cross-linker	Photothermal-therapy	Kefayat et al. (2019)
Salmonella typhimurium	FlaB	Genetic modification	Immunotherapy	Zheng et al. (2017)

Reproduced with permission from Cao and Liu (2020)

Gram-negative and Gram-positive bacteria. Using this property, OMVs can be made into antibiotic carriers to kill pathogenic bacteria. HER2 is a transmembrane receptor with tyrosine kinase activity that is overexpressed in breast, ovarian, and gastric cancer cells. Using its HER2 targeting, Gujrati et al. (2019) delivered siRNA-loaded OMVs to tumor cells, successfully silencing the target gene and inhibiting tumor cell growth.

8.7 Dendritic Cells

Dendritic cells (DCs) emerged as a carrier, mainly used for anti-tumor cell therapy (Moreau et al. 2017). Immature DCs have strong migration ability. Mature DCs can effectively activate the initial T cells and play a central role in initiating, regulating, and maintaining the immune response (Na et al. 2016). Patients with numbers of DC infiltrates in most solid tumors have a good prognosis.

The US Food and Drug Administration (FDA) has approved sipuleucel-T (Provenge), an autologous cell immunotherapy drug that uses DC as its main effector cell, for the treatment of asymptomatic or mild metastatic caste-resistant prostate cancer (Gardner et al. 2020). The core of effective anti-tumor immune response is to generate cellular immune response dominated by CD8 + T cells, which can produce anti-tumor effect through activation of cytotoxic T cells (CTL), which is also the basis of assisted activation of T cells as immunotherapy (Le Gall et al. 2018). As long as the obstacles of in vitro isolation and proliferation of DC can be overcome, DC will be a good choice for immunotherapy (Mukhopadhyay 2020). Creusot et al. (Firdessa-Fite and Creusot 2020) have developed a liposome nanoparticle platform targeting endogenous antigen presentation in vivo and in vitro, delivering mRNAs electroporated dendritic cells. After intraperitoneal injection, the nanoparticle-platform promoted effective mRNA entry into a variety of endogenous antigen-presenting cells, including lymph node stromal cells, and evoked robust T cell responses in a broader lymphoid tissue network compared to dendritic cells. After intravenous injection, mRNA electroporation dendritic cells and nanoparticle platforms were mainly located in lung and spleen, respectively. When administered locally through the intradermal route, both platforms resulted in injection site mRNA expression and robust T cell response in the draining lymph nodes.

Acknowledgements This study was supported by the Hunan Provincial Natural Science Foundation of China (No. 2023JJ30124) (X. Hu), Fujian Provincial Natural Science Foundation of China (No. 2023J01247) (X. Hu), the Natural Science Foundation of Chongqing, China (No. CSTB2022NSCQ-MSX1551) (X. Hu), Guangdong Basic and Applied Basic Research Foundation (No. 2024A1515012771) (X. Hu), the Postdoctoral Fellowship Program of CPSF (No. GZB20240215) (Y. Liu).

References

Alapan Y, Yasa O, Schauer O et al (2018) Soft erythrocyte-based bacterial microswimmers for cargo delivery. Sci Robot 3(17):1–10. https://doi.org/10.1126/scirobotics.aar4423

Altaner C, Altanerova V, Cihova M et al (2014) Complete regression of glioblastoma by mesenchymal stem cells mediated prodrug gene therapy simulating clinical therapeutic scenario. Int J Cancer 134(6):1458–1465. https://doi.org/10.1002/ijc.28455

An Y, Yang Q (2020) Tumor-associated macrophage-targeted therapeutics in ovarian cancer. Int J Cancer 149(1):21–30. https://doi.org/10.1002/ijc.33408

Biagiotti S, Paoletti MF, Fraternale A et al (2011) Drug delivery by red blood cells. IUBMB Life 2011(63):621–631. https://doi.org/10.1002/iub.478

Bielamowicz K, Fousek K, Byrd TT et al (2018) Trivalent CAR T-cells overcome interpatient antigenic variability in glioblastoma. Neuro-Oncology 20(4):506–518. https://doi.org/10.1093/neuonc/nox182

Cao Z, Liu J (2020) Bacteria and bacterial derivatives as drug carriers for cancer therapy. J Control Release 326:396–407. https://doi.org/10.1016/j.jconrel.2020.07.009

Chandra D, Selvanesan BC, Yuan Z et al (2017) 32-Phosphorus selectively delivered by listeria to pancreatic cancer demonstrates a strong therapeutic effect. Oncotarget. https://doi.org/10.18632/oncotarget.15117

Chen W, Wang Y, Qin M et al (2018) Bacteria-driven hypoxia targeting for combined biotherapy and photothermal therapy. ACS Nano 12(6):5995–6005. https://doi.org/10.1021/acsnano.8b02235

Chen W, Zhang F, Ju Y et al (2021) Gold nanomaterial engineering for macrophage-mediated inflammation and tumor treatment. Adv Healthc Mater 10(5):e2000818–ee200834. https://doi.org/10.1002/adhm.202000818

Chinnasamy D, Yu Z, Kerkar SP et al (2012) Local delivery of interleukin-12 using T cells targeting VEGF receptor-2 eradicates multiple vascularized tumors in mice. Clin Cancer Res 18(6):1672–1683. https://doi.org/10.1158/1078-0432.CCR-11-3050

Chmielewski M, Kopecky C, Hombach AA et al (2011) IL-12 release by engineered T cells expressing chimeric antigen receptors can effectively Muster an antigen-independent macrophage response on tumor cells that have shut down tumor antigen expression. Cancer Res 71(17):5697–5706. https://doi.org/10.1158/0008-5472.CAN-11-0103

Cihova M, Altanerova V, Altaner C (2011) Stem cell based cancer gene therapy. Mol Pharm 8(5):1480–1487. https://doi.org/10.1021/mp200151a

Combes F, Meyer E, Sanders NN (2020) Immune cells as tumor drug delivery vehicles. J Control Release 327:70–87. https://doi.org/10.1016/j.jconrel.2020.07.043

Ektate K, Munteanu MC, Ashar H et al (2018) Chemo-immunotherapy of colon cancer with focused ultrasound and Salmonella-laden temperature sensitive liposomes (thermobots). Sci Rep-UK 8(1):13062. https://doi.org/10.1038/s41598-018-30106-4

Everson EM, Olzsko ME, Leap DJ et al (2016) ASGCT 19th annual meeting: abstracts. Mol Ther 24(Suppl 1):S1–S304. https://doi.org/10.1038/mt.2016.78

Ferguson L, Hood E, Shuvaeva T et al (2020) Dual targeted red blood cell hitchhiking: a platform liposomal drug delivery system to improve circulation time and increase delivery to target organs. American Thoracic Society 2020 International Conference, Philadelphia, PA, 15–20 May 2020

Firdessa-Fite R, Creusot RJ (2020) Nanoparticles versus dendritic cells as vehicles to deliver mRNA encoding multiple epitopes for immunotherapy. Mol Ther Methods Clin Dev 16:50–62. https://doi.org/10.1016/j.omtm.2019.10.015

Gardner A, de Mingo PÁ, Ruffell B (2020) Dendritic cells and their role in immunotherapy. Front Immunol 11(924):1–14. https://doi.org/10.3389/fimmu.2020.00924

Glassman PM, Villa CH, Ukidve A et al (2020) Vascular drug delivery using carrier red blood cells: focus on RBC surface loading and pharmacokinetics. Pharmaceutics 12(5):1–21. https://doi.org/10.3390/pharmaceutics12050440

Golinelli G, Mastrolia I, Aramini B et al (2020) Arming mesenchymal stromal/stem cells against cancer: has the time come? Front Pharmacol 11:1–14. https://doi.org/10.3389/fphar.2020.529921

Goradel NH, Baker AT, Arashkia A et al (2021) Oncolytic virotherapy: challenges and solutions. Curr Probl Cancer 45(1):100639–100656. https://doi.org/10.1016/j.currproblcancer.2020.100639

Grisendi G, Bussolari R, Cafarelli L et al (2010) Adipose-derived mesenchymal stem cells as stable source of tumor necrosis factor-related apoptosis-inducing ligand delivery for cancer therapy. Cancer Res 70(9):3718–3729. https://doi.org/10.1158/0008-5472.CAN-09-1865

Guiho R, Biteau K, Grisendi G et al (2016) TRAIL delivered by mesenchymal stromal/stem cells counteracts tumor development in orthotopic Ewing sarcoma models. Int J Cancer 139(12):2802–2811. https://doi.org/10.1002/ijc.30402

Gujrati V, Prakash J, Malekzadeh-Najafabadi J et al (2019) Bioengineered bacterial vesicles as biological nano-heaters for optoacoustic imaging. Nat Commun 10(1):1114–1123. https://doi.org/10.1038/s41467-019-09034-y

Guo L, Zhang Y, Yang ZP et al (2019) Tunneling nanotubular expressways for ultrafast and accurate M1 macrophage delivery of anticancer drugs to metastatic ovarian carcinoma. ACS Nano 13(2):1078–1096. https://doi.org/10.1021/acsnano.8b08872

Guo L, Zhang Y, Wei R et al (2020) Proinflammatory macrophage-derived microvesicles exhibit tumor tropism dependent on CCL2/CCR2 signaling axis and promote drug delivery via SNARE-mediated membrane fusion. Theranostics 10(15):6581–6598. https://doi.org/10.7150/thno.45528

Gutova M, Goldstein L, Metz M et al (2017) Optimization of a neural stem-cell-mediated carboxylesterase/irinotecan gene therapy for metastatic neuroblastoma. Mol Ther Oncolytics 4:67–76. https://doi.org/10.1016/j.omto.2016.11.004

Han X, Wang L, Li T et al (2020) Beyond blocking: engineering RNAi-mediated targeted immune checkpoint nanoblocker enables T-cell-independent cancer treatment. ACS Nano 14(12):17524–17534. https://doi.org/10.1021/acsnano.0c08022

Hao NB, Lü MH, Fan YH et al (2012) Macrophages in tumor microenvironments and the progression of tumors. Clin Dev Immunol. 2012:2012:948098. https://doi.org/10.1155/2012/948098. Epub 2012 Jun 19. PMID: 22778768; PMCID: PMC3385963

Hayashi K, Yamada S, Sakamoto W et al (2018) Red blood cell-shaped microparticles with a red blood cell membrane demonstrate prolonged circulation time in blood. ACS Biomater Sci Eng 4(8):2729–2732. https://doi.org/10.1021/acsbiomaterials.8b00197

Ho YJ, Chiang YJ, Kang ST et al (2018) Camptothecin-loaded fusogenic nanodroplets as ultrasound theranostic agent in stem cell-mediated drug-delivery system. J Control Release 278:100–109. https://doi.org/10.1016/j.jconrel.2018.04.001

Hoyos V, Savoldo B, Quintarelli C et al (2010) Engineering CD19-specific T lymphocytes with interleukin-15 and a suicide gene to enhance their anti-lymphoma/leukemia effects and safety. Leukemia 24(6):1160–1170. https://doi.org/10.1038/leu.2010.75

Hu B, Ren J, Luo Y et al (2017) Augmentation of antitumor immunity by human and mouse CAR T cells secreting IL-18. Cell Rep 20(13):3025–3033. https://doi.org/10.1016/j.celrep.2017.09.002

Huang B, Abraham WD, Zheng Y et al (2015) Active targeting of chemotherapy to disseminated tumors using nanoparticle-carrying T cells. Sci Transl Med 7(291):1–21. https://doi.org/10.1126/scitranslmed.aaa5447

Jan AT (2017) Outer membrane vesicles (OMVs) of Gram-negative bacteria: a perspective update. Front Microbiol 8:1–11. https://doi.org/10.3389/fmicb.2017.01053

Jones RB, Mueller S, Kumari S et al (2017) Antigen recognition-triggered drug delivery mediated by nanocapsule-functionalized cytotoxic T-cells. Biomaterials 117:44–53. https://doi.org/10.1016/j.biomaterials.2016.11.048

Kang SR, Nguyen DH et al (2022) Bacteria and bacterial derivatives as delivery carriers for immunotherapy. Adv Drug Deliver Rev 181(2022):14085. https://doi.org/10.1016/j.addr.2021.114085

Kefayat A, Ghahremani F, Motaghi H et al (2019) Alive attenuated Salmonella as a cargo shuttle for smart carrying of gold nanoparticles to tumour hypoxic regions. J Drug Target 27(3):315–324. https://doi.org/10.1080/1061186X.2018.1523417

Klyachko NL, Polak R, Haney MJ et al (2017) Macrophages with cellular backpacks for targeted drug delivery to the brain. Biomaterials 140:79–87. https://doi.org/10.1016/j.biomaterials.2017.06.017

Krishnamurthy S, Gnanasammandhan MK, Xie C et al (2016) Monocyte cell membrane-derived nanoghosts for targeted cancer therapy. Nanoscale 8(13):6981–6985. https://doi.org/10.1039/c5nr07588b

Le Gall CM, Weiden J, Eggermont LJ et al (2018) Dendritic cells in cancer immunotherapy. Nat Mater 17(6):474–475. https://doi.org/10.1038/s41563-018-0093-6

Lee S, Kivimae S, Dolor A et al (2016) Macrophage-based cell therapies: the long and winding road. J Control Release 240:527–540. https://doi.org/10.1016/j.jconrel.2016.07.018

Lejeune A, Moorjani M, Gicquaud C et al (1994) Nanoerythrosome, a new derivative of erythrocyte ghost: preparation and antineoplastic potential as drug carrier for daunorubicin. Anticancer Res 14(3A):915–919

Li L, Li F, Tian H et al (2014) Human mesenchymal stem cells with adenovirus-mediated TRAIL gene transduction have antitumor effects on esophageal cancer cell line Eca-109. Acta Biochim Biophys Sin (Shanghai) 46(6):471–476. https://doi.org/10.1093/abbs/gmu024

Li Y, Liu X, Xu X et al (2019) Red-blood-cell waveguide as a living biosensor and micromotor. Adv Funct Mater 29. https://doi.org/10.1002/adfm.201905568

Liesveld JL, Sharma N, Aljitawi OS (2020) Stem cell homing: from physiology to therapeutics. Stem Cells 38(10):1241–1253. https://doi.org/10.1002/stem.3242

Liu X, Zhang Y, Cheng C et al (2017) CRISPR-Cas9-mediated multiplex gene editing in CAR-T cells. Cell Res 27(1):154–157. https://doi.org/10.1038/cr.2016.142

Liu CG, Wang Y, Liu P et al (2020) Aptamer-T cell targeted therapy for tumor treatment using sugar metabolism and click chemistry. ACS Chem Biol 15(6):1554–1565. https://doi.org/10.1021/acschembio.0c00164

Luo YL, Xu CF, Li HJ et al (2018) Macrophage-specific in vivo gene editing using cationic lipid-assisted polymeric nanoparticles. ACS Nano 12(2):994–1005. https://doi.org/10.1021/acsnano.7b07874

Lye KL, Nordin N, Vidyadaran S et al (2016) Mesenchymal stem cells: from stem cells to sarcomas. Cell Biol Int 40(6):610–618. https://doi.org/10.1002/cbin.10603

Mader EK, Maeyama Y, Lin Y et al (2009) Mesenchymal stem cell carriers protect oncolytic measles viruses from antibody neutralization in an orthotopic ovarian cancer therapy model. Clin Cancer Res 15(23):7246–7255. https://doi.org/10.1158/1078-0432.CCR-09-1292

Markley JC, Sadelain M (2010) IL-7 and IL-21 are superior to IL-2 and IL-15 in promoting human T cell-mediated rejection of systemic lymphoma in immunodeficient mice. Blood 115(17):3508–3519. https://doi.org/10.1182/blood-2009-09-241398

Melzer MK, Zeitlinger L, Mall S et al (2019) Enhanced safety and efficacy of oncolytic VSV therapy by combination with T cell receptor transgenic T cells as carriers. Mol Ther Oncolytics 12:26–40. https://doi.org/10.1016/j.omto.2018.12.001

Metruccio MM, Evans DJ, Gabriel MM et al (2016) Pseudomonas aeruginosa outer membrane vesicles triggered by human mucosal fluid and lysozyme can prime host tissue surfaces for bacterial adhesion. Front Microbiol 7(871):1–19. https://doi.org/10.3389/fmicb.2016.00871

Mooney R, Abdul Majid A, Batalla J et al (2017) Cell-mediated enzyme prodrug cancer therapies. Adv Drug Deliv Rev 118:35–51. https://doi.org/10.1016/j.addr.2017.09.003

Moreau A, Alliot-Licht B, Cuturi MC et al (2017) Tolerogenic dendritic cell therapy in organ transplantation. Transpl Int 30(8):754–764. https://doi.org/10.1111/tri.12889

Mostaghaci B, Yasa O, Zhuang J et al (2017) Bioadhesive bacterial microswimmers for targeted drug delivery in the urinary and gastrointestinal tracts. Adv Sci 4(6):1–9. https://doi.org/10.1002/advs.201700058

Mukhopadhyay M (2020) Sugar-coating dendritic cells. Nat Methods 17(7):651. https://doi.org/10.1038/s41592-020-0892-y

Na H, Cho M, Chung Y (2016) Regulation of Th2 cell immunity by dendritic cells. Immune Netw 16(1):1–12. https://doi.org/10.4110/in.2016.16.1.1

Nakamura K, Ito Y, Kawano Y et al (2004) Antitumor effect of genetically engineered mesenchymal stem cells in a rat glioma model. Gene Ther 11(14):1155–1164. https://doi.org/10.1038/sj.gt.3302276

Nguyen VD, Han JW, Choi YJ et al (2016) Active tumor-therapeutic liposomal bacteriobot combining a drug (paclitaxel)-encapsulated liposome with targeting bacteria (Salmonella Typhimurium). Sens Actuators B Chem 224:217–224. https://doi.org/10.1016/j.snb.2015.09.034

Nguyen VD, Min HK, Kim DH et al (2020) Macrophage-mediated delivery of multifunctional nanotherapeutics for synergistic chemo-photothermal therapy of solid tumors. ACS Appl Mater Interfaces 12(9):10130–10141. https://doi.org/10.1021/acsami.9b23632

Ouyang X, Wang X, Kraatz HB et al (2020) Correction: a Trojan horse biomimetic delivery strategy using mesenchymal stem cells for PDT/PTT therapy against lung melanoma metastasis. Biomater Sci 8(4):1160–1170. https://doi.org/10.1039/d0bm90006k

Pang L, Qin J, Han L et al (2016) Exploiting macrophages as targeted carrier to guide nanoparticles into glioma. Oncotarget 7(24):37081–37091. https://doi.org/10.18632/oncotarget.9464

Pang L, Zhu Y, Qin J et al (2018) Primary M1 macrophages as multifunctional carrier combined with PLGA nanoparticle delivering anticancer drug for efficient glioma therapy. Drug Deliv 25(1):1922–1931. https://doi.org/10.1080/10717544.2018.1502839

Park B-W, Zhuang J, Yasa O et al (2017) Multifunctional bacteria-driven microswimmers for targeted active drug delivery. ACS Nano 11(9):8910–8923. https://doi.org/10.1021/acsnano.7b03207

Pegram HJ, Purdon TJ, van Leeuwen DG et al (2015) IL-12-secreting CD19-targeted cord blood-derived T cells for the immunotherapy of B-cell acute lymphoblastic leukemia. Leukemia 29(2):415–422. https://doi.org/10.1038/leu.2014.215

Qiu Y, Ren K, Zhao W et al (2021) A "dual-guide" bioinspired drug delivery strategy of a macrophage-based carrier against postoperative triple-negative breast cancer recurrence. J Control Release 329:191–204. https://doi.org/10.1016/j.jconrel.2020.11.039

Quispe-Tintaya W, Chandra D, Jahangir A et al (2013) Nontoxic radioactive Listeriaat is a highly effective therapy against metastatic pancreatic cancer. Proc Natl Acad Sci U S A 110(21):8668–8673. https://doi.org/10.1073/pnas.1211287110

Ren C, Kumar S, Chanda D et al (2008) Therapeutic potential of mesenchymal stem cells producing interferon-alpha in a mouse melanoma lung metastasis model. Stem Cells 26(9):2332–2338. https://doi.org/10.1634/stemcells.2008-0084

Rossi L, Fraternale A, Bianchi M, Magnani M (2019) Red blood cell membrane processing for biomedical applications. Front Physiol 10(1070):1–8. https://doi.org/10.3389/fphys.2019.01070

Rossignoli F, Spano C, Grisendi G et al (2019) MSC-delivered soluble TRAIL and paclitaxel as novel combinatory treatment for pancreatic adenocarcinoma. Theranostics 9(2):436–448. https://doi.org/10.7150/thno.27576

Sengupta S, Balla VK (2018) A review on the use of magnetic fields and ultrasound for non-invasive cancer treatment. J Adv Res 14:97–111. https://doi.org/10.1016/j.jare.2018.06.003

Shum T, Omer B, Tashiro H et al (2017) Constitutive signaling from an engineered IL7 receptor promotes durable tumor elimination by tumor-redirected T cells. Cancer Discov 7(11):1238–1247. https://doi.org/10.1158/2159-8290.CD-17-0538

Su Y, Xie Z, Kim GB et al (2004) Design strategies and applications of circulating cell-mediated drug delivery systems. ACS Biomater Sci Eng 1(4):201–217. https://doi.org/10.1021/ab500179h

Su YX, Xie ZW, Kim GB et al (2015) Design strategies and applications of circulating cell-mediated drug delivery systems. ACS Biomater Sci Eng 1 (4):201-217. https://doi.org/10.1021/ab500179h

Suh S, Jo A, Traore MA et al (2019) Nanoscale bacteria-enabled autonomous drug delivery system (NanoBEADS) enhances intratumoral transport of nanomedicine. Adv Sci 6(3):1–13. https://doi.org/10.1002/advs.201801309

Tahmasebi S, Elahi R, Khosh E et al (2020) Programmable and multi-targeted CARs: a new breakthrough in cancer CAR-T cell therapy. Clin Transl Oncol 23(6):1003–1019. https://doi.org/10.1007/s12094-020-02490-9

Toyofuku M, Nomura N, Eberl L (2019) Types and origins of bacterial membrane vesicles. Nat Rev Microbiol 17(1):13–24. https://doi.org/10.1038/s41579-018-0112-2

Tran E, Turcotte S, Gros A et al (2014) Cancer immunotherapy based on mutation-specific CD4+ T cells in a patient with epithelial cancer. Science 344(6184):641–645. https://doi.org/10.1126/science.1251102

Tran E, Robbins PF, Lu YC et al (2016) T-cell transfer therapy targeting mutant KRAS in cancer. N Engl J Med 375(23):2255–2262. https://doi.org/10.1056/NEJMoa1609279

Usman WM, Pham TC, Kwok YY et al (2018) Efficient RNA drug delivery using red blood cell extracellular vesicles. Nat Commun 9, 2359. https://doi.org/10.1038/s41467-018-04791-8

Villa CH, Anselmo AC, Mitragotri S, Muzykantov V (2016) Red blood cells: supercarriers for drugs, biologicals, and nanoparticles and inspiration for advanced delivery systems. Adv Drug Deliv Rev 2016:88–103. https://doi.org/10.1016/j.addr.2016.02.007

Wang Z, Wu Z, Liu Y et al (2017) New development in CAR-T cell therapy. J Hematol Oncol 10(1):1–11. https://doi.org/10.1186/s13045-017-0423-1

Wang HF, Liu Y, Wang T et al (2020a) Tumor-microenvironment-on-a-chip for evaluating nanoparticle-loaded macrophages for drug delivery. ACS Biomater Sci Eng 6(9):5040–5050. https://doi.org/10.1021/acsbiomaterials.0c00650

Wang X, Wu Z, Qiu W et al (2020b) Programming CAR T cells to enhance anti-tumor efficacy through remodeling of the immune system. Front Med 14(6):726–745. https://doi.org/10.1007/s11684-020-0746-0

Wu HH, Zhou Y, Tabata Y et al (2019) Mesenchymal stem cell-based drug delivery strategy: from cells to biomimetic. J Control Release 294:102–113. https://doi.org/10.1016/j.jconrel.2018.12.019

Wynn RF, Hart CA, Corradi-Perini C et al (2004) A small proportion of mesenchymal stem cells strongly expresses functionally active CXCR4 receptor capable of promoting migration to bone marrow. Blood 104(9):2643–2645. https://doi.org/10.1182/blood-2004-02-0526

Xia Q, Zhang Y, Li Z et al (2019) Red blood cell membrane-camouflaged nanoparticles: a novel drug delivery system for antitumor application. Acta Pharm Sin B. https://doi.org/10.1016/j.apsb.2019.01.011

Xia Y, Rao L, Yao H et al (2020) Engineering macrophages for cancer immunotherapy and drug delivery. Adv Mater 32(40):e2002054–e2002e73. https://doi.org/10.1002/adma.202002054

Xu Z, Liu H, Tian H et al (2020) Real-time imaging tracking of engineered macrophages as ultrasound-triggered cell bombs for cancer treatment. Adv Funct Mater 30(14):1910304–1910313. https://doi.org/10.1002/adfm.201910304

Xue J, Zhao Z, Zhang L et al (2017) Neutrophil-mediated anticancer drug delivery for suppression of postoperative malignant glioma recurrence. Nat Nanotechnol 12(7):692–700. https://doi.org/10.1038/nnano.2017.54

Yan J, Yu J, Wang C et al (2017) Red blood cells for drug delivery. Small Methods 1(12):1–18. https://doi.org/10.1002/smtd.201700270

Yong T, Zhang X, Bie N et al (2019) Tumor exosome-based nanoparticles are efficient drug carriers for chemotherapy. Nat Commun 10(1):3838–3853. https://doi.org/10.1038/s41467-019-11718-4

Yong-Chen L, Parker LL, Tangying L et al (2017) Treatment of patients with metastatic cancer using a major histocompatibility complex class II–restricted T-cell receptor targeting the cancer germline antigen MAGE-A3. J Clin Oncol 35(29):3322–3329. https://doi.org/10.1200/jco.2017.74.5463

Yu H, Yang Z, Li F et al (2020) Cell-mediated targeting drugs delivery systems. Drug Deliv 27(1):1425–1437. https://doi.org/10.1080/10717544.2020.1831103

Zendedel E, Atkin SL, Sahebkar A (2019) Use of stem cells as carriers of oncolytic viruses for cancer treatment. J Cell Physiol 234:14906–14913. https://doi.org/10.1002/jcp.28320

Zhang W, Wang M, Tang W et al (2018a) Nanoparticle-laden macrophages for tumor-tropic drug delivery. Adv Mater 30(50):e1805557–e1805e67. https://doi.org/10.1002/adma.201805557

Zhang Q, Xiang W, Yi DY et al (2018b) Current status and potential challenges of mesenchymal stem cell-based therapy for malignant gliomas. Stem Cell Res Ther 9(1):228–236. https://doi.org/10.1186/s13287-018-0977-z

Zhang Y, Zhang Y, Guo Q et al (2019) Trained macrophage bioreactor for penetrating delivery of fused antitumor protein. ACS Appl Mater Interfaces 11(26):23018–23025. https://doi.org/10.1021/acsami.9b06097

Zhang D, Zhou C, Liu F et al (2020) Red blood cells membrane vehicle co-delivering DOX and IR780 for effective prostate cancer therapy. J Mater Res 35:3116–3123. https://doi.org/10.1557/jmr.2020.273

Zheng JH, Nguyen VH, Jiang SN et al (2017) Two-step enhanced cancer immunotherapy with engineered Salmonella typhimurium secreting heterologous flagellin. Sci Transl Med 9(376):eaak9537.1–eaaeaak.10. https://doi.org/10.1126/scitranslmed.aak9537

Zheng H, Li J, Luo X et al (2019) Murine RAW264.7 cells as cellular drug delivery carriers for tumor therapy: a good idea? Cancer Chemother Pharmacol 83(2):361–374. https://doi.org/10.1007/s00280-018-3735-0

Zhou S (2016) Bacteria synchronized for drug delivery. Nature 536(7614):33–34. https://doi.org/10.1038/nature18915

Chapter 9
Nanocrystal Technology

Shudan Huang, Ting Zhang, Cuiping Shi, Ye Zi, Jiamin Xu, Jing Su, Daixu Wei, Xichang Wang, and Jian Zhong

Abstract The targeting of drug nanocrystals to the malignant and metastatic components is a potential strategy for tumor therapy. Compared with nanocarrier drugs, nanocrystals have many excellent advantages such as relatively simple preparation process, high drug loading, and good physical stability. In this chapter, the application of nanocrystal technology in the treatment of tumors is reviewed and discussed. First, the preparation methods of nanocrystals are summarized. Second, administration routes of drug nanocrystals are reviewed. Third, the status of drug nanocrystals in the market and clinical trials is summarized. Fourth, recent researches on drug nanocrystals in the tumor target delivery are discussed. Finally, the application of drug nanocrystals for the treatment of tumors is summarized and reviewed. This chapter provides useful information to understand the research and development of drug nanocrystals for the treatment of tumors.

Keywords Administration routes · Bottom-up techniques · Nanocrystal products · Top-down techniques · Tumor target delivery

S. Huang · C. Shi · Y. Zi
Shanghai Key Laboratory of Pediatric Gastroenterology and Nutrition, Shanghai Institute for Pediatric Research, Xinhua Hospital, Shanghai Jiao Tong University School of Medicine, Shanghai, China

T. Zhang · J. Xu · X. Wang
National R&D Branch Center for Freshwater Aquatic Products Processing Technology (Shanghai), Integrated Scientific Research Base on Comprehensive Utilization Technology for By-Products of Aquatic Product Processing, Ministry of Agriculture and Rural Affairs of the People's Republic of China, Shanghai Engineering Research Center of Aquatic-Product Processing & Preservation, College of Food Science & Technology, Shanghai Ocean University, Shanghai, China

J. Su
School of Pharmacy, Shanghai Jiao Tong University, Shanghai, China

L. Teng et al. (eds.), *Drug Delivery to Tumors*,
https://doi.org/10.1007/978-981-19-8930-8_9

9.1 Introduction

Nowadays, noncommunicable diseases are the main cause of death for the global population. Almost in every country, cancer is becoming the first reason of death in the twenty-first century. The malignant tumor has become the first cause of death in China. Both the malignant cells and other cells recruited to the tumor area composed the tumor tissue, and together, these cells create the tumor microenvironment (TME). A variety of barriers consist the TME, including the extracellular matrix and associated cells such as endothelia cells, pericytes, and cancer-associated fibroblasts. Due to the complexity of TME, the anticancer therapies struggle to be more effective (Balkwill et al. 2012).

Different tumors have different TMEs, which are different in several ways, like acidosis, hypoxia, and high interstitial fluid pressure (IFP) (Yang and Gao 2017). Acidosis, which is created by the development of tumor, is an important feature in the TME. In this process, the glucoses are degraded by a large amount of enzyme-released acidic substances (Martinez et al. 2006; Moellering et al. 2008; Romero-Garcia et al. 2011). These acidic substances will affect the pH of TME and lower the efficiency of drug delivery (Bae et al. 2012). Hypoxia is caused by the rapid tumor development. Up to 50–60% of locally advanced solid tumors may show hypoxic and/or anoxic tissue areas that are heterogeneously distributed within the tumor mass (Vaupel and Mayer 2007). Regulating the hypoxia can promote the efficacy of tumor therapy (Hockel and Vaupel 2001; Brown and Giaccia 1998; Brizel et al. 1996). High IFP is another remarkable feature in TME. Generally, IFP in normal tissue is only 0–3 mmHg. However, IFP in tumors is 5–40 mmHg, and in some areas, it can even reach 5–130 mmHg (Milosevic et al. 2004; Provenzano et al. 2012). If IFP is lower than microvascular pressure, the drug nanocrystals would not participate in the therapies. Hence, modulating the IFP can improve the efficiency of drug delivery (Khawar et al. 2015). However, despite the differences in the TME,

D. Wei
Key Laboratory of Resource Biology and Biotechnology in Western China, Ministry of Education, School of Medicine, Department of Life Sciences and Medicine, Northwest University, Xi'an, Shaanxi, China

J. Zhong (✉)
Shanghai Key Laboratory of Pediatric Gastroenterology and Nutrition, Shanghai Institute for Pediatric Research, Xinhua Hospital, Shanghai Jiao Tong University School of Medicine, Shanghai, China

National R&D Branch Center for Freshwater Aquatic Products Processing Technology (Shanghai), Integrated Scientific Research Base on Comprehensive Utilization Technology for By-Products of Aquatic Product Processing, Ministry of Agriculture and Rural Affairs of the People's Republic of China, Shanghai Engineering Research Center of Aquatic-Product Processing & Preservation, College of Food Science & Technology, Shanghai Ocean University, Shanghai, China
e-mail: jzhong@shsmu.edu.cn

there are common features among several types of tumors (Li and Burgess 2020). Therefore, targeting malignant and metastatic components around the TME is potential strategy for tumor therapies (Cheng et al. 2016; Bahrami et al. 2018; Roma-Rodrigues et al. 2019).

About 40% of developed and marketed chemicals are poorly soluble compounds, and they result in lots of problems in the application such as low bioavailability, nonfocused distribution, and poor solubility (Ranjita 2013). The nonspecific distribution of most drugs throughout the body will bring various side effects and reduce their use in clinic (Trapani et al. 2012). Therefore, nanocrystal technology is a good potential solution to solve these problems. In 1991, Elan Nanosystem first proposed the concept of nanocrystals, registered the trade name of NanoCrystal®, propagated the use of nanocrystals instead of microcrystals for oral bioavailability enhancement, and used nanocrystal suspensions for intravenous or pulmonary drug delivery (Junghanns and Müller 2008).

Nanomedicine is derived from the combination of nanotechnology with pharmacy. According to different forms, nanomedicine can be divided into nanocarrier drugs (with the help of carrier materials, drugs can be dispersed in the carrier nanomaterial such as nanoparticles, nanolipid carriers, and nanoemulsions) and nanocrystal drug (the drug itself becomes nano-sized). The preparation process of nanocarrier drugs is complicated and instable, which greatly limited the drug loading (Trapani et al. 2012). Only several nanocarrier-based preparations has been published, e.g., Doxil®, DaunoXome®, and Abraxane®. Nanocrystal is a pure drug nano-colloidal dispersion system without carrier material. It has many excellent advantages such as relatively simple preparation process, high drug loading, and good physical stability (Lu et al. 2016). Compared with other nanoparticles (e.g., polymeric nanoparticles, liposomes, lipid nanoparticles, and nanoemulsions), nanocrystals have high drug loading (as high as 100%). They will deliver the drugs more efficiently and accurately. Therefore, nanocrystals have become potentially helpful in the treatment of various cancers (Lu et al. 2015). They may produce desirable pharmacological actions such as the enhancement of oral bioavailability (Zhang et al. 2013). According to the pharmacokinetics data obtained in rats, the bioavailability was improved a lot.

In this chapter, we mainly introduce the preparation methods, administration routes, status in the market and clinical trials, and recent research advances of drug nanocrystals in the tumor therapy.

9.2 Preparation of Nanocrystals

As shown in Fig. 9.1, the nanocrystal preparation methods can be divided into three main categories: bottom-up techniques, top-down techniques, and combination techniques (Miao et al. 2018).

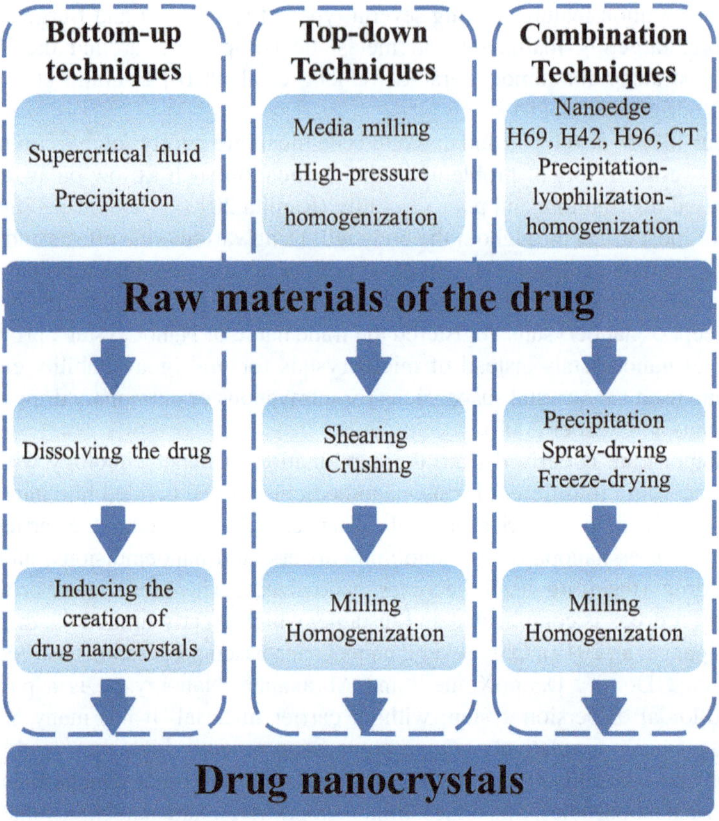

Fig. 9.1 Preparation methods of nanocrystals

9.2.1 Bottom-Up Techniques

The bottom-up technique refers to an approach that makes the drug dissolve in a solvent, then an agent will be added and/or a condition will be adjusted to induce the creation of drug nanocrystals. This process may be called "a classical precipitation process" (Junyaprasert and Morakul 2015). Typical bottom-up techniques include supercritical fluid technique and precipitation technique.

9.2.1.1 Supercritical Fluid Technique

Based on the roles that supercritical fluid plays in the crystallization process, the supercritical fluid technique can be divided into rapid expansion of supercritical solution (RESS) method and supercritical antisolvent (SAS) method (Guisasola et al. 2018).

The RESS method refers to dissolving the poorly water-soluble drugs in the supercritical fluid (like CO_2), when the fluid flows through the micro-aperture nozzle and becomes atomized. When the supercritical fluid vaporizes rapidly, it will produce the nanocrystal solid particles. Preparation of nanocrystals can be achieved using supercritical carbon dioxide due to its special advantages such as its physical properties and diffusivity characteristics like gas and solubilization behavior like liquid (Tozuka et al. 2010). The advantages of RESS method are simple procedure, low cost, and use of environmental-friendly solvents. It is possible to produce nanocrystal solid particles without any organic solvent residues. However, the RESS uses large amounts of supercritical fluids. Therefore, it is not suitable for some drugs that are not soluble in supercritical fluid (Möschwitzer and Müller 2006; Salazar et al. 2012).

The SAS method refers to dissolving poorly water-soluble drugs in an "antisolvent" that is miscible with a supercritical fluid and atomizing them at the same time. Under high pressure, the supercritical fluid can absorb the "antisolvent" completely and precipitate nanocrystal particles. The main advantage of the SAS method is its suitability for some drugs which are not soluble in supercritical fluid. Therefore, in other words, it has wide range of applications in this field (Gao et al. 2011; Sinha et al. 2013). The process of the supercritical antisolvent will not change the crystal structure and will only reduce the particle size.

9.2.1.2 Precipitation Technique

The precipitation technique is a traditional bottom-up procedure to prepare nanocrystals (Mohammad et al. 2019). In a typical precipitation process, a good solvent solution for the poorly water-soluble drugs is added to a miscible poor solvent under agitation to produce nanocrystals by controlling the conditions of crystallization. There are two key steps in the process of precipitation: one is crystal nucleation process and the other is crystal growth process. The nucleation process is very important because it produces uniform and small nanocrystals. When the nucleation rate becomes higher, more nuclei will precipitate from the supersaturated solution. Moreover, in the nucleation process, large amounts of nuclei will be obtained with a narrow particle size distribution (Sinha et al. 2013; Kakran et al. 2012). The advantages of precipitation are simple procedure, immediate removal of fluids without extra drying of nanocrystals, low requirements to instrument, and low cost. Besides, the particle size is small and the distribution is narrow. The major disadvantage of precipitation technique is that it is not suitable for many drugs. Due to uncontrolled particle growth, large amounts of poorly water-soluble drugs are not suitable for use. Meanwhile, it is also not suitable for some drugs which are either soluble in water or in nonaqueous solvents (Yue et al. 2012). Also, some precipitation procedures are costly because of the removal of organic solvents. Finally, for the drugs with low solubility in water and organic solvents, the experimental procedure requires a large amount of water and organic solvents. Therefore, in pharmaceutical industry, only several products are produced by the precipitation technique.

9.2.2 *Top-Down Techniques*

Top-down techniques refer to the application of mechanical forces such as grinding or homogenization to reduce the particle size of poorly water-soluble drugs to nanometer range, including media milling and high-pressure homogenization (HPH) (Van Eerdenbrugh et al. 2008). The flexibility in scale is one of the main advantages of top-down techniques. Therefore, it is considered the universal method for preparation of nanocrystals. Due to its universal and robust nature, various commercial nanocrystal products have been developed based on this technique. But there are also some disadvantages of this technique, including time-consuming (from several hours to days), high energy requirement (even to 1700 bar), and potential contamination to final products such as unexpected side effects with traces of grinding media (Chen et al. 2011).

9.2.2.1 Media Milling

Media milling uses rigid grinding media (glass, zircon oxide, hard polystyrene derivatives, etc.) in a milling chamber to prepare nanocrystals (Beeravelli et al. 2014). This technique was first developed to prepare nanodrugs in 2003 (Merisko-Liversidge et al. 2003). Recirculating chamber for milling process, coolant for controlling temperature rise, and milling media for grinding the particles to nanometer range consist of the major components. Pressure, shearing, impact, and attrition are four main kinds of forces in the media milling process (Chen et al. 2011; Juhnke et al. 2012). Shearing was generated by the movement of milling media to reduce the particle size. Besides, the size and dosage of milling media, time and speed of grinding, and concentration of drugs have obvious effects on the particle size of drug nanocrystals (Bitterlich et al. 2015). Dry milling such as jet milling is generally not efficient to obtain nanocrystals with the ideal nanometer size. Wet milling has been adopted as an efficient method to prepare nanocrystals (Junyaprasert and Morakul 2015). In a typical wet milling process, the drug is first dispersed in a surfactant/stabilizer solution, and then the obtained macrosuspension was exposed under the milling energy.

The nanocrystals produced by media milling have a well-defined shape and a narrow size distribution. There are several crucial advantages such as ease to scale up when it can be finished in a close proximity, simple process, and capability for various drugs that are both insoluble in water and nonaqueous solvents. Compared with precipitation technique, there is no need of organic solvent in media milling. Therefore, it is an environmental-friendly fabrication method to obtain nanocrystals. When under a much higher power density, the operation time was significantly reduced from several days (low energy ball milling) to 30–120 min (Merisko-Liversidge and Liversidge 2011).

In this procedure, heat generation is the main disadvantage. The formation and shape of nanodrugs may be affected by the temperature change. Ostwald ripening may occur to decrease the stability during this process (Juhnke et al. 2012). According to Ostwald-Freundlich theory, large particles have lower saturation solubility than smaller ones (Salazar et al. 2011). Hence, due to the concentration gradient between the small and large particles, small particles will diffuse and accumulate on the surface of larger particles, resulting in unexpected growth or agglomeration, even precipitation or crystallization (Kanthamneni et al. 2016). During the storage, the Ostwald ripening will result in the increase of polydispersity index (PDI) of nanocrystals (wide the distribution of particle size) (Wei et al. 2015). These changes are important obstacles in the application of drug nanocrystals, especially in the drug delivery via different administration routes such as pulmonary or intravenous (Chang et al. 2015). The media milling has two different modes: discontinuous or continuous types. In low size range, the continuous mode is preferred. During the milling process, the temperature will rise and affect the stability of particles. To improve this situation, some coolants are used to control the temperature in the chamber (Monteiro et al. 2013; George and Ghosh 2013).

The erosion of milling materials during the process is a normal problem in this technique. Obviously, the milling media technique needs special milling media materials. Choosing a proper media plays an important role in this technique. If glass beads or zirconium oxide milling beads is applied, the nanocrystals will be contaminated. Both the erosion of milling beads and milling chamber will result in the contamination of the obtained nanodrugs (George and Ghosh 2013; Juhnke et al. 2012). In early times, impurities produced by abrasion were one of the main reasons of stopping the application of milling media technique. Luckily, with the introduction of highly crosslinked polystyrene beads as milling media, the milling media has made great progress (Kesisoglou et al. 2007). The difference between highly crosslinked polystyrene beads and other milling materials is the elastic deformation of polystyrene beads, which can reduce the formation of cracks and abrasion from beads. There are dozens of poorly water-soluble drugs produced by this technique such as Rapamune®, Emend®, Tricor®, and Megace® (Liu et al. 2010b).

9.2.2.2 High-Pressure Homogenization (HPH)

Among several methods used to prepare drug nanocrystals, HPH is considered the second most important technology (Möschwitzer 2013). In HPH, under very high pressure, the crude drug suspensions were forced through highly narrow homogenization channels. The sizes of the crude drugs were reduced to nanometer sizes due to the cavitation, collisions, and high-shear forces among particles. In the pharmaceutical industry and the food industry, HPH has been used to produce liposomes and emulsion systems (Brandl et al. 1990). One of the significant advantages of HPH is the possibility to employ the production equipment for various formulation (multipurpose production lines). According to differences of equipment and media, the HPH can be divided into three types: piston-gap homogenization in water (Dissocubes® technology), piston-gap homogenization in water mixtures or in non-aqueous media (Nanopure® technology), and Microfluidizer technology (IDD-P™

technology). One of the main disadvantages of HPH is its high cost. It is very expensive to establish production lines in-house.

Dissocubes® technology was developed by Müller and coworkers (Kipp 2004). In a typical preparation process, nanosuspension is produced in the aqueous media at room temperature (Möschwitzer 2013). First, the drug particles are dispersed in an aqueous media with proper stabilizer. Then, with the help of a piston, at an extremely high pressure (1500–4000 bar), the dispersion is forced to go through a very narrow homogenization gap to produce the ideal nanocrystals. The width of homogenization gap can be adjusted from 5 to 20 μm to meet different requirements (e.g., particle size and nanosuspension viscosity) of different drugs. The suspension has high streaming velocity, which will increase the dynamic pressure. According to the Bernoulli's law, this pressure will be compensated by a reduction in the static pressure below the vapor pressure of the aqueous phase. The simplified form of Bernoulli's law is shown below:

$$p + q = p_0 \tag{9.1}$$

where p_0 is total pressure, p is static pressure, q is dynamic pressure.

The gas bubbles will occur when the water starts boiling. When the liquid leaves the homogenization gap under normal pressure of 1 bar, the gas bubbles will collapse. This phenomenon of the formation and implosion of the bubbles is called as cavitation, and this will cause shockwaves. There are several forces to reduce the drug particles such as turbulent flow, high-shear forces, and the power of shockwaves. Since the suspension is aqueous, there are many disadvantages of this technology such as stability issues caused by the hydrolysis of aqueous sensitive drugs, problems during subsequent drying steps, and employment of expensive lyophilization techniques (Möschwitzer 2013; Shegokar and Müller 2010).

Compared with the piston-gap homogenization in water (Dissocubes® technology), the piston-gap homogenization in water mixtures or in nonaqueous media (Nanopure® technology) has several advantages. This technology applies the nonaqueous media such as oils, polyethylene glycol 400 (PEG), and polyethylene glycol 600. Therefore, it is suitable for aqueous unstable drugs. However, the oils or fatty acids have lower vapor pressures at room temperature. Thus, the boiling points are higher, resulting in the very little or nonexistent cavitation (the drop of the static pressure is not enough to produce cavitation in the homogenization). It is not as efficient in the particle size reduction as compared with aqueous dispersions. But even without cavitation, it is enough to reduce the particle size through shear forces, particle collisions, and turbulences. As this process is prepared under low temperature, it is proper for temperature labile drugs. Moreover, it is possible to produce the oil-dispersed nanocrystals, which is filled in soft gelatin capsules or into hard gelatin or HPMC capsules (Möschwitzer 2013; Müller et al. 2011).

Different from the piston-gap homogenization in water, piston-gap homogenization in water mixtures or in nonaqueous media, the microfluidization technique is based on the jet stream principle. The microfluidization technique refers to dispersed large particles in an aqueous solution or other solvents to form coarse suspension, which passes through a "Z shaped" or "Y shaped" homogenization chamber under high pressure at a high speed. The forces used in the microfluidization technique are similar to the piston-gap homogenization methods. These forces include impact force, shear force, and cavitation, which reduce the drug particle size. However, in the "Z shaped" chamber, the suspension changes the direction of liquid, and the shear force is produced due to the collision. In the "Y shaped" chamber, the shear force is generated due to the collisions in the two pipes at the intersection. To obtain ideal particle size, this method needs a number of cycles (50–100 passes). This also leads to the time consumption, which is one of the major problems of this technique for preparing the drug nanocrystals (Yang et al. 2012).

High-pressure homogenization is a simple technique. In this technique, the pressure, number of homogenization cycles, and temperature are three main important parameters. With the increasing of pressure and cycles, the particle size will decrease. In addition, high pressure processing increases the temperature, which is not good for temperature-sensitive drugs. Thus, a heat exchanger is suggested to be placed ahead of the homogenizer valve to control the increase of temperature. In addition, compared with the media milling, high-pressure homogenization has a significant advantage: lower contamination due to low erosion of the homogenizer (Müller et al. 2011; Keck and Muller 2006; Möschwitzer 2013).

9.2.3 Combination Techniques

Although bottom-up techniques and top-down techniques have been applied in the nanocrystals production, there are still some disadvantages for each of them. It is better for the equipment to use micronized materials. If drug particles are too coarse, it may lead to some equipment troubles. Moreover, it is too much time-consuming. When the process is performed in a big-scale mill, the process will cost a relatively long time. Therefore, to overcome these disadvantages, the particles are suggested to be pretreated before this technique. The combination of the bottom-up techniques and top-down techniques is a good approach to dissolve these disadvantages for preparing drug nanocrystals. The combination technologies generally combine a precipitation pretreatment process with a high energy process (Chen et al. 2015).

Today, several combination techniques have been successfully developed such as Nanoedge™, H69, H42, H96, CT, and precipitation-lyophilization-homogenization technologies. Nanoedge™ technology was first developed by Baxter Healthcare (MőSchwitzer and Müller 2006). It combined a bottom-up step (microprecipitation) with a top-down step (high-pressure homogenization). H69 technology is an alternative method of Nanoedge™ technology and was first developed by Möschwitzer and colleagues (Müller and Möschwitzer 2008). This method was designed to minimize the time between two steps in order to produce even smaller nanocrystals by making the microprecipitation step immediately followed by high-pressure homogenization. H42 technology (spray-drying and high-pressure homogenization) and H96 technology (freeze-drying and high-pressure homogenization) were developed by Möschwitzer and colleagues. These two technologies can minimize the particle aggregation and improve the efficiency of particle size reduction. CT technology refers to the method of media milling followed by high-pressure homogenization.

Recently, precipitation-lyophilization-homogenization (PLH) method was developed by Junyaprasert group (Morakul et al. 2013). In this method, the drug is first dissolved in an organic solvent. Second, it was added to the aqueous phase to precipitate small crystals. Third, the precipitation material is treated by freeze drying for removing the aqueous and organic solvents. Finally, it is homogenized under high pressure. It can drastically reduce the fabrication time by reducing the homogenization cycles. Only after the fifth cycles, the clarithromycin nanocrystals can reduce to the 400 nm by this method. On the other hand, if the clarithromycin nanocrystals are produced by the high-pressure homogenization technique without any pretreatment, it will need around 30 cycles to reduce them to the same size.

Obviously, the combination techniques can improve the fabrication methods and shorten the fabrication time. The disadvantages of the combination techniques are expensive and complex. Only if the normal methods (top-down and bottom-up techniques) cannot be used to obtain the ideal results, the combined methods will be considered.

9.3 Administration Routes of Drug Nanocrystals

Due to their high drug loading, relatively easy scale up for production, lower cost for manufacture, and applicability to many administration routes, drug nanocrystals have been applied in clinical therapy. The absorption of drug nanocrystals is influenced by several factors such as particle size, Zeta potential, crystalline state, shape, stabilizers, and administration routes. There are various administration routes such as oral, dermal, pulmonary, ocular, parenteral, and targeted delivery (Table 9.1).

9.3.1 Oral Drug Delivery

Oral drug delivery is the main administration route in drug delivery due to its advantages such as the administration convenience, relatively low cost, and relatively high safety. In oral administration, poorly water-soluble drugs have several significant problems such as low bioavailability, and easy susceptibility by eating speed and dosage. Luckily, the creation of drug nanocrystals solves these problems and offers benefits for oral drug delivery.

Nanocrystal technologies can enhance the bioavailability of poorly water-soluble drugs by undergoing complex absorption mechanisms for the transport of drugs or intact nanocrystals from the gastrointestinal tract to the blood or lymphatic system (Liu et al. 2020). The absorption mechanisms of drug nanocrystals mainly include four aspects: (1) Enhanced saturation solubility and dissolution rate. Owing to their

Table 9.1 Administration routes of drug nanocrystals

Routes of administration	Advantages to conventional drugs
Oral	Enhanced saturation solubility and dissolution rate Enhanced interaction with mucus layer Enhanced transport across the intestinal membrane Enhanced absorption by stabilizers
Dermal	Increased adhesiveness to surfaces/cell membranes due to increased contact areas Improved dissolution rate Increased flux caused by increased saturation solubility Application potential for hair follicle targeting Increased dermal bioavailability Simulated local action
Pulmonary	Local administration within the respiratory tract Rapid drug action Reduced dose Reduction in systemic side effects Reduced evasion of first-pass hepatic metabolism by absorbed drug Potential for pulmonary administration of systemically active materials
Ocular	Improved ocular safety Increased retention of formulation in cul-de-sac Enhanced corneal permeation Enhanced ocular bioavailability Dual drug release behavior in the eye Improved tolerability
Injection	Safe Reduced dosage Suitable for some formulations which need the high dosage and high concentration Some stabilizers can improve the dissolution rate in vitro, improve the drug loading, and improve biological safety of the preparation
Targeted	A high drug concentration to the target site is assured The drug therapeutical level in the disease site can be more easily reached Surface properties and in vivo behavior of drug nanocrystals can be easily altered Reduce the side effect to minimum

small particle size and large surface area, drug nanocrystals can gain a higher dissolution rate and water solubility than coarse drugs. (2) Enhanced interaction with mucus layer. Mucosa is the moist surface for protection and lubrication in the gastrointestinal tract. The moistness is usually caused by the presence of a mucus layer. Mucus layer can block the contact of drugs to epithelial cell, thus hindering the absorption of drugs. The formation of mucoadhesion can be explained by following general theories: electronic, wetting, adsorption, mechanical, and fracture theories. Due to the small particle size and potential use of functional stabilizers, drug nanocrystals have higher electrostatic attraction forces, van der Waals' forces, and surface area. Therefore, enhanced mucoadhesion to mucosa and prolonged retention time of drugs are obtained. In addition, compared with coarse drugs, nanocrystals can generate a higher penetration ability across the mucus layer by forming a reservoir to release the drugs with a smaller effective diffusion thickness and by using surface-altering agents such as stabilizers. Therefore, faster diffusion through mucus layer leads to improved drug contact with epithelial cells and enhanced absorption. (3) Enhanced transport across the intestinal membrane. Compared with coarse drugs, drug nanocrystals can be up-taken by epithelial cells via caveoin-mediated, caveolae-mediated, lipid raft-mediated, and macrophage-mediated endocytosis. This process can be modulated by cav-1, dynamin, actin filaments, etc. The entered drug nanocrystals can be transport across the epithelial cell with the aid of lipid raft, Golgi complexes, lysosomes, and endosomes. In addition, drug nanocrystals can also be up-taken by M cells in Peyer's patches and be transported to lymph vessel. (4) Enhanced absorption by stabilizers. With the aid of stabilizers, drug nanocrystals can inhibit P-glycoprotein efflux, which can increase the bioavailability of drug nanocrystals. In addition, tight junctions can be opened by functional stabilizers, which may construct a paracellular pathway to transport drug nanocrystals and improve the bioavailability.

There are several important factors to influence the oral absorption of drug nanocrystals: particle size, Zeta potential, crystalline state, shape of nanocrystals, and stabilizers. They have different effects on different absorption mechanisms (Liu et al. 2020). Rational design and preparation to optimize these factors should be taken into consideration during the development of drug nanocrystals for oral delivery.

9.3.2 Dermal Drug Delivery

The skin is the largest organ in the human body and acts as a principal barrier. Dermal drug delivery is a good way of administration by the skin surface to achieve local treatment or systemic treatment without first-pass effect (Liu et al. 2018). Dermal delivery can be achieved through stratum corneum and skin appendages. There are three ways to transfer drugs inside the stratum corneum: intercellular, transcellular, and follicular ways (Romero et al. 2015).

In the dermal drug delivery, compared with coarse drugs, the fundamental mechanism of drug nanocrystals is the improvement of the solubility of poorly water-soluble drugs in aqueous phase, which increases the concentration gradient between the formulation and skin. Therefore, the drug nanocrystals can transdermally penetrate the skin. Although the adhesion of drug nanocrystals is a significant property, the fast solubility and increased penetration are very important to dermal drug delivery.

The first application of this technology was in the cosmetics industry, and then it was applied in drug delivery (Shegokar and Müller 2010). The best dermal nanocrystal formulations are supposed to have various characteristics such as high kinetic saturation solubility and high concentration gradient (Vidlářová et al. 2016). Moreover, the density of nanocrystals on the skin is low, but the nanocrystals are able to cover the skin densely. Therefore, drug nanocrystals can get a large area to directly contact with the skin, which can increase the drug adsorption. It should be noted that some chemicals such as tretinoin used in the nanosuspension can improve the dermal delivery and stability of the drug nanocrystals (Zhu et al. 2013). Therefore, the dermal delivery efficacy of nanocrystal is affected by particle size, particle concentration, saturation solubility, and stabilizers.

Dermal delivery of drug nanocrystals has several important advantages (Kumar et al. 2019): increased adhesiveness to surfaces/cell membranes due to increased contact areas, improved dissolution rate, increased flux caused by increased saturation solubility, application potential for hair follicle targeting, increased dermal bioavailability, and simulated local action.

9.3.3 Pulmonary Drug Delivery

In the human body, the lungs are highly perfused organs. Therefore, it can efficiently and rapidly deliver the drug to the systemic circulation. It is generally achieved by an aerosol generated with either an inhaler or a nebulizer. Pulmonary drug delivery has been generally considered an attractive noninvasive way for both local and systemic treatments (Patil and Sarasija 2012).

There are two kinds of pulmonary formulations containing drug nanocrystals (Gao et al. 2012): (1) Aqueous nanosuspensions packaged and administrated by a nebulizer. In this method, drug nanocrystals can be collected and transported into the lung by the small aerosol droplets generated by the nebulizer. (2) Inhaled powders containing dried nanocrystals dispersed in some inhalable carriers such as inhalable lactose. In this method, the powders can be inhaled by commercial dry powder inhalers.

Compared with coarse drugs, drug nanocrystals offer a more homogenous distribution of drugs in the nebulized droplets or among the inhalable carriers. This is beneficial for pulmonary drug delivery. Drug nanocrystals can spread more evenly on the lung surface. Moreover, drug nanocrystals dissolve rapidly in the lung lining fluid. The induced high concentration of drugs can raise the driving force for permeation through the alveoli membrane. Therefore, the absorption rate of the poorly water-soluble drugs increases (Tolman and Williams 2010).

Pulmonary delivery of drug nanocrystals has several important advantages (Shah et al. 2012; Kumar et al. 2020): local administration within the respiratory tract, rapid drug action, reduced dose, reduction in systemic side effects, reduced evasion of first-pass hepatic metabolism by absorbed drug, and potential for pulmonary administration of systemically active materials.

9.3.4 Ocular Drug Delivery

Ocular drug delivery is generally used for the treatment of both intraocular (e.g., age-related macular degeneration, diabetes retinopathy, glaucoma, uveitis, keratoconus, and proliferative vitreoretinopathy) and periocular (e.g., dry eyes, conjunctivitis, and blepharitis) diseases. There are several membranes and barriers that control the transport of materials into and out of the eye: physiological barrier, anatomical barrier, corneal barrier, conjunctival barrier, blood-ocular barrier, and drug efflux transporters. It may not suitable for systemic administration of drugs because blood-aqueous and blood-retinal barriers limit the transportation of drugs into extravascular retinal space, as well as the aqueous and vitreous humor of the eye. Due to normal eye physiology, such as lacrimation or continuous flow of tear fluid, nasolachrymal drainage, conventional ocular drug formulations fail to achieve effective concentration in both intraocular and periocular regions. Therefore, repeated administration is generally required for effective treatment.

Nanocrystal technologies help in tailoring the property of poorly water-soluble drugs such as particle size, particle shape, solubility, and net charge on surface. Therefore, the ocular bioavailability is increased. In addition, mucoadhesive nature of drug nanocrystals helps in minimizing the drainage of formulation with tear fluid and increases the retention time in cul-de-sac, thereby reducing the administration frequency and improving the drug ocular bioavailability. In all, nanocrystals can be used for both intraocular drug delivery and periocular drug delivery due to their significant advantages (Sharma et al. 2016): improved ocular safety, increased retention of formulation in cul-de-sac, enhanced corneal permeation, enhanced ocular bioavailability, dual drug release behavior in the eye, and improved tolerability.

9.3.5 Parenteral Drug Delivery

The parenteral drug delivery includes subcutaneous delivery, intramuscular delivery, intravenous (IV) delivery, intradermal delivery, and intra-arterial injection delivery. This method delivers the drug to a blood vessel, organ, tissue, or lesion directly. It has many advantages such as immediate effects, overcoming the first-pass effect, targeting effect, high drug loading, and conventional sterilization techniques (gamma irradiation, filtration, and thermal sterilization) to sterilize the formulations.

The poorly water-soluble drugs have difficulty in the injection delivery such as poor compliance and many side effects, which have greatly restricted their clinical use. Compared with the conventional injection delivery, the nanocrystals have various advantages including safe and reduced dosage. This technology is suitable for some formulations which need the high dosage and high concentration. Besides, some stabilizers can improve the dissolution rate in vitro, improve the drug loading, and improve biological safety of the preparation. Therefore, drug nanocrystals can be applied in the intravenous delivery.

Due to the sufficiently small size and safe aqueous composition of nanosuspension, it can be injected intravenously and achieve 100% bioavailability (Gao et al. 2012). In the intravenous delivery, the stability of nanocrystals during the storage is of great importance. The nanosuspensions are supposed to control the content of particle (the size >5 mm) in case of avoiding the capillary blockade and embolism.

Compared with other carrier-based nanoparticles, drug nanocrystals experience a much faster particle size reduction during the dissolution process, and this may lead to a distinct pharmacokinetic progress after intravenous injection administration. In addition, other natures of drug nanocrystals such as the intrinsic drug solubility and surface properties may also play important roles. The pharmacokinetic performances of nanosuspensions after intravenous injection administration can be classified into two types: (1) Pharmacokinetic properties of drug nanocrystals with small particle size (generally <200 nm) or relatively high intrinsic solubility in the blood circulation are similar to those of solution. (2) Pharmacokinetic properties of drug nanocrystals with large particle size or very poor solubility in the blood circulation are similar to those of carrier nanoparticles.

9.3.6 *Targeted Drug Delivery*

One of the major problems associated with drug delivery is the inability to deliver drugs to a specific disease site of the body, resulting in nonspecific toxicity. Nanoscaled drug delivery systems have been developed to improve targeting drug delivery. Compared with other nanocarrier delivery systems, drug nanocrystals have high drug loading of 100%. A high drug concentration to the target site is assured and the drug therapeutical level in the disease site can be more easily reached. In addition, surface properties and in vivo behavior of drug nanocrystals can be easily altered for the targeting drug delivery.

The targeted delivery of drug nanocrystals can be divided into three types, including passive targeting, specific organs targeting, and cell-based targeting. There are two pathways for passive targeting. One is the recognition as foreign matters and the rapid clearance by the abundant phagocytic cells in special tissues and organs such as liver, lung, and spleen (Ye et al. 2008). The other is the accumulation of drug nanocrystals in solid tumors due to the enhanced permeability and retention effect (Matsumura and Maeda 1986). For specific organs targeting, some targeting moieties such as transferrin and folate are attached to drug nanocrystal surface via some polymer coating for realizing exact affinity with specific organs (Hilgenbrink and Low 2005). For cell-based targeting, some nanocrystal-loaded cell systems are developed to release the drug content in circulation or at selected sites or other relevant cells in the body. The common cell systems used in cell-based targeting include macrophages and red blood cells (Dou et al. 2006; Staedtke et al. 2010).

The targeted delivery can reduce the side effect to minimum level. Due to the controlled particle size and the surface modification, drug nanocrystals can selectively target specific parts, like lung, brain, or bone marrow. However, the cell-specific ligands cannot always improve the transport across the cell membrane (Wu et al. 2020). Recently, though the research of targeted delivery is a hot issue, the strategies applied to overcome the physiological barrier of tumors, make the tumor issues penetrate deeply, improve the tumor specificity of chemotherapeutics, release in the tumor cell controlledly are urgently required.

9.4 Status of Drug Nanocrystals in the Market and Clinical Trials

As shown in Table 9.2, about 12 drug nanocrystal products are currently approved in clinics. The market for drug nanocrystal is enormous. These commercial products are mainly produced by wet milling or high-pressure homogenization. The administration routes are mainly oral and injection routes. The path of oral administration to commercialization is easier than other administration routes. There are several drug nanocrystal products currently in clinical trials, as shown in Table 9.3. These clinical trials are updated on the clinical trials website (clinicaltrial.gov). Interested readers can search nanocrystal products in this website for more detailed information. Though some progresses on the commercial nanocrystals have already made for the treatment of human diseases in this field, drug nanocrystals have great potential in drug pharmaceutics and are worthy to be paid more attention in the future.

9.5 Researches on Drug Nanocrystals in the Tumor Target Delivery

The treatment of tumor in the body generally involves target drug delivery. The targeted delivery of drug nanocrystals can be divided into three types: passive targeting, specific organs targeting, and cell-based targeting. Some typical examples are shown in Table 9.4.

9.5.1 Passive Targeting

There are two pathways for passive targeting. One is generally based on the enhanced permeability and retention effect of tumor tissues. In tumor tissues, tumor cells proliferate rapidly and metabolize vigorously. Intratumoral neovascularization proliferates abnormally and the endothelial cells of the neovascularization are

Table 9.2 Typical products based on nanocrystals approved by FDA (Liu et al. 2020)

Product	Company	Drug compound	Clinical purpose	Production approach	Administration route	Approval year
Rapamune®	Wyeth	Sirolimus	Immunosuppressive	Wet milling	Oral	2000
Focalin XR®	Novartis	Dexmethylphenidate HCl	Anti-psychotic	Wet milling	Oral	2001
Avinza®	King Pharma	Morphine sulfate	Anti-chronic pain	Wet milling	Oral	2002
Ritalin LA®	Novartis	Methylphenidate HCl	Anti-psychotic	Wet milling	Oral	2002
Zanaflex Capsules™	Acorda	Tizanidine HCl	Muscle relaxant	Wet milling	Oral	2002
Emend®	Merck	Aprepitant	Antiemetic	Wet milling	Oral	2003
Tricor®	Abbott	Fenofibrate	Hypercholesterolemia	Wet milling	Oral	2004
Megace® ES	Par Pharma	Megestrol acetate	Appetite stimulant	Wet milling	Oral	2005
Triglide™	Skye Pharma	Fenofibrate	Hypercholesterolemia	HPH	Oral	2005
Invega Sustenna®	Johnson	Tizanidine HCl	Anti-depression	HPH	Injection (i.m.)	2009
Invega Trinza®	Johnson	Tizanidine HCl	Anti-depression	HPH	Injection (i.m.)	2015
Aristada®	Alkermes	Tizanidine HCl	Schizophrenia	HPH	Injection (i.m.)	2015

Table 9.3 Nanocrystal drug products in clinical trials by FDA (Jarvis et al. 2019)

Product	Company	Drug compound	Clinical purpose	Production approach	Administration route	Clinical status
Semapimod	Cytokine Phamasciences	Guanylhydrazone	TNF-α Inhibitor	Self-developed	Injection (i.m.)	II
Paxceed®	Angiotech	Paclitaxel	Anti-inflammatory	Unknown	Injection (i.m.)	III
Theralux	Celmed	Themectacin	Autoimmune diseases and cancer	Media milling	Injection (i.m.)	II
Nucryst®	Nucryst Pharmaceuticals	Silver	Atopic dermatitis	Self-eveloped	Topical	II
PanzemNCD	EntreMed	2-Methoxy estradiol	Ovarian cancer	Media milling	Oral	II

Table 9.4 Typical research examples on drug nanocrystals in the tumor target delivery

Target strategy	Drug compound	Modification	Administration route	Tumor model	Mechanism	Advantages	Reference
Passive targeting	Indomethacin paclitaxel	IDM/PTX Assembly	Injection	MDA-MB-231 tumor	EPR effect, the IDM/PTX assemblies improved synergetic antitumor effects	1. Appropriate nanoscale size 2. A total of 100% drug loading 3. Perfect encapsulation efficiency 4. High drug loading of PTX	Zhang et al. (2019)
Passive targeting	Hydroxycamptothecin (HCPT)	Precipitation-combined ultra-sonication prepares HCPT nanosuspensions	Injection	H22-tumor	Recognition and rapid clear of the abundant phagocytic cells in special tissues and organs	1. Higher antitumor efficacy 2. Lower systemic side effects	Han et al. (2013)
Specific organs targeting	Doxorubicin hydrochloride, calcium carbonate nanocrystal	DOX-loaded $CaCO_3$ nanocrystals	/	MG 63 osteosarcoma cell	The cellular uptake of $CaCO_3$/DOX nanocrystals is through nonspecific endocytosis, this mechanism of $CaCO_3$/DOX nanocrystal delivery to tumors may circumvent the effect of multidrug resistant proteins which are always present in cancer cells	1. Do not induce specific toxicity 2. Achieving the effective osteosarcoma cancer cell death	Kamba et al. (2013)

(continued)

Table 9.4 (continued)

Target strategy	Drug compound	Modification	Administration route	Tumor model	Mechanism	Advantages	Reference
Specific organs targeting	Paclitaxel (PTX), camptothecin (CPT), Pluronic F127 (F127) polymer	Folate ligand-modified	Injection	Human lung cancer, murine breast cancer	The ligands can recognize the specific receptor in the special organs, the nanocrystals are transported to the specific lesion site by receptor-mediated phagocytosis	1. Excellent antitumor activity 2. Low toxicity 3. Ease of scale-up manufacture	Liu et al. (2010a)
Specific organs targeting	PTX, TPGS, citrate acid, dopamine	Polyethylene glycol and RGD peptide modification	Injection	A549 model tumor	The ligands can recognize the specific receptor in the special organs, the nanocrystals are transported to the specific lesion site by receptor-mediated phagocytosis	1. Enhancing intratumor distribution 2. Enhancing tumor growth inhibition efficacy	Huang et al. (2019)
Cell-based targeting	Anaerobic *Bifidobacterium breve*, *Clostridium difficile*	Near-infrared light excitation for two forms of the anaerobic *Bifidobacterium breve* and *Clostridium difficile*	Injection	A549 model tumor	Au NRs to attach on the bacteria, cargo-carrying approach brings NPs with bacteria to tumor, the quantity of the NPs would be diluted	1. Stronger imaging 2. Longer retention period 3. Effective therapy to completely remove tumors	Luo et al. (2016)

Cell-based targeting	HS-PEG-COOH, AuNRs	Various thiolated poly(ethylene glycol)s (HS-PEG) modification	Injection	4T1 tumor	HS-PEG activate the macrophages, macrophages loaded more Anionic-AuNRs and achieve more endocytosis and aggregation	1. Simultaneously achieve high phagocytosis quantity of theranostics agent and visually observe hypoxic regions 2. High drug loading and enhanced tumor inhibition ability	An et al. (2019)
Cell-based targeting	Synthesize NC(DTX), RBC-NC(DTX), multivalent avidin–biotin	The red blood cell (RBC) membrane modification by targeting peptides	Injection	U87 tumor	Apoptosis and angiogenesis inhibition	1. Superior tumor accumulation and therapeutic efficacy 2. High drug loading, long-term stability 3. Excellent biocompatibility and prolonged retention time	Chai et al. (2019)

loosely connected. Therefore, the vascular endothelial cell spaces with leaky vessels and pores (100 nm to 2 μm in diameter) are formed, which makes the tumor tissue produce enhanced permeability and retention effect (Kalyane et al. 2019). This effect promotes the selective distribution of drugs in tumor tissues, which can increase drug efficacy and reduce systemic side effects. The drug nanocrystals with particle size of 10–500 nm can enter the tumor tissues through the vascular endothelial cell space and achieve specific distribution and accumulation in tumor tissue (Prabhakar et al. 2013). The particle size and surface properties of drug nanocrystals play important roles in the passive targeting of nanocrystal delivery. Therefore, the optimization of the particle size and surface modification can inhibit the scavenging effect of the reticuloendothelial system effectively, prolong the blood circulation time of the drug nanocrystals, and increase the clinical efficacy of drug nanocrystals for the treatment of tumors (Nehoff et al. 2014). Recently, based on the strong intermolecular interactions, paclitaxel-induced indomethacin nanocrystal aggregates were prepared via one-pot self-assembly without any nonactive excipient (Zhang et al. 2019). In serum, these nanoassemblies could rapidly collapse into a great number of smaller nanoparticles, thus targeting the tumor site through the enhanced permeability and retention effect.

The other pathway for passive targeting is based on the recognition and rapid clearance by the abundant phagocytic cells in special tissues and organs such as liver, lung, and spleen. When drug nanocrystals are intravenous injected into body, drug nanocrystals are easily absorbed by the opsonic protein in the plasma. Then, they are recognized and swallowed by macrophages. In the end, they will be eliminated by macrophages from the blood circulation. Generally, about 5 minutes after the drug nanocrystals are intravenously injected into body, approximately 90% of the drug nanocrystals are swallowed by macrophages in the liver, and only approximately 5% of the drug nanocrystals are swallowed by macrophages in the spleen (Fuhrmann et al. 2014). Compared with coarse drugs, hydroxycamptothecin nanocrystals increased their bioavailability in the liver, spleen, and lungs by 410, 46, and 40 folds, respectively (Han et al. 2013). In all, drug nanocrystals with particle size of 100–1000 nm can be swallowed by macrophages, which cause drug accumulation, cytotoxic reactions, and the delivery of anticancer or chemotherapeutic drugs into special tissues and organs (Lu et al. 2015).

9.5.2 Specific Organs Targeting

Specific organs targeting is generally based on specific molecular markers in tumor tissues. The polymer stabilizers absorbed on the surface of drug nanocrystals play an important role in the stability and inhibition of the coalescence and growth of the drug nanocrystals. Physical and chemical modifications of the polymer on the surface of nanocrystals can contribute the nanocrystals with the properties of responding to environmental stimuli such as temperature, pH, and enzymes. The modification can not only make the nanocrystals act as a stabilizer but also increase the targeted ability to special tissues. For example, acid pH-sensitive calcium carbonate nanocrystals were developed, and they released slowly in the normal physiological environment with pH 7.4. In contrast, they released fast in the acidic environment (under the tumor tissues with pH 4.8) to achieve the specific organ targeting (Kamba et al. 2013).

To further improve the tumor specificity, the cell-binding ability, and cell up taking, some targeting moieties such as small molecule ligands (e.g., folic acid, mannose, and galactose), aptamers, peptides, antibody fragments, or monomers can be used to modify the surface of the drug nanocrystals, which make the drug nanocrystals have the ability of active targeted delivery. In the ligand-targeted delivery, the ligands can be absorbed onto the surface of the drug nanocrystals directly. The ligands can recognize the specific receptor in the special organs, then the nanocrystals are transported to the specific lesion site by receptor-mediated phagocytosis. For example, the folate ligand-modified nanocrystals showed excellent tumor targeting ability, significant therapeutic effects, low toxicity, and ease of scale-up manufacture (Liu et al. 2010a). Recently, paclitaxel nanocrystals were modified by polyethylene glycol for stabilization and RGD peptide for tumor targeting (Huang et al. 2019). The results suggested the RGD-modified PEGylated paclitaxel nanocrystals could be an effective lung cancer-targeted delivery system.

This technology offers a new and efficient strategy to transport the anticancer drugs to tumor tissues. These targeting moieties are generally used to attach to the polymer chains on the drug nanocrystal surface. However, polymer chains are commonly physically (not chemically) anchored on nanocrystal surface. Therefore, the polymers and attached targeting moieties can be reversibly adsorbed onto and desorbed from nanocrystal surface, which may decrease the targeting efficacy. The adsorption stability of the polymers and attached targeting moieties on the nanocrystal surface should be carefully considered during the preparation process. Chemical crosslinking methods may be helpful to solve this issue (Kim and Lee 2011).

There are three main key factors that affect the property of nanocrystals: the particle size and morphology of the drug nanocrystals, the zeta potential of the drug nanocrystals, and the hydrophobic properties on the surface of the drug nanocrystals. The affinity of adsorbed proteins on the surface of nanocrystals is different for the different ordinary raw materials. Besides, with the changing of the particle size, the targeting property of nanocrystals also changes (Jansch et al. 2013). Compared with the nanocrystals with charged surface, the neutral surface of nanocrystals absorbs protein obviously slowly, which shows that the zeta potential of nanocrystals has the relation with the ability of adsorbing proteins. Moreover, the hydrophobic properties of the surface of polymer-modified drug nanocrystals can not only determine the content of adsorbed protein but also significantly affect the type of adsorbed protein (Lu et al. 2017).

9.5.3 Cell-Based Targeting

Excellent dissolution velocity of drug nanocrystals is one of the main advantages of drug nanocrystals. However, it should be noted that drug nanocrystals may dissolve in the transportation process before reaching the target site. Cell-based drug delivery targeting can be employed to deal with this problem. Cell-based delivery systems use cell carriers such as bacteria and animal cells to load drugs for the treatment of human diseases. These systems release the loaded drugs in blood circulation or at target sites.

The hypoxia region in a solid tumor is a complex microenvironment with very low oxygen concentration and deficient nutrients. The hypoxic environment reduces the susceptibility of the cancer cells to anticancer drugs, less proliferation of cancer cells, and low response of free radicals. However, the hypoxic environment provides an appreciable habitat for anaerobic bacteria to colonize. Therefore, bacteria-based drug delivery system may be a good strategy to deliver drugs to tumor sites. Recently, nanoparticle-loaded cargo-carrying and antibody-directed bacteria-mediated targeting hypoxia methods were developed for diagnosis and therapy in cancer diseases (Luo et al. 2016). Upconversion nanorods and Au nanorods were loaded for imaging and photothermal ablation upon near-infrared light excitation, respectively. This work took the advantage of anaerobic bacterial's habit to combine with functional nanoparticles to improve the perplexed treatment in tumor hypoxia, which provided an alternative to enhance targeting diagnosis and therapy.

Typical animal cells used in this method are macrophages and red blood cells. Macrophages are differentiated cells of the immune system and are able to phagocytize microorganisms and nanoparticulate materials. Macrophages can avoid interception by the immune system and cross the blood vessel barriers to reach the hypoxic regions of tumors. Therefore, macrophages are good potential cell-based drug delivery systems for delivering drug nanocrystals (Andón et al. 2017). Macrophages are first loaded with drug nanocrystals ex vivo and then re-infused into the target sites that favor homing of macrophages. Red blood cells have unique properties to be used as cell-based drug delivery systems such as suitability for nanocrystal encapsulation, biodegradability, biocompatibility, and long-term drug releasing ability (Rossi et al. 2005). They can be easily treated to load nanocrystals ex vivo by several techniques (Serafini et al. 2004). Recently, small gold nanorod-loaded macrophage-mediated delivery system was developed for tumor hypoxia photoacoustic imaging and enhanced photothermal therapy (An et al. 2019). The system had high drug loading and enhanced tumor inhibition ability. In addition, a nanocrystal-loaded red blood cell membrane system was modified with targeting peptides through a facile insertion method involving avidin-biotin interactions (Chai et al. 2019). The complex systems showed high drug loading, long-term stability, excellent biocompatibility, and prolonged retention time. These advantages proved that they are suitable for effective drug targeting delivery.

9.6 Summary and Future Perspectives

Although huge efforts and resources have been applied in the area of cancer therapy, cancer remains the leading cause of death worldwide. Not only the malignant cell masses but also the surrounding tumor microenvironment, which includes both cellular and noncellular components, consists of the tumor. There are common characters among tumors despite of the differences between the tumor microenvironments. Therefore, the targeting of drug nanocrystals to the malignant and metastatic components is a potential strategy for tumor therapy.

Compared with the nanocarrier drugs, nanocrystals have many excellent advantages such as relatively simple preparation process, high drug loading, and good physical stability. The preparation methods of nanocrystals can be classified into bottom-up, top-down, and combination techniques. The bottom-up techniques include the supercritical fluid technique and the precipitation technique. The top-down techniques include the media milling and HPH. The combination techniques combine the bottom-up technique and top-down technique. Typical combination techniques include Nanoedge™, H69, H42, H96, CT, and precipitation-lyophilization-homogenization technologies.

Due to the various advantages, the drug nanocrystals have been widely applied and explored in the cancer therapy. There are various administration routes, such as oral, dermal, pulmonary, ocular, parenteral, and targeted delivery. The main administration route is oral drug delivery. The drug nanocrystals still face great numbers of challenges such as the optimum dose, the optimum release, and the targeted delivery of the drug nanocrystals. In the future, it is highly expected that a great deal of drug nanocrystals will be produced, and more targeted delivery methods will be applied in the tumor therapy.

References

An L, Wang Y, Lin J et al (2019) Macrophages-mediated delivery of small gold nanorods for tumor hypoxia photoacoustic imaging and enhanced photothermal therapy. ACS Appl Mater Interfaces 11:15251–15261. https://doi.org/10.1021/acsami.9b00495

Andón FT, Digifico E, Maeda A et al (2017) Targeting tumor associated macrophages: the new challenge for nanomedicine. Semin Immunol 34:103–113. https://doi.org/10.1016/j.smim.2017.09.004

Bae YM, Park YI, Nam SH et al (2012) Endocytosis, intracellular transport, and exocytosis of lanthanide-doped upconverting nanoparticles in single living cells. Biomaterials 33:9080–9086. https://doi.org/10.1016/j.biomaterials.2012.08.039

Bahrami A, Hassanian SM, Khazaei M et al (2018) The therapeutic potential of targeting tumor microenvironment in breast cancer: rational strategies and recent progress. J Cell Biochem 119:111–122. https://doi.org/10.1002/jcb.26183

Balkwill FR, Capasso M, Hagemann T (2012) The tumor microenvironment at a glance. J Cell Sci 125:5591. https://doi.org/10.1242/jcs.116392

Beeravelli S, Kothagunda N, Murthy KVR (2014) Nanosuspensions as a versatile carrier based drug delivery system – an overview. Curr Drug Deliv 11:299–305. https://doi.org/10.2174/1567201811666140323131342

Bitterlich A, Laabs C, Krautstrunk I et al (2015) Process parameter dependent growth phenomena of naproxen nanosuspension manufactured by wet media milling. Eur J Pharm Biopharm 92:171–179. https://doi.org/10.1016/j.ejpb.2015.02.031

Brandl M, Bachmann D, Drechsler M et al (1990) Liposome preparation by a new high pressure homogenizer Gaulin Micron Lab 40. Drug Dev Ind Pharm 16:2167–2191. https://doi.org/10.3109/03639049009023648

Brizel DM, Scully SP, Harrelson JM et al (1996) Tumor oxygenation predicts for the likelihood of distant metastases in human soft tissue sarcoma. Cancer Res 56:941–943

Brown JM, Giaccia AJ (1998) The unique physiology of solid tumors: opportunities (and problems) for cancer therapy. Cancer Res 58:1408–1416

Chai Z, Ran D, Lu L et al (2019) Ligand-modified cell membrane enables the targeted delivery of drug nanocrystals to glioma. ACS Nano 13:5591–5601. https://doi.org/10.1021/acsnano.9b00661

Chang T-L, Zhan H, Liang D et al (2015) Nanocrystal technology for drug formulation and delivery. Front Chem Sci Eng 9:1–14. https://doi.org/10.1007/s11705-015-1509-3

Chen H, Khemtong C, Yang X et al (2011) Nanonization strategies for poorly water-soluble drugs. Drug Discov Today 16:354–360. https://doi.org/10.1016/j.drudis.2010.02.009

Chen A, Shi Y, Yan Z et al (2015) Dosage form developments of nanosuspension drug delivery system for oral administration route. Curr Pharm Des 21:4355–4365. https://doi.org/10.2174/1381612821666150901105026

Cheng YT, Yang CC, Shyur LF (2016) Phytomedicine-modulating oxidative stress and the tumor microenvironment for cancer therapy. Pharmacol Res 114:128–143. https://doi.org/10.1016/j.phrs.2016.10.022

Dou H, Destache CJ, Morehead JR et al (2006) Development of a macrophage-based nanoparticle platform for antiretroviral drug delivery. Blood 108:2827–2835. https://doi.org/10.1182/blood-2006-03-012534

Fuhrmann K, Gauthier MA, Leroux JC (2014) Targeting of injectable drug nanocrystals. Mol Pharm 11:1762–1771. https://doi.org/10.1021/mp5001247

Gao L, Liu G, Wang X et al (2011) Preparation of a chemically stable quercetin formulation using nanosuspension technology. Int J Pharm 404:231–237. https://doi.org/10.1016/j.ijpharm.2010.11.009

Gao L, Liu G, Ma J et al (2012) Drug nanocrystals: in vivo performances. J Control Release 160:418–430. https://doi.org/10.1016/j.jconrel.2012.03.013

George M, Ghosh I (2013) Identifying the correlation between drug/stabilizer properties and critical quality attributes (CQAs) of nanosuspension formulation prepared by wet media milling technology. Eur J Pharm Sci 48:142–152. https://doi.org/10.1016/j.ejps.2012.10.004

Guisasola E, Asín L, Beola L et al (2018) Beyond traditional hyperthermia: in vivo cancer treatment with magnetic-responsive mesoporous silica nanocarriers. ACS Appl Mater Interfaces 10:12518–12525. https://doi.org/10.1021/acsami.8b02398

Han M, Liu X, Guo Y et al (2013) Preparation, characterization, biodistribution and antitumor efficacy of hydroxycamptothecin nanosuspensions. Int J Pharm 455:85–92. https://doi.org/10.1016/j.ijpharm.2013.07.056

Hilgenbrink AR, Low PS (2005) Folate receptor-mediated drug targeting: from therapeutics to diagnostics. J Pharm Sci 94:2135–2146. https://doi.org/10.1002/jps.20457

Hockel M, Vaupel P (2001) Tumor hypoxia: definitions and current clinical, biologic, and molecular aspects. J Natl Cancer Inst 93:266–276. https://doi.org/10.1093/jnci/93.4.266

Huang Z-G, Lv F-M, Wang J et al (2019) RGD-modified PEGylated paclitaxel nanocrystals with enhanced stability and tumor-targeting capability. Int J Pharm 556:217–225. https://doi.org/10.1016/j.ijpharm.2018.12.023

Jansch M, Jindal AB, Sharmila BM et al (2013) Influence of particle shape on plasma protein adsorption and macrophage uptake. Pharmazie 68:27–33. https://doi.org/10.1691/ph.2013.2088
Jarvis M, Krishnan V, Mitragotri S (2019) Nanocrystals: a perspective on translational research and clinical studies. Bioeng Transl Med 4:5–16. https://doi.org/10.1002/btm2.10122
Juhnke M, Martin D, John E (2012) Generation of wear during the production of drug nanosuspensions by wet media milling. Eur J Pharm Biopharm 81:214–222. https://doi.org/10.1016/j.ejpb.2012.01.005
Junghanns J-UAH, Müller RH (2008) Nanocrystal technology, drug delivery and clinical applications. Int J Nanomedicine 3:295–309. https://doi.org/10.2147/ijn.s595
Junyaprasert VB, Morakul B (2015) Nanocrystals for enhancement of oral bioavailability of poorly water-soluble drugs. Asian J Pharm Sci 10:13–23. https://doi.org/10.1016/j.ajps.2014.08.005
Kakran M, Shegokar R, Sahoo NG et al (2012) Fabrication of quercetin nanocrystals: comparison of different methods. Eur J Pharm Biopharm 80:113–121. https://doi.org/10.1016/j.ejpb.2011.08.006
Kalyane D, Raval N, Maheshwari R et al (2019) Employment of enhanced permeability and retention effect (EPR): nanoparticle-based precision tools for targeting of therapeutic and diagnostic agent in cancer. ACS Biomater Sci Eng 98:1252–1276. https://doi.org/10.1016/j.msec.2019.01.066
Kamba SA, Ismail M, Hussein-Al-Ali SH et al (2013) In vitro delivery and controlled release of Doxorubicin for targeting osteosarcoma bone cancer. Molecules 18:10580–10598. https://doi.org/10.3390/molecules180910580
Kanthamneni N, Valiveti S, Patel M et al (2016) Enhanced bioavailability of danazol nanosuspensions by wet milling and high-pressure homogenization. Int J Pharm Investig 6:218–224. https://doi.org/10.4103/2230-973X.195931
Keck CM, Muller RH (2006) Drug nanocrystals of poorly soluble drugs produced by high pressure homogenisation. Eur J Pharm Biopharm 62:3–16. https://doi.org/10.1016/j.ejpb.2005.05.009
Kesisoglou F, Panmai S, Wu Y (2007) Nanosizing – oral formulation development and biopharmaceutical evaluation. Adv Drug Deliv Rev 59:631–644. https://doi.org/10.1016/j.addr.2007.05.003
Khawar IA, Kim JH, Kuh HJ (2015) Improving drug delivery to solid tumors: priming the tumor microenvironment. J Control Release 201:78–89. https://doi.org/10.1016/j.jconrel.2014.12.018
Kim S, Lee J (2011) Folate-targeted drug-delivery systems prepared by nano-comminution. Drug Dev Ind Pharm 37:131–138. https://doi.org/10.3109/03639045.2010.496788
Kipp JE (2004) The role of solid nanoparticle technology in the parenteral delivery of poorly water-soluble drugs. Int J Pharm 284:109–122. https://doi.org/10.1016/j.ijpharm.2004.07.019
Kumar M, Shanthi N, Mahato AK (2019) Pharmaceutical drug nanocrystals: role in dermal delivery. Nanosci Nanotechnol Asia 9:300–310. https://doi.org/10.2174/2210681208666180516093855
Kumar M, Jha A, Dr M et al (2020) Targeted drug nanocrystals for pulmonary delivery: a potential strategy for lung cancer therapy. Expert Opin Drug Deliv 17:1459–1472. https://doi.org/10.1080/17425247.2020.1798401
Li J, Burgess DJ (2020) Nanomedicine-based drug delivery towards tumor biological and immunological microenvironment. Acta Pharm Sin B 10:2110. https://doi.org/10.1016/j.apsb.2020.05.008
Liu F, Park J-Y, Zhang Y et al (2010a) Targeted cancer therapy with novel high drug-loading nanocrystals. J Pharm Sci 99:3542–3551. https://doi.org/10.1002/jps.22112
Liu Y, Sun C, Hao Y et al (2010b) Mechanism of dissolution enhancement and bioavailability of poorly water soluble celecoxib by preparing stable amorphous nanoparticles. J Pharm Pharm Sci 13:589–606. https://doi.org/10.18433/J3530J
Liu Y, Zhao J, Wang L et al (2018) Nanocrystals technology for transdermal delivery of water-insoluble drugs. Curr Drug Deliv 15:1221–1229. https://doi.org/10.2174/1567201815666180518124345
Liu J, Tu L, Cheng M et al (2020) Mechanisms for oral absorption enhancement of drugs by nanocrystals. J Drug Deliv Sci Technol 56:101607. https://doi.org/10.1016/j.jddst.2020.101607

Lu Y, Chen Y, Gemeinhart RA et al (2015) Developing nanocrystals for cancer treatment. Nanomedicine (Lond) 10:2537–2552. https://doi.org/10.2217/nnm.15.73

Lu Y, Li Y, Wu W (2016) Injected nanocrystals for targeted drug delivery. Acta Pharm Sin B 6:106–113. https://doi.org/10.1016/j.apsb.2015.11.005

Lu Y, Qi J, Dong X et al (2017) The in vivo fate of nanocrystals. Drug Discov Today 22:744–750. https://doi.org/10.1016/j.drudis.2017.01.003

Luo C-H, Huang C-T, Su C-H et al (2016) Bacteria-mediated hypoxia-specific delivery of nanoparticles for tumors imaging and therapy. Nano Lett 16:3493–3499. https://doi.org/10.1021/acs.nanolett.6b00262

Martinez D, Vermeulen M, Trevani A et al (2006) Extracellular acidosis induces neutrophil activation by a mechanism dependent on activation of phosphatidylinositol 3-kinase/Akt and ERK pathways. J Immunol 176:1163–1171. https://doi.org/10.4049/jimmunol.176.2.1163

Matsumura Y, Maeda H (1986) A new concept for macromolecular therapeutics in cancer chemotherapy: mechanism of tumoritropic accumulation of proteins and the antitumor agent smancs. Cancer Res 46:6387–6392

Merisko-Liversidge E, Liversidge GG (2011) Nanosizing for oral and parenteral drug delivery: a perspective on formulating poorly-water soluble compounds using wet media milling technology. Adv Drug Deliv Rev 63:427–440. https://doi.org/10.1016/j.addr.2010.12.007

Merisko-Liversidge E, Liversidge GG, Cooper ER (2003) Nanosizing: a formulation approach for poorly-water-soluble compounds. Eur J Pharm Sci 18:113–120. https://doi.org/10.1016/S0928-0987(02)00251-8

Miao X, Yang W, Feng T et al (2018) Drug nanocrystals for cancer therapy. Wiley Interdiscip Rev Nanomed Nanobiotechnol 10:e1499. https://doi.org/10.1002/wnan.1499

Milosevic M, Fyles A, Hedley D et al (2004) The human tumor microenvironment: invasive (needle) measurement of oxygen and interstitial fluid pressure. Semin Radiat Oncol 14:249–258. https://doi.org/10.1016/j.semradonc.2004.04.006

Moellering RE, Black KC, Krishnamurty C et al (2008) Acid treatment of melanoma cells selects for invasive phenotypes. Clin Exp Metastasis 25:411–425. https://doi.org/10.1007/s10585-008-9145-7

Mohammad IS, Hu H, Yin L et al (2019) Drug nanocrystals: fabrication methods and promising therapeutic applications. Int J Pharm 562:187–202. https://doi.org/10.1016/j.ijpharm.2019.02.045

Monteiro A, Afolabi A, Bilgili E (2013) Continuous production of drug nanoparticle suspensions via wet stirred media milling: a fresh look at the Rehbinder effect. Drug Dev Ind Pharm 39:266–283. https://doi.org/10.3109/03639045.2012.676048

Morakul B, Suksiriworapong J, Leanpolchareanchai J et al (2013) Precipitation-lyophilization-homogenization (PLH) for preparation of clarithromycin nanocrystals: influencing factors on physicochemical properties and stability. Int J Pharm 457:187–196. https://doi.org/10.1016/j.ijpharm.2013.09.022

Möschwitzer JP (2013) Drug nanocrystals in the commercial pharmaceutical development process. Int J Pharm 453:142–156. https://doi.org/10.1016/j.ijpharm.2012.09.034

Möschwitzer J, Müller RH (2006) New method for the effective production of ultrafine drug nanocrystals. J Nanosci Nanotechnol 6:3145–3153. https://doi.org/10.1166/jnn.2006.480

Müller RH, Möschwitzer J (2008) Method and device for producing very fine particles and coating such particles. US Patent, 0297565 A1

Müller RH, Gohla S, Keck CM (2011) State of the art of nanocrystals – special features, production, nanotoxicology aspects and intracellular delivery. Eur J Pharm Biopharm 78:1–9. https://doi.org/10.1016/j.ejpb.2011.01.007

Nehoff H, Parayath NN, Domanovitch L et al (2014) Nanomedicine for drug targeting: strategies beyond the enhanced permeability and retention effect. Int J Nanomedicine 9:2539–2555. https://doi.org/10.2147/IJN.S47129

Patil JS, Sarasija S (2012) Pulmonary drug delivery strategies: a concise, systematic review. Lung India 29:44–49. https://doi.org/10.4103/0970-2113.92361

Prabhakar U, Maeda H, Jain RK et al (2013) Challenges and key considerations of the enhanced permeability and retention effect for nanomedicine drug delivery in oncology. Cancer Res 73:2412–2417. https://doi.org/10.1158/0008-5472.CAN-12-4561

Provenzano PP, Cuevas C, Chang AE et al (2012) Enzymatic targeting of the stroma ablates physical barriers to treatment of pancreatic ductal adenocarcinoma. Cancer Cell 21:418–429. https://doi.org/10.1016/j.ccr.2012.01.007

Ranjita S (2013) Nanosuspensions: a new approach for organ and cellular targeting in infectious diseases. J Pharm Investig 43:1–26. https://doi.org/10.1007/s40005-013-0051-x

Roma-Rodrigues C, Mendes R, Baptista PV et al (2019) Targeting tumor microenvironment for cancer therapy. Int J Mol Sci 20:840. https://doi.org/10.3390/ijms20040840

Romero GB, Chen R, Keck CM et al (2015) Industrial concentrates of dermal hesperidin smartCrystals® – production, characterization & long-term stability. Int J Pharm 482:54–60. https://doi.org/10.1016/j.ijpharm.2014.11.039

Romero-Garcia S, Lopez-Gonzalez JS, Baez-Viveros JL et al (2011) Tumor cell metabolism: an integral view. Cancer Biol Ther 12:939–948. https://doi.org/10.4161/cbt.12.11.18140

Rossi L, Serafini S, Pierigé F et al (2005) Erythrocyte-based drug delivery. Expert Opin Drug Deliv 2:311–322. https://doi.org/10.1517/17425247.2.2.311

Salazar J, Heinzerling O, Müller RH et al (2011) Process optimization of a novel production method for nanosuspensions using design of experiments (DoE). Int J Pharm 420:395–403. https://doi.org/10.1016/j.ijpharm.2011.09.003

Salazar J, Ghanem A, Muller RH et al (2012) Nanocrystals: comparison of the size reduction effectiveness of a novel combinative method with conventional top-down approaches. Eur J Pharm Biopharm 81:82–90. https://doi.org/10.1016/j.ejpb.2011.12.015

Serafini S, Rossi L, Antonelli A et al (2004) Drug delivery through phagocytosis of red blood cells. Transfus Med Hemother 31:92–101. https://doi.org/10.1159/000078042

Shah N, Shah V, Chivate N (2012) Pulmonary drug delivery: a promising approach. J Appl Pharm Sci 2:33–37. https://doi.org/10.7324/JAPS.2012.2632

Sharma OP, Patel V, Mehta T (2016) Nanocrystal for ocular drug delivery: hope or hype. Drug Deliv Transl Res 6:399–413. https://doi.org/10.1007/s13346-016-0292-0

Shegokar R, Müller RH (2010) Nanocrystals: industrially feasible multifunctional formulation technology for poorly soluble actives. Int J Pharm 399:129–139. https://doi.org/10.1016/j.ijpharm.2010.07.044

Sinha B, Muller RH, Moschwitzer JP (2013) Bottom-up approaches for preparing drug nanocrystals: formulations and factors affecting particle size. Int J Pharm 453:126–141. https://doi.org/10.1016/j.ijpharm.2013.01.019

Staedtke V, Brähler M, Müller A et al (2010) In vitro inhibition of fungal activity by macrophage-mediated sequestration and release of encapsulated amphotericin B nanosupension in red blood cells. Small 6:96–103. https://doi.org/10.1002/smll.200900919

Tolman JA, Williams RO (2010) Advances in the pulmonary delivery of poorly water-soluble drugs: influence of solubilization on pharmacokinetic properties. Drug Dev Ind Pharm 36:1–30. https://doi.org/10.3109/03639040903092319

Tozuka Y, Miyazaki Y, Takeuchi H (2010) A combinational supercritical CO2 system for nanoparticle preparation of indomethacin. Int J Pharm 386:243–248. https://doi.org/10.1016/j.ijpharm.2009.10.044

Trapani G, Denora N, Trapani A et al (2012) Recent advances in ligand targeted therapy. J Drug Target 20:1–22. https://doi.org/10.3109/1061186X.2011.611518

Van Eerdenbrugh B, Van den Mooter G, Augustijns P (2008) Top-down production of drug nanocrystals: nanosuspension stabilization, miniaturization and transformation into solid products. Int J Pharm 364:64–75. https://doi.org/10.1016/j.ijpharm.2008.07.023

Vaupel P, Mayer A (2007) Hypoxia in cancer: significance and impact on clinical outcome. Cancer Metastasis Rev 26:225–239. https://doi.org/10.1007/s10555-007-9055-1

Vidlářová L, Romero GB, Hanuš J et al (2016) Nanocrystals for dermal penetration enhancement – effect of concentration and underlying mechanisms using curcumin as model. Eur J Pharm Biopharm 104:216–225. https://doi.org/10.1016/j.ejpb.2016.05.004

Wei L, Ji Y, Gong W et al (2015) Preparation, physical characterization and pharmacokinetic study of paclitaxel nanocrystals. Drug Dev Ind Pharm 41:1343–1352. https://doi.org/10.3109/03639045.2014.950272

Wu F, Xue H, Li X et al (2020) Enhanced targeted delivery of adenine to hepatocellular carcinoma using glycyrrhetinic acid-functionalized nanoparticles in vivo and in vitro. Biomed Pharmacother 131:110682. https://doi.org/10.1016/j.biopha.2020.110682

Yang S, Gao H (2017) Nanoparticles for modulating tumor microenvironment to improve drug delivery and tumor therapy. Pharmacol Res 126:97–108. https://doi.org/10.1016/j.phrs.2017.05.004

Yang Y, Marshall-Breton C, Leser ME et al (2012) Fabrication of ultrafine edible emulsions: comparison of high-energy and low-energy homogenization methods. Food Hydrocoll 29:398–406. https://doi.org/10.1016/j.foodhyd.2012.04.009

Ye J, Wang Q, Zhou X et al (2008) Injectable actarit-loaded solid lipid nanoparticles as passive targeting therapeutic agents for rheumatoid arthritis. Int J Pharm 352:273–279. https://doi.org/10.1016/j.ijpharm.2007.10.014

Yue PF, Wang Y, Wan J et al (2012) The research progress of preparation methods of solid nanocrystal delivery system. Yao Xue Xue Bao 47:1120–1127

Zhang J, Huang Y, Liu D et al (2013) Preparation of apigenin nanocrystals using supercritical antisolvent process for dissolution and bioavailability enhancement. Eur J Pharm Sci 48:740–747. https://doi.org/10.1016/j.ejps.2012.12.026

Zhang C, Long L, Xiong Y et al (2019) Facile engineering of indomethacin-induced paclitaxel nanocrystal aggregates as carrier-free nanomedicine with improved synergetic antitumor activity. ACS Appl Mater Interfaces 11:9872–9883. https://doi.org/10.1021/acsami.8b22336

Zhu L, Wang D, Wei X et al (2013) Multifunctional pH-sensitive superparamagnetic iron-oxide nanocomposites for targeted drug delivery and MR imaging. J Control Release 169:228–238. https://doi.org/10.1016/j.jconrel.2013.02.015

Chapter 10
Application of Microneedles in Antitumor Therapy

Guilan Quan and Xin Pan

Abstract As an essential alternative for antitumor application, transdermal drug delivery system (TDDS) can directly deliver drugs into the cutaneous lesion site to exert topical or systemic therapeutic effect. Unfortunately, the skin's stratum corneum barrier has taken a toll on the efficacy of transdermal drug penetration, ultimately diminishing the overall curative effect.

To break down the limitation of stratum corneum, microneedles (MNs), which consist of hundreds of micron-scale needles attached to a supporting base, have emerged to remarkably improve the transdermal drug permeability and further bioavailability. Currently, the types of MNs mainly include solid MNs, coated MNs, hollow MNs, dissolving MNs, and hydrogel MNs. The use of MNs for antitumor therapy has the following advantages: (1) Improved drug penetration, especially biomacromolecules and nanoparticles, through the created microchannels after skin puncture. (2) Realizing spatiotemporally controlled drug release by encapsulating multiple chambers. (3) Noninvasive drug delivery with minimal pain. (4) Increased drug accumulation at superficial tumor site with minimal systemic side effects. (5) Strong controllability of the length, size, and shape of the needles for personalized treatment needs.

Keywords Microneedles · Transdermal drug delivery system · Antitumor therapy

10.1 Introduction

Skin is the largest organ in the body, which consists of three layers: epidermis, dermis, and hypodermis. The epidermis can be divided into nonviable epidermis and viable epidermis. The stratum corneum is the outmost layer of nonviable epidermis with a thickness of 10–15 μm, and acts as the major barrier of transdermal drug

G. Quan
College of Pharmacy, Jinan University, Guangzhou, China

X. Pan (✉)
School of Pharmaceutical Sciences, Sun Yat-sen University, Guangzhou, China
e-mail: panxin2@mail.sysu.edu.cn

L. Teng et al. (eds.), *Drug Delivery to Tumors*,
https://doi.org/10.1007/978-981-19-8930-8_10

delivery (TDD). The viable epidermis is an avascular tissue under the stratum corneum, containing a small quantity of immunocompetent cells and nerve tissue. The dermis and hypodermis are below the viable epidermis, and contain abundant nerve fibers, blood vessels, and lymphatic vessels that have a primordial role in TDD (Dragicevic and Maibach 2018; Sala et al. 2018). Skin with this type of structure has evolved to protect the body against invasion of toxic chemicals and pathogens as well as to prevent the excessive loss of water. In addition, it can also be a safe and efficient administration route.

Transdermal drug delivery system (TDDS) is an essential alternative to deliver drugs to exert topical or systemic therapeutic effect, which has been widely adopted for clinical application (MCP, Juárez et al. 2021). TDDS can deliver drugs directly into the lesion site at the epidermis or dermis to treat localized diseases. TDDS can avoid the first-pass liver effect and gastrointestinal enzyme destruction in systemic drug delivery when compared to oral administration, and it also has advantages of reduced pain, increased convenience for self-administration, and enhanced patient compliance compared to hypodermic injection (Prausnitz and Langer 2008). However, the skin stratum corneum barrier that significantly restricts drug transdermal delivery efficacy remains a remarkable concern (Seah and Teo 2018). How to break down the limitations of the stratum corneum, thereby increasing drug penetration and further improving drug bioavailability and therapeutic effect, is the key challenge in the development of TDDS.

Therefore, multitudinous chemical and physical technologies have been utilized to enhance the drug permeability through the stratum corneum (Marwah et al. 2016). The chemical penetration enhancers are commonly used to reduce the barrier resistance for increased drug delivery by changing the skin physiological characteristics or disorganizing the tightly packed lipid layer arrangement. But the long-term utilization of chemical enhancers may cause adverse reactions such as skin irritation and allergies due to its nonspecificity (Lane 2013). Physical penetration enhancement approaches, such as iontophoresis, electroporation, jet injection, ultrasound, microwave, thermal ablation, and laser ablation, are also widely used to alter the physical state of the skin stratum or increase the molecular energy to improve drug penetration (Zhang et al. 2019). The traditional physical process usually relies on professional instruments, which impedes its practical application. Even though these chemical and physical methods could efficiently facilitate the transdermal penetration of small molecule drugs, limitations remain on the overall successfulness of drug delivery, especially for high molecular weight drugs.

Microneedles (MNs), as a novel technique to increase the permeability of drugs, have been widely used in recent years. The concept of MNs was first reported in 1971 by Martin S. Gerstel and Virgil A. Place in a United States patent (Gerstel 1976), and until 1998 the first proof-of-concept study using MNs for drug delivery has been reported by Henry et al. (1998). MNs are micro-sized percutaneous drug delivery devices, consisting of hundreds of micron-scale needles attached as an array to a supporting base. MNs integrate the advantages of a hypodermic syringe and a transdermal patch. They can be used to penetrate into the stratum corneum noninvasively and painlessly to create reversible microchannels, increasing drug permeability and improving safety and comfort (Waghule et al. 2019). Additionally,

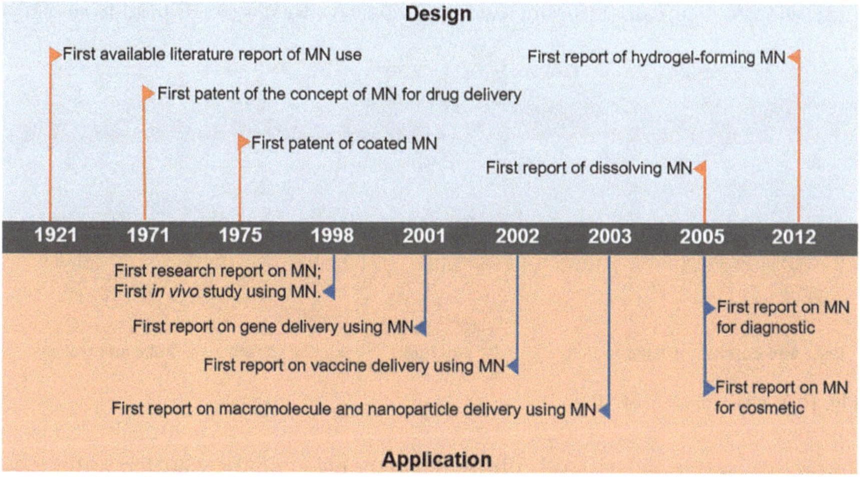

Fig. 10.1 Chronological timeline of MNs development

MNs can deliver drugs ranging in sizes from small molecules to macromolecules and even nanoparticles (McAllister et al. 2003; He et al. 2019). The length, size, and shape of the needles can also be customized for the specific treatment needs (Luzuriaga et al. 2018). Since the first MNs were proposed, the MNs-based drug delivery system has been developed over several decades. The chronological timeline of MNs development is shown in Fig. 10.1. Currently, various properties of MNs have been explored, such as different drug delivery mechanisms (solid MNs, coated MNs, hollow MNs, dissolving MNs, and hydrogel MNs), materials (silicon, metal, ceramic, glass, and polymer), and geometries (octagonal, cylindrical, rectangular, pyramidal, conical, and quadrangular) (Waghule et al. 2019; Li et al. 2021; Alimardani et al. 2020; Sharma et al. 2019). Taken together, this will reshape the landscape of TDDS.

10.2 Classification and Characteristics of MNs

Based on the needle structure, composition, and fabrication, MNs can be classified into five types: solid MNs, coated MNs, hollow MNs, dissolving MNs, and hydrogel MNs, respectively (Fig. 10.2).

10.2.1 Solid MNs

Solid MNs are mostly utilized for pretreating the skin to enhance drug penetration in a "Poke and Patch" manner (Sharma et al. 2019). Solid MNs are pierced into the skin and subsequently removed to form microchannels, then drugs are spread out on

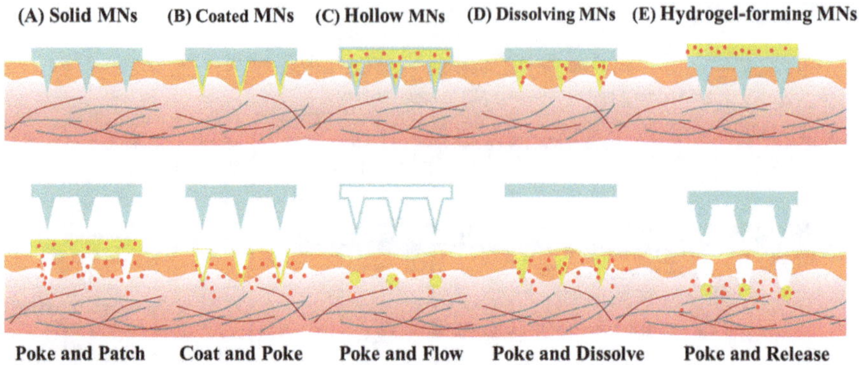

Fig. 10.2 Classification of MNs

the pretreated skin and passively diffused into the epidermis or even dermis for topical or systemic therapy (Prausnitz 2017). In this two-step approach, the pre-created micro-sized pores in the skin are under dynamic recovery, which might affect the quantity of diffused drug resulting in an inaccurate dose being delivered. In addition, the mechanical strength of solid MNs is hard enough to pierce into the stratum corneum, which can be affected by the needle length (Li et al. 2017b). The needles will be weaker and easier to break when the length is extended, which might increase the risks of allergy and infection since metal materials are usually used for solid MNs preparation.

10.2.2 Coated MNs

Coated MNs are used for transdermal administration based on the "Coat and Poke" approach, in which drugs are applied as a coating layer on the surface of needles. After insertion into the skin, the drug is dissolved and then diffused into the deep tissue. This one-step approach simplifies the administration procedure and provides more precise drug dosing compared to solid MNs (Alimardani et al. 2020). The drug loading amount can be adjusted by the thickness of the coating layer (Li et al. 2017a). Moreover, changing the needle size can also influence the loading capacity. However, the total quantity of drug is still relatively low due to the limited superficial area of MNs, which results in decreased drug delivery efficiency and narrow clinical application.

10.2.3 Hollow MNs

Hollow MNs are identified as micro-syringes used for the delivery of liquids (Maaden et al. 2018). These MNs are hollow inside and have holes at the needle tips, operating through the "Poke and Flow" approach. The drug solution can continually flow out from the holes and be directly deposited into the epidermis or the upper dermis after MN is inserted into the skin. These MNs are capable of administrating a large amount of drug compared to solid MNs, and mostly used for the delivery of high molecular weight compounds such as proteins, vaccines, and oligonucleotides (Ita 2015). The drug flow rate can be adjusted by the driving force by using a pump, pressurized gas, or a microfluidic chip (Miller et al. 2016). However, hollow MNs suffer from some limitations due to the empty structure, such as possible breakage and blockage during the skin insertion and infusion (Alimardani et al. 2020). Furthermore, the high cost and complex fabrication process also limit its application.

10.2.4 Dissolving MNs

Dissolving MNs are manufactured using water-soluble biodegradable polymers with drugs dispersed in the only tip or whole body of needles, and even in the base of MN patches (Du et al. 2019; Qin et al. 2020). The MNs are inserted into the skin where needles are dissolved by the interstitial fluid and the preloaded drugs are released. In this "Poke and Dissolve" approach, the MN patches do not need to be removed until the needles have dissolved, providing convenient, noninvasive, and painless drug self-administration (Arya et al. 2017). The dissolving MNs have unique properties that are superior to other kinds of MNs due to these characteristics: (1) The whole structure of the MN matrix can be used to encapsulate drugs, which will overcome the problem of low drug loading comparing to coated MNs. (2) The drug release rate can be accurately controlled by selecting different polymers with different water solubility. (3) The biocompatible and biodegradable polymers can reduce the production cost and improve safety, while avoiding needle waste. (4) The solid-state storage condition contributes to maintaining the drug activity and stability, especially for the biomacromolecule drugs (Ling and Chen 2013; Lee et al. 2008).

10.2.5 Hydrogel-Forming MNs

Hydrogel-forming MNs were first reported in 2012 and are produced by chemically or physically cross-linked polymers capable of super-swelling without dissolving (Turner et al. 2020). These MNs operate via the "Poke and Release" approach. The polymers have a hydrophilic structure that has the ability to absorb a greater amount of water into their three-dimensional polymeric network (Courtenay et al. 2020). When pierced into the skin, hydrogel polymers absorb the interstitial fluid and then swell, leading to the diffusion of preloaded drugs. These MNs are flexible in size and shape, and can provide a controlled release profile by adjusting the cross-link density (Waghule et al. 2019; Lutton et al. 2015). They can also be removed without any residual polymer residues remaining in the skin, which minimizes the risk of irritation at the site. Despite such attractive features, they have not yet shown great potential in cancer treatment.

10.3 Preparation Methods of MNs

As a result of the advancement in material science and of modern preparation equipment, various technologies have been developed to construct MNs based on their material, structure, geometry, and size, which facilitate large-scale production of MNs. Numerous techniques have been used for different types of MNs (Table 10.1).

10.3.1 Solid MNs

Solid MNs are mainly fabricated by silicon or metal to ensure good mechanical strength to pierce the skin. The silicon solid MNs are produced using microelectronics-based construction techniques such as wet etching, dry etching, reactive ion etching, and deep reactive ion etching (He et al. 2019; Deng and Juang 2014; Gittard et al. 2010). Another microelectronics-based technique that has been used for preparing solid MNs involves photolithography in combination with patterned photoresist masks. In this technique, a photoresist is patterned onto the surface and unprotected material is subsequently removed using X-ray exposure. This technique allows MNs to be created using a wide variety of materials such as polymers and ceramics. The geometries of MNs are limited to radially symmetrical structures, excluding complex shapes, as these etching processes are line-of-sight methods. The metallic solid MNs are commonly prepared through infrared laser ablation. In this technique, an infrared laser is used to ablate a two-dimensional pattern of the needle shape on a stainless steel sheet, and then the sheet is bent to create a three-dimensional structure. Electroplating and micromolding have also been utilized for fabricating metallic solid MNs. The commercial viability of solid MNs prepared by the aforementioned fabrication techniques is still limited by numerous factors, such as expensive equipment, costly cleanroom facilities, and professional hazardous waste services.

10.3.2 Coated MNs

The fabrication of coated MNs includes two steps; first, the solid needle part is prepared based on the solid MNs, and second, a drug-containing coating layer is loaded on the surface of the needles. Multiple coating approaches have been developed, including dip-coating, gas-jet drying, spray coating, ink-jet printing, and

Table 10.1 Properties, materials, and preparation methods of different MNs

	Solid MNs	Coated MNs	Hollow MNs	Dissolving MNs	Hydrogel MNs
Drug delivery approaches	Poke and Patch	Coat and Poke	Poke and Flow	Poke and Dissolve	Poke and Release
Drug loading	/	37 ng–3.8 mg (Ingrole and Gill 2019)	/	0.06 μg–1.00 mg (Leone et al. 2017)	0.8–50 mg
Advantages	High mechanical strength	Single-step process, precise drug dosage	Large dose administration, continuous drug delivery	Single-step process, accurate drug dosage, controlled drug delivery, easy fabrication process, low cost, environmental friendly	Single-step process, large drug loading, controlled drug delivery, intact removal of MN patches
Drawbacks	Two-step process, limited dose administration, impeded drug delivery caused by reversible microchannels, metal allergy, risk of needle breakage, needle waste	Limited drug loading, difficult control of the coating thickness, metal allergy, risk of needle breakage	Risk of needle blockage and breakage, less mechanical strength, complex fabrication, costly, needle waste	Possibility of polymer deposition, poor mechanical strength	Possibility of polymer deposition, poor mechanical strength, complex preparation of cross-linked polymers
Materials	Silicon, titanium, nickel, stainless steel, ceramic, bioceramic, PMMA, PEGDA	Silicon, titanium, stainless steel, PLLA, PLA, liquid crystal polymer	Silicon, titanium, nickel, ceramic, glass, polycarbonate	PVA, PVP, PGA, CMC, HA, chitosan, gelatin, dextran, and sugars	Cross-linked polymers, PEGDA, MeHA, PMVE/MA-PEG, and pHEMA
Technique	Wet etching (Wilke et al. 2005b), dry etching (Izumi and Aoyagi 2007), reactive ion etching, deep reactive ion etching (Held et al. 2007), photolithography (Rodriguez et al. 2005), laser cutting (Mina et al. 2016), electroplating (Wilke et al. 2005a), micromolding (Cai et al. 2015), hot embossing (Cahill et al. 2019)	Coating approaches: dip-coating (Caudill et al. 2018), gas-jet drying (Chen et al. 2009), spray coating (Mcgrath et al. 2011), ink-jet printing (Uddin et al. 2015), electro-hydrodynamic atomization-based process (Mehta et al. 2017)	Wet etching (Vinayakumar et al. 2014), dry etching (Vinayakumar et al. 2014), deep reactive ion etching (Li et al. 2019b), sputter deposition, electrodeposition (Norman et al. 2013), micromolding (Rui et al. 2020), laser writing (Juríek et al. 2013)	Hot embossing (Jin et al. 2009), solvent casting, injection (Ono et al. 2017), 3-dimensional printing (Boehm et al. 2013), heat imprint lithography (Tomomo 2018)	Hot embossing (Lutton et al. 2015), solvent casting (McAlister et al. 2020)

electro-hydrodynamic atomization-based process, to overcome the drawback of drug wastage and loss, variable coating thickness, and inaccurate drug dosage (Haj-Ahmad et al. 2015). The dip-coating process is a simple method for fabricating the coating layer. Briefly, the MNs are dipped into the drug solution and then removed to form a liquid film on the MNs. Thereafter, the liquid layer is dried to finally form a solid film coating (Kapoor et al. 2019). The drying process of the dip-coating technique is the rate-limiting step, while gas-jet drying technique efficiently speeds up the drying process by means of continuously wetting the MNs with a small amount of drug solution. A uniform drug coating layer is the key point during the coated MN preparation, which is crucial to ensure the accurate dosage. Some surfactants or solution viscosity enhancers have been added into the drug solution to make a homogenous drug coating layer.

10.3.3 Hollow MNs

The production process of hollow MNs is more complex that requires precise control to form hollow structure. Several solid MN fabrication techniques are also appropriate for hollow MN fabrication, such as wet etching, dry etching, reactive ion etching, deep reactive ion etching, X-ray photolithography, laser casting, and electrodeposition (Li et al. 2019b). The materials used in the production include silicon, metal, ceramic, glass, stainless steel, and some polymers. The design of the needle structure can be exploited to deal with drawbacks associated with hollow MNs such as needle breakage and blockage. For example, beveled tip hollow MNs were designed to improve their mechanical strength and diffusion capacity. Additionally, side bore-opening hollow MNs are designed to overcome clogging during tissue insertion.

10.3.4 Dissolving MNs

Dissolving MNs are mainly generated using biocompatible and biodegradable polymeric materials with good water solubility, such as polyvinyl alcohol (PVA), polyvinyl pyrrolidone (PVP), polyglutamic acid (PGA), carboxymethyl cellulose (CMC), hyaluronic acid (HA), chitosan, gelatin, dextran, and sugars. A wide variety of techniques, such as droplet-born air blowing, drawing lithography, continuous liquid interface production, and micromolding, have been employed in the fabrication of dissolving MNs. The micromolding technique is the most common method for the dissolving MN production, which includes methods such as hot embossing and solvent casting. Hot embossing is a thermal micromolding process requiring a force gauge and heat controller (Andersen et al. 2018). In brief, high temperature is used to form a liquid polymer flow, after which the polymer is squeezed into a mold under a designed pressure for a few minutes. The pressure is then released and the system is cooled down to room temperature. Finally, the dissolving polymeric MNs are peeled off the mold. Hot embossing allows for fast replication of MNs, but the high processing temperature reduces their use since a considerable range of drugs have poor thermal stability. It also prevents some active pharmaceutical ingredients to be loaded into dissolving MNs. Solvent casting is a mild micromolding process that uses an aqueous drug formulation. The drug solution is applied into the mold, after which centrifugation and vacuum drying follows to remove the solvent. Finally, a dry MN patch is formed and taken down from the mold. Since the solvent casting process is done under the room temperature, this technique is more suitable for preparing biomolecule-loaded MNs such as protein, peptide, and vaccine.

10.3.5 Hydrogel MNs

Since hydrogel MNs and dissolving MNs are both polymer-based, their fabrication processes are similar, but differ in the polymeric materials used. The special cross-linked hydrophilic polymers, such as polyethylene glycol diacrylate (PEGDA), methacrylated hyaluronic acid (MeHA), and poly 2-hydroxyethyl methacrylate (pHEMA), have been applied to produce hydrogel MNs (Turner et al. 2020). The polymer cross-linking process is an important step that ensures the swellable property. Various chemical or physical techniques have been used, including esterification reaction or covalent linkage, heating, UV exposure, and lyophilization (Maitra and Shukla 2014).

10.4 Characteristics of MNs

10.4.1 MNs for Drug Loading

As an emerging TDD system, MNs are equipped with several unique characteristics: adjustable stability, noninvasiveness, and convenience of self-administration. MNs have been widely explored to enhance the transdermal permeability of small molecules, biomacromolecules, and even nanoparticles. In particular, the drug loading dosage of dissolving MNs and hydrogel MNs is summarized in Table 10.2.

10.4.1.1 Small Molecular Drugs

Drugs with a molecular weight lower than 500 Da are able to passively diffuse across the skin. However, for those that require high doses, it is still difficult to achieve effective drug concentration above the therapeutic window just through passive diffusion. The formation of microchannels in the skin created by MNs is

Table 10.2 Drug loading dosage of dissolving MNs and hydrogel MNs

	Material and structure	Active drug	Total loaded amount	Ref
Dissolving MNs	PVA/two-layered conical structure with a needle height of around 900 μm and a base diameter of 300 μm	Curcumin	10.9 ± 1.1 μg	Abdelghany et al. (2019)
	Chondroitin sulfate (CS) + PVP/pyramid with 380 μm in base length and 680 μm in height	Interferon-α-2b	4.94 ± 0.64–23.79 ± 2.48 μg	Chen et al. (2016a)
	HA/MN patch consisted of 100 (10 × 10) needles on a baseplate with a height of 650 ± 19 μm, base width of 220 ± 7 μm	Methotrexate	6.5 ± 1.3, 13.8 ± 1.8, 30.6 ± 2.3, and 65.3 ± 2.9 μg	Du et al. (2019)
	HA + PVP/MN patch with 19 × 19 needles and 500 μm in height	Gentamicin	4.8 mg	González-Vázquez et al. (2017)
	Dextran, HA/100 microneedle arrays in a 10 × 10 matrix with 500 μm length	Sumatriptan	31.6 ± 4.5, 24.1 ± 0.9 μg	Ito et al. (2011a)
	CS/100 needles with length of 469.8 ± 4.7 μm and basement diameter of 284.5 ± 9.8 μm	Leuprolide acetate	14.3 ± 1.6 μg	Ito et al. (2011b)
	CS + Trehalose/MN patch containing 196 needles	Staphylococcal enterotoxin B	13 ± 1.4 μg	Liu et al. (2019)
	HA/220 needles with 800 μm in height	Human parathyroid hormone (1–34)	20.5 ± 3.0 μg	Naito et al. (2018)
	PVP + PVA/256 needles of a square-based pyramid with side-length of 300 μm at the baseplate and needle-needle interspacing of 100 μm	Methotrexate	2.48 mg	Tekko et al. (2020b)
	HA/25 (5 × 5) pyramidal needles with the height, tip diameter, width of base, and needle pitch of 907 ± 20 μm, 12 ± 5 μm, 309 ± 16 μm, and 1123 ± 2 μm	5-Aminolevulinic acid	122 μg	Zhao et al. (2018)

Hydrogel MNs	CA and PVP matrix/0.5 cm^2 patch consisted of 11 × 11 needle arrays with 729.5 ± 50.2 μm in height and 300 μm in width at the baseplate	Methotrexate	18.2 ± 1.5, 20.2 ± 2.0 mg	Tekko et al. (2020a)
	Gantrez® S-97/MN arrays (19 × 19 over 0.5 cm^2) with 550 μm in height	OVA	2.5 mg	Courtenay et al. (2019)
	Gantrez® S-97/11 × 11 needle array density, 600 μm height, 300 μm width at the base, and 150 μm interspacing	Metformin hydrochloride	75 mg	Migdadi et al. (2018)
	Gantrez® AN-139 and S-97/MN arrays (11 × 11 needle density, 600 μm height, 300 μm width at base, and 150 μm interspacing)	Donepezil hydrochloride	854.71 ± 122.71 μg	Kearney et al. (2016)
	2-Hydroxyethyl methacrylate and ethylene glycol dimethacrylate/121 (11 × 11) needles of conical shape and 600 μm in depth, with base width of 300 μm and interspacing of 300 μm	Ibuprofen	50 mg	Hardy et al. (2016)

conducive to improve the permeability and therapeutic effect of small molecular drugs. The employment of a MN roller was reported as a pretreatment of the application area of the skin of mice with B16-F10 melanoma tumor model, after which topical 5-fluorouracil (5-FU) cream was applied (Naguib et al. 2014). As revealed by an in vitro permeation assay, the flux of 5-FU across the mouse skin was enhanced by up to 4.5 times for the pretreated group, and the tumor growth was almost completely suppressed.

Depending on the type of MN, the loading amount of small molecular drugs is mainly in the range of 50–600 μg per patch. For coated MNs, the drug loading capacity is easily affected by the shape and number of MNs in the array, but for dissolving or hydrogel MNs, it is mainly related to the properties of the polymer materials used. Hydrophilic drugs can be directly dissolved in their preparation solution to achieve effective drug loading for dissolving MNs. Zhao and colleagues (Zhao et al. 2018) constructed tip-concentrated dissolving MNs based on HA by using the two-casting method, in which the loading amount of the drug 5-aminolevulinic acid (ALA) reached up to 122 μg per MN patch. Attractively, even with a lower dose, the antitumor photodynamic therapy (PDT) efficacy of ALA-encapsulated MNs was superior to that of the ALA injection group. In another study (Bhatnagar et al. 2019), dissolving MNs co-loaded with 533.5 μg of the drugs doxorubicin (DOX) and 227.0 μg of docetaxel (DTX) were prepared by directly dissolving the drug molecules in PVPK360 matrix solution.

Hydrophobic drugs have been widely used in the treatment of various diseases, but most preparation materials of dissolving MNs are water-soluble polymers, resulting in limited solubility of hydrophobic drugs in the needle body solution followed by insufficient drug loading. Intermolecular interactions between drugs and polymer materials, such as hydrophobic interaction, electrostatic adsorption, and host-guest interaction, have been investigated to increase the solubility and insoluble drug loading (Dangol et al. 2016). The introduction of hydroxypropyl-β-cyclodextrin (HP-β-CD) to form an inclusion compound with triamcinolone resulted in a significantly improved drug payload of 92% (increased by nearly five times, Fig. 10.3a) (Lin et al. 2019). Yang et al. (2020) innovatively designed core-shell MNs composed of a charged shell and a hydrophilic core. Polymers rich in hydrogen bond donors served as the core materials to form hydrogen bonds with the indoleamine 2, 3-dioxygenase (IDO) inhibitor 1-methyl-D,L-tryptophan (1-MT) to produce a supersaturated tip solution, thereby obtaining an improved 1-MT payload (Fig. 10.3b).

10.4.1.2 Macromolecular Drugs

For decades, protein and peptide drugs have offered ample potential for biomedicine and diagnosis. These macromolecular drugs are often used clinically as subcutaneous injections and require cold-chain transportation and storage, which come with several drawbacks, including poor stability, easy inactivation, short half-life, and frequent administration. The rise of the MN technology provides a novel avenue

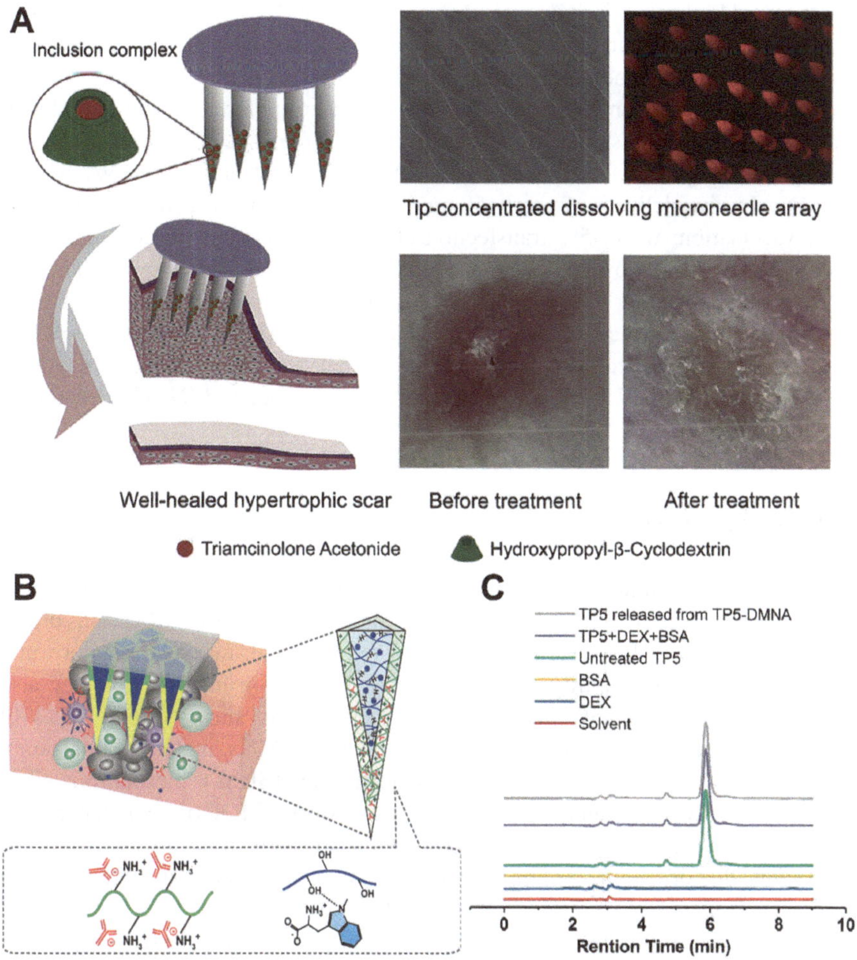

Fig. 10.3 (**a**) Schematic illustration of microneedle-assisted delivery of triamcinolone employing HP-β-CD for treating a hypertrophic scar. (Reproduced with permission. Copyright 2019, Elsevier). (**b**) The co-delivery of anti-programmed cell death receptor ligand 1 (aPD-L1) antibody and 1-MT in core-shell microneedles for melanoma immunotherapy. (Reproduced with permission. Copyright 2020, Elsevier). (**c**) Chromatograms of different samples evaluating the stability of the encapsulated thymopentin peptide. (Reproduced with permission. Copyright 2017, Elsevier)

for efficient and safe delivery of macromolecular drugs, by breaking the stratum corneum barrier and increasing their skin penetration amount.

The bullet-shaped hydrogel MNs were designed by Yang and colleagues (Seong et al. 2017) for insulin transdermal delivery. The tips absorbed the body fluid and swelled reversibly after insertion into the skin, resulting in a constant and sustained protein release. Lin et al. (2017) adopted a modified two-step molding technology to construct thymopentin-loaded dissolving MNs, which maintained the structural integrity and biological activity after drug loading (Fig. 10.3c). The in vivo pharmacodynamics experiment indicated that thymopentin-containing MNs displayed a comparable immunomodulatory effect with intravenous thymopentin injection.

Apart from protein and peptide, MNs have also been explored as delivery vehicles for nucleic acids. For instance, Cole and coworkers (Cole et al. 2018) took advantage of lyophilization to enhance the plasmid DNA (pDNA) payload within PVA dissolving MNs. In vivo delivery of nearly 50 μg pDNA per patch was achieved for robust DNA vaccination, with 95% transfection efficacy. A nanoparticle-embedded MN system was employed by Wang and colleagues (Wang et al. 2020) to transdermally deliver small interfering RNAs (siRNAs), composed of a hydrophilic HA matrix and mesoporous silica-coated upconversion nanoparticles (UCNPs@mSiO_2). siRNAs were successfully loaded into the porous mSiO_2 shell and protected from enzymatic degradation, while the UCNP core enabled the visible trace of MN skin penetration and drug diffusion.

10.4.1.3 Nanoparticles

As a result of the rapid development of nanotechnology, multifunctional nanoparticles have been widely employed to improve the therapeutic effect of various drugs. Not only can they tackle several limitations faced by raw drugs, such as poor water solubility, low bioavailability, and nonspecific distribution in vivo, but they also help overcome biological obstacles and target specific sites. Several types of nanoparticles have been successfully encapsulated into MNs, including liposomes, micelles, gold nanoparticles, solid lipid nanoparticles (SLNs), mesoporous silica nanoparticles (MSNs), and metal-organic framework (MOFs), to name a few. Combining the strengths of both, nanoparticle-encapsulated MNs are able to implement controlled drug delivery and multi-mode synergistic treatment strategies. Four different nanoparticles were selected by Du et al. (2017) to investigate the potential of hollow MNs for intradermal vaccination. Ovalbumin (OVA, model antigen) and immunoadjuvant poly(I:C) were co-encapsulated into nanoparticles, which led to stronger IgG2a response as compared with OVA/poly(I:C) solution because the nanoparticles could be better taken up by dendritic cells and then transferred to lymph nodes more quickly. Furthermore, OVA/poly(I:C)-loaded liposomes were found to induce the highest $CD8^+$ and $CD4^+$ T cell responses, suggesting its superiority in Th1/CTL immune response.

Thermal-sensitive SLNs containing paclitaxel (PTX) and photothermal agent IR-780 were concentrated in the tips of dissolving MNs to achieve dual-modal chemo-photothermal therapy against melanoma (Qin et al. 2020). In another study from the same group, PTX and indocyanine green (ICG), co-loaded in functionalized poly(lactic-co-glycolic acid) (PLGA) nanoparticles, were also introduced into dissolving MNs for local tumor treatment (Peng et al. 2020). Pei and coworkers (Pei et al. 2018) constructed PVP MN patches co-loaded with DOX and ICG, in which MSNs were employed to improve the stability of ICG and prevent its aggregation-caused quenching.

Lan et al. (2018) took advantage of dissolving MNs for the transdermal delivery of lipid-coated cisplatin nanoparticles (LCC-NPs). With the help of MN patches, LCC-NPs were able to effectively pass through the stratum corneum barrier and then be released from the MNs. Moreover, dissolving MNs were employed to transdermally deliver core-shell dual MOFs for effective synergistic antitumor therapy, which were composed of an ICG-loaded MIL-88 core and a chemotherapeutic DOX-containing ZIF-8 shell (Wu et al. 2020).

10.4.2 Mechanical Strength

The sufficient mechanical strength of MNs plays a decisive role in breaking through the stratum corneum barrier to effectively deliver therapeutic drugs, which is mainly determined by the inherent mechanical characteristics of needle materials. Silicon and metal MNs are often equipped with an adequate mechanical strength for skin penetration, while polymeric MNs generally display a weak mechanical strength. Researchers have adopted various strategies to construct reinforced MNs, for example, by constructing composite materials and by introducing nanomaterials.

Several studies have successfully fabricated reinforced MNs with sufficient mechanical strength by introducing small-molecule carbohydrates, medium or high molecular weight polymers, etc., to construct composites with needle tip materials. Sucrose, trehalose, and maltose were confirmed to enhance the elastic modulus of CMC MNs, with a stronger mechanical property of about 1.5-fold when compared to CMC alone (Raphael et al. 2016; Loizidou et al. 2015). Lin and coworkers (Lin et al. 2019) revealed that carbohydrates and polyols with abundant hydroxyl groups could be connected by hydrogen bonds, which led to a relatively high-ordered molecular lattice arrangement, thereby increasing the mechanical strength of MNs. The authors selected a HA polymer with good toughness for the fabrication of dissolving MNs, while medium molecular HP-β-CD as a reinforcer was employed to form a composite with HA. The results showed that the mechanical strength of MNs was gradually enhanced with the increased amount of HP-β-CD because of the intermolecular hydrogen bonds that were formed between HA and HP-β-CD (Fig. 10.4a). In another study (Vora et al. 2017), a PVA polymer served to adjust the brittleness of PVP, successfully constructing bilayer MN arrays with good mechanical properties. The incorporation of bovine serum albumin (BSA), a kind of biomacromolecule, was also able to obtain MNs with a higher Young's modulus (Lee et al. 2008), which was consistent with the results reported by another study (Fig. 10.4b) (Lin et al. 2017).

In addition to the abovementioned carbohydrates and polymers, nanoparticles have also been investigated to strengthen the mechanical characteristics of dissolving MNs. For instance, by incorporating 5% (w/w) of layered double hydroxide (LDH) nanoparticles into CMC, nanocomposite-reinforced MN arrays were successfully manufactured as reported by Yan et al. (2014). The obtained CMC–LDH MNs were confirmed to reliably pierce into pig and human skin and then rapidly

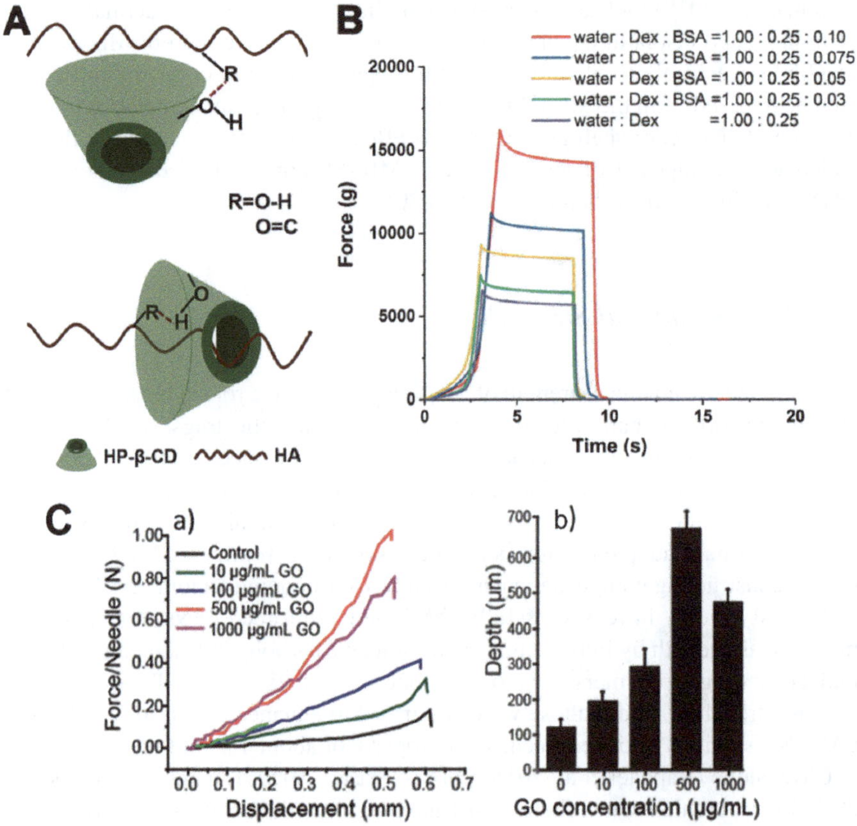

Fig. 10.4 (**a**) Schematic illustration of interactions between HA and HP-β-CD. (Reproduced with permission. Copyright 2019, Elsevier). (**b**) Mechanical strength of the drug-free dissolving microneedles. (Reproduced with permission. Copyright 2017, Elsevier). (**c**) (a) Mechanical behaviors of PVP MNs and (b) Skin insertion depth of PVP MNs containing different concentrations of GO ($n = 3$). (Reproduced with permission. Copyright 2019, ACS Publications)

release the encapsulated cargo. Chen and colleagues (Chen et al. 2020c) incorporated graphene oxide (GO) into biocompatible polymers for MN construction. As shown in Fig. 10.4c, GO-doped MNs displayed an increased mechanical strength (10–17-fold at 500 μg/mL GO) and an improved skin insertion depth (6 times). MSNs were also demonstrated to help fabricate MN patches with more robust toughness and mechanical strength (Xu et al. 2017). The failure force was strengthened to 0.28 N per needle by MSNs, while only 0.12 N per needle for pure PVP MNs was obtained. Similarly, gold nanocage (AuNC) was explored to act as a mechanical reinforcer of MNs (Dong et al. 2018). AuNC-containing MNs possessed a distinctively improved Young's moduli of 224.9 MPa, compared to 68.9 MPa for raw HA ones, which was due to the reinforcement effect of the Au nanocage on the polymeric materials.

10.4.3 *Drug Release and Distribution*

10.4.3.1 Rapid Drug Release

Based on excellent hydrophilicity of the applied polymers, MNs for rapid drug release can promptly dissolve once the needle body enters the skin so as to realize drug release. They are suitable for drugs that need to reach the therapeutic concentration in a short time, or that exert a local effect. The common materials for constructing these quick-dissolving MNs include natural carbohydrate polymers and synthetic soluble polymers. With good biocompatibility, natural carbohydrate materials mainly include maltose, dextran, and HA, to name a few. Therein, HA is one of the most widely used polymers for rapid-release MNs due to its good mechanical strength and the lack of heating requirement during the molding process. Xu et al. (2020) selected HA (about 10 kDa) as the matrix to construct rapidly dissolving MNs for the co-encapsulation of p53 DNA and IR820 (Fig. 10.5a). The obtained HA MN patch could effectively penetrate the stratum corneum and rapidly dissolve in the interstitial fluid within only 5 minutes. HA-based dissolving MNs have also been applied for the fast release of anti-inflammatory diclofenac sodium (Hou et al. 2019) and different kinds of nanoparticles (Peng et al. 2020; Wu et al. 2020).

In comparison, artificially synthesized soluble polymers possess the advantage of adjustable structure and performance. Chen and coworkers (Chen et al. 2020a) fabricated core-shell dissolving MNs for co-delivery of ICG-loaded chitosan nanoparticles as the tip shell and 1-MT, which was loaded into the cross-linked PVP K30 and PVA composite gel as the MN core. Upon insertion into the skin of the tumor site, the core-shell MNs would readily dissolve and thus enable the loaded drugs to release and exert a robust synergistic antitumor immunotherapy.

10.4.3.2 Sustained Drug Release

Numerous therapeutic drugs require sustained release over a long time to realize satisfactory therapeutic effects. Accordingly, biodegradable or swellable polymers have been adopted to fabricate transdermal MNs for the sustained release of the drugs, which leads to maintenance of the blood drug concentration to a certain extent and reduction of side effects. The drug release profile can be modulated through appropriate selection of polymers with varied molecular weights and degradability or expandability.

For biodegradable MNs, the drug release profile mainly depends on the hydrolysis and degradation of the polymeric matrix and the payload diffusion rate (Ye et al. 2018). Synthetic polymers (e.g., polylactic acid (PLA), PGA, PLGA) and natural polymers (e.g., chitosan, silk fibroin) are commonly used biodegradable materials acting as MN matrices. For instance, DeMuth et al. (2013) incorporated antigen and adjuvant into PLGA microparticles, followed by encapsulation into a highly hydrophilic poly(acrylic acid) (PAA) MN matrix. Upon administration, the outer PAA

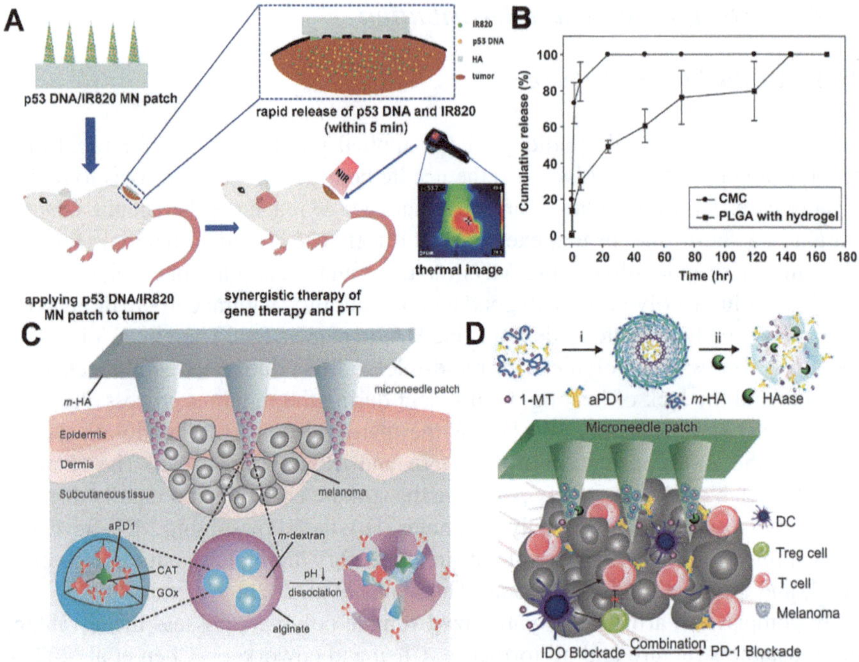

Fig. 10.5 (**a**) Illustration of p53 DNA/IR820 MN patch and the synergistic combination of gene therapy and PTT for tumor treatment. (Reproduced with permission. Copyright 2020, the Royal Society of Chemistry). (**b**) Release profile of PLGA microneedles containing hydrogel and rhodamine and CMC microneedles with rhodamine as a function of time. (Reproduced with permission. Copyright 2011, Elsevier). (**c**) Schematic drawing of the anti-PD-1 delivered by an MN patch loaded with physiologically self-dissociated NPs triggered by pH. (Reproduced with permission. Copyright 2016, ACS Publications). (**d**) Schematic illustration of encapsulation and release of IDO inhibitor 1-MT and anti-PD-1 from self-assembled 1-MT-conjugated HA NPs, which would be dissociated by HAase. (Reproduced with permission. Copyright 2016, ACS Publications)

layer would rapidly dissolve so that the inner PLGA microspheres could be released where they remained as drug depots inside the skin for weeks. Likewise, transdermal MN arrays were employed to deliver chicken OVA-containing PLGA nanoparticles for evoking robust immune responses, which was reported by Zaric and coworkers (Zaric et al. 2013).

Hydrogel MNs composed of cross-linked polymers can also realize sustained drug release, in which the loaded drugs are often gradually released through the microchannels created by swellable hydrogel matrices. Kim and colleagues (Kim et al. 2012) innovatively constructed swelling MNs by micromolding PLGA with drug-encapsulated hydrogel microparticles. It was demonstrated that after MN application, hydrogel expansion could well mediate the needle separation from skin for topical drug delivery over several days to months (Fig. 10.5b).

10.4.3.3 Stimulus-Responsive Drug Release

Stimulus-responsive MNs can be triggered to precisely release drugs as required according to changes in the physiological environment or external signals, such as variation of pH, glucose concentration, and enzyme concentration, so as to improve therapeutic efficiency while reducing side effects.

The skin environment is weakly acidic, with a pH range of about 4.0–7.0 in the epithelium, which enables TDD using pH-sensitive MNs. PVP MNs containing pH-sensitive PLGA microspheres were fabricated by Sung and colleagues (Ke et al. 2012), in which the fluorescent dye Cy5 and $NaHCO_3$, functioning as a pH-responsive element, were encapsulated. After piercing into mildly acidic skin, the PVP matrix rapidly dissolved. Hydrogen ions in the interstitial fluid gradually permeated into the drug-loaded PLGA microspheres, leading to release of $NaHCO_3$-generated carbon dioxide bubbles that created micropores on the surface of the PLGA microspheres, thereby promoting Cy5 release. Wang et al. (2016) employed self-degradable HA MNs to assist the transdermal delivery of anti-programmed cell death receptor-1 (anti-PD-1) antibody by promoting the generation of an acidic environment (Fig. 10.5c). The integrated pH-sensitive dextran nanoparticles served as the reservoir of both anti-PD-1 antibody and glucose oxidase (GOx), the latter of which allowed for the conversion of blood glucose to gluconic acid and facilitated gradual dissociation of nanoparticles, thus leading to the triggered release of embedded anti-PD-1 for robust immunotherapy against melanoma. Additionally, other researchers also successfully constructed pH-responsive MNs for the delivery of both hydrophilic and hydrophobic therapeutic agents, such as anti-cytotoxic T lymphocyte-associated antigen 4 (anti-CTLA-4) antibody, zinc phthalocyanine (Chen et al. 2020b), and cisplatin (Lan et al. 2018).

Moreover, the utilization of disease-related enzymes for intelligent drug delivery has attracted extensive attention. Given that overexpressed hyaluronidase (HAase) was often presented at the tumor site, Ye and coworkers (Ye et al. 2016) designed a self-assembled immunotherapeutic nanocapsule using 1-MT-modified HA for the delivery of anti-PD-1 antibody, which was then made into HAase-responsive MNs. Upon reaching the tumor microenvironment, drug release was triggered by the enzymatic degradation of HA polymers (Fig. 10.5d).

External stimuli are also widely utilized for realizing on-demand drug release. An example of this was reported in a study by Chen et al. (2016b), in which near-infrared (NIR) light-activatable MNs were produced by combining chemotherapy and PTT against superficial tumors. Polycaprolactone served as the MN matrix for encapsulating chemotherapeutic DOX and photosensitive lanthanum hexaboride, and dissolvable PVA/PVP composite was employed as the supporting array. After insertion into the skin, the MN supporting array dissolved and left the embedded polycaprolactone within the tissue lesion. When irradiated with NIR laser, lanthanum hexaboride converted light into heat and induced thermal ablation to release DOX, which synergistically destroyed tumor cells.

10.4.3.4 Drug Penetration and Distribution

After being released from MNs, drug penetration and distribution also play an important part in effective topical therapy. Therefore, Zhang and colleagues (Zhang et al. 2010) investigated the penetration and distribution of PLGA nanoparticles in human skin pretreated with MNs. The simulation study of Franz diffusion cell showed that transdermal permeation of nanoparticles could be enhanced by MNs. Furthermore, their distribution was visualized through confocal laser scanning microscopy, which revealed that PLGA nanoparticles were delivered via microchannels created by MNs. It was also shown that they penetrated into the epidermis and dermis, displaying a particle size-dependent distribution manner.

In another study, hydrophilic rhodamine B (RhB) was selected as a model dye to evaluate the transdermal biodistribution of the free drug and of nanoparticle-encapsulated drug in dissolving MNs, respectively (Kennedy et al. 2017). Following MN administration, RhB was detected in several main organs and superficial lymphatic capillaries at a slightly faster distribution rate for free RhB than that of nanoparticle-encapsulated RhB. Notably, the nanoparticle uptake in the draining lymph nodes highlighted the potential of noninvasive MN delivery in transdermal vaccination.

10.5 The Application of MNs for Cancer Therapy

Cancer is ranked as the first or second leading cause of death before the age of 70 years in 112 of 183 countries (Kennedy et al. 2017). According to the latest report provided by the Global Cancer Statistics 2020, there were estimated 1.20 million people being diagnosed with non-melanoma skin cancer, 0.30 million people with melanoma, and 2.26 million people being diagnosed with breast cancer worldwide in 2020 (Sung et al. 2021). These cancers are obviously a great global health problem that causes continuous psychosocial impacts and excessive medical expenses. Unfortunately, the anticancer efficacy of conventional therapy via systemic administration is lower than desired and often accompanied by severe side effects, particularly for skin cancers. This is a result from insufficient drug targeting and accumulation at the tumor sites after systemic circulation (Hare et al. 2017). Recently, MN-based cancer therapy has aroused great attention of researchers due to the significant advantages of good efficacy, convenience, and compliance (Li et al. 2021). Among these different kinds of MNs, dissolving MNs is considered a promising approach, since it could either provide spatially controlled drug delivery through well-designed array structure and well-chosen needle materials, or by achieving multifunctional combined therapy by integrating them with nanotechnology. Although there are currently no commercially available anticancer dissolving MNs, pre-clinical research in cancer therapy is ongoing (Qin et al. 2020; Pei et al. 2018; Tian et al. 2019; Hao et al. 2020).

10.5.1 The Application of MNs for Chemotherapy

Chemotherapy has been widely used in clinic as cancer treatment. However, the benefits of chemotherapy are limited by its nonspecific drug accumulation under systemic administration, resulting in inadequate drug concentration at the targeted tumor site and severe side effects. MN-based chemotherapeutics rely on local administration, which can efficiently improve the drug targeting and reduce the side effects caused by the systemic exposure to the drug (Ingrole and Gill 2019; Ma and Wu 2017; Lan et al. 2018). MNs can provide a better spatial and temporal control by forming multipoint drug deposits at the tumor site in order to decrease the drug leakage to the adjacent healthy tissues (Park et al. 2006). Bhatnagar and coworkers designed dissolving MNs using PVP/PVA to co-deliver DOX and DTX (DOX/DTX-PVP/PVA@MN) for breast cancer treatment in mouse model (Bhatnagar et al. 2019). The two drugs were both incorporated into the needles along with a polymer matrix in this work. This DOX/DTX-PVP/PVA@MN had enough mechanical strength to insert it into the skin, and compared to intravenous injection, MN-mediated chemotherapy mitigated nonspecific toxicity in the mouse breast cancer model and led to an increase in survival (Bhatnagar et al. 2019).

10.5.2 The Application of MNs for Phototherapy

Phototherapy has emerged as a potential novel cancer therapy strategy and is considered a promising approach to improve the treatment efficacy of superficial tumors in recent years. Phototherapeutic methods include photothermal therapy (PTT) and photodynamic therapy (PDT). PTT can cause irreversible tumor cell damage by using photothermal agents to convert external NIR light energy into hyperthermia to induce tumor cell thermal ablation (Chu and Dupuy 2014; Carcamo-Martinez et al. 2020). In contrast, PDT promotes intracellular oxidative stress caused by an excessive reactive oxygen species (ROS) generation, leading to cell damage (Huang et al. 2019). Phototherapy shows high antitumor efficiency with low dose requirements, since the photothermal agents have great photothermal conversion properties. Simultaneously, due to the flexible and accurate control of laser irradiation, this therapy has minimal invasiveness to the surrounding healthy tissues, which lowers the risk of side effects. In the phototherapy process, photosensitizers need to be previously delivered to the tumor lesion before laser treatment. Therefore, accurate drug targeting and enough drug accumulation are the key points related to the therapeutic efficiency. Using MNs to delivery photosensitizers can be an efficient method for superficial cancer treatment (Chen et al. 2016b; Song et al. 2020; Pei et al. 2018). Tseng et al. developed a polymeric MN patch with gold nanorods incorporated at the needle tips (Tseng et al. 2009). This MN patch could successfully deliver gold nanorods into the skin. Upon NIR light irradiation, the surface plasmon resonance in the gold nanorods could efficiently convert to heat to induce tumor thermal ablation.

10.5.3 The Application of MNs for Immunotherapy

Since there are abundant immune cells populating the skin dermis and hypodermis, such as antigen-presenting cells (APCs), macrophages, Langerhans cells, and dendritic cells (DCs), the skin plays a crucial role in the immune system. Transdermal delivery of cancer vaccines or other immunotherapy drugs, such as monoclonal antibodies, immune adjuvants, and genetic materials, has been a promising strategy for cancer therapy.

10.5.3.1 Cancer Vaccine

The stepwise process of the antitumor immune response is referred to as the "cancer immunity" circle (Chen and Mellman 2013), which include the following: (1) tumor neoantigen exposure; (2) antigen capture and presentation; (3) effector T cell activation and proliferation; (4) activated effector T cell migration and infiltration to tumor sites; and (5) T cells recognize tumor cells and kill them. The recognition of tumor-associated neoantigens is crucial for antitumor immunotherapy. For that purpose, researchers have been exploring the application of MN patches that contain immunostimulatory adjuvants and/or antigens for immunotherapy (Lee et al. 2019; Shakya et al. 2018). Cancer vaccines can induce durable antitumor immunity by exposure of neoantigens, subsequently initiating the cancer immunity circle to activate a systemic immune response that contributes to the protection against tumor recurrence and metastasis (Banchereau and Palucka 2018). Zaric et al. reported using polymeric MNs for delivering OVA-loaded PLGA nanoparticles to trigger the antigen-specific immune response against OVA-expressing B16 melanoma tumors (Zaric et al. 2013). The ex vivo results demonstrated that these MNs were able to penetrate the skin and successfully deposit OVA-nanoparticles into the skin dermal layer, subsequently stimulating DC-related antigen capture and recognition. Furthermore, the resident DCs migrated into the regional lymph nodes, triggering antigen-specific T cell proliferation and cytokine production. In the in vivo assays, significant antitumor efficacy was observed in the mice model after MN treatment. In addition, DNA-based cancer vaccines also present tremendous potential for cancer immunotherapy (Seok et al. 2014; Bernelin-Cottet et al. 2019; Duong et al. 2018). Duong and coworkers fabricated nanoengineered DNA vaccine-loaded MNs with layer-by-layer coating of pH-responsive polymers and immunostimulatory adjuvants to achieve efficacious vaccine delivery (Duong et al. 2018). These well-designed MN patches could stimulate both humoral and cellular immune responses. When perforating the stratum corneum, the MNs disassembled at the immune-cell-rich dermis, followed by the release of the DNA vaccine and adjuvants, which could induce the maturation of dendritic cells. Additionally, type I interferons and antigen-specific antibodies were produced. Transcutaneous application of MNs in mice elicited greater numbers of antibodies and cytotoxic CD8+ T cells compared to subcutaneous injection. Consequently, the in vivo studies showed that the DNA cancer vaccine-loaded MNs rejected the B16/OVA melanoma tumors in mice model and efficiently inhibited metastasis to the lungs (Duong et al. 2018).

10.5.3.2 Antibody Immunotherapy

The complicated tumor microenvironment frequently results in immune escape. CTLA-4 and PD-1 are major immune checkpoints, which can generate immune suppressive signals to impede T cell-mediated immune responses (Nair and Elkord 2018). Therefore, MNs can be engineered to deliver antibody-based immune therapies aimed to revert/by-pass the immune suppressive signals of tumor cells (Yang et al. 2020; Lopez-Ramirez et al. 2020). Wang and coworkers produced self-degradable MNs for mediating the delivery of anti-PD-1 to B16F10 melanoma tumors (Wang et al. 2016). In this approach, anti-PD-1 and GOx were encapsulated into pH-responsive dextran nanoparticles, which were integrated into biocompatible MNs. The GOx could create a local acidic environment by catalyzing oxidation of glucose, which promoted the dissociation of nanoparticles and the release of anti-PD-1. The anti-PD-1 targets the PD-1 receptors, therefore avoiding the inhibitory regulatory signals from the tumors. Animal studies indicated that the administration of MNs could effectively facilitate activated T cell infiltration into the tumor site and stimulate a strong antigen-specific immune response, leading to significant inhibition of tumor progression (Wang et al. 2016). Compared to traditional hypodermic needles, MNs can avoid the pain and discomfort of needle insertion, reduce the risk of bleeding or infection, and reduce needle waste as well. Moreover, MNs contain drugs in a dried solid form, which contributes to maintaining the long-term bioactivity of proteins and peptides as well as improving their thermal stability.

10.5.4 The Application of MNs for Combination Cancer Therapy

Because of the limitation of a single cancer therapy, the development of combination cancer therapies has reached an important inflection point in the history of cancer treatment. MNs can mediate simultaneous delivery of several drugs, provided that the drug formulations are well designed. Especially with the advance in nanotechnology, drug-nanoparticle-loaded MNs have been widely studied for improving the efficacy of anticancer treatments (Wang et al. 2020; Peng et al. 2020; Lan et al. 2018, 2020). Therefore, the MN-based combined cancer therapy has recently been widely explored.

10.5.4.1 Chemo-photothermal Therapy

Qian and coworkers fabricated a chemo-photothermal synergistic system to treat epidermoid cancer, in which NIR-responsive 5-FU and ICG co-loaded nanoparticles were integrated with HA dissolving MNs (Hao et al. 2017). Under the NIR irradiation, the MN patch presented good photothermal conversion effect due to the function of ICG to kill the tumor cells. The high temperature further promoted the release of 5-FU in the nanoparticles. Moreover, this MN system exhibited excellent antitumor efficacy in the A431 epidermoid tumor at a low dosage of the chemotherapeutic drug.

In a similar strategy, Qin et al. constructed a MN-based spatiotemporally controlled pulsatile release system for chemo-photothermal therapy to treat aggressive melanoma (Qin et al. 2020). In this design, PTX and photothermal agent IR-780 were co-encapsulated into thermal-sensitive SLNs, which could undergo phase transition after NIR irradiation, thereby leading to the release of PTX. The results of in vivo studies showed that this MN-based chemo-photothermal system had significant antitumor efficacy with a curable rate of 100% in 30 days, which was obviously more efficient than the single therapy approach. Moreover, compared to intravenous and intratumoral injection, MNs could be directly inserted into the tumor site and provide stable, uniform drug deposits, resulting in enhanced bioavailability and safety.

10.5.4.2 Photothermal-immunotherapy

As reported, the hyperthermic effect or ROS caused by photosensitive agents can elicit antitumor immunity by inducing immunogenic cell death, leading to calreticulin expression on the surface of tumor cells, which is regarded as an "eat me"-signal that stimulates phagocytosis of tumor cells by dendritic cells (Li et al. 2019a). Hence, photothermal therapy can contribute to enhanced immunotherapy with the application of immunotherapeutic drugs in a synergistic way.

Chen and colleagues designed a core-shell structured dissolving MN system to co-deliver photosensitizer ICG and IDO blockade 1-MT to achieve combined photothermal-immunotherapy (Chen et al. 2016a). Under NIR irradiation, ICG could convert light energy into hyperthermia, inducing the generation of cell death–associated antigens that further facilitated the maturation of dendritic cells. Furthermore, the 1-MT released from the cores of the MN patches subsequently relieved the immunosuppression in the tumor microenvironment, stimulating a robust systematic immune response. The in vivo studies indicated that this MN-based combined therapy could efficiently inhibit the growth and recurrence of the primary tumor, and also significantly impede the distal tumor in the melanoma lung metastasis model.

Similarly, Chen et al. produced a self-degradable MN-assisted system for combining PDT and immunotherapy, which generated a synergistic antitumor outcome while reducing the side effects through a controlled co-delivery of the photosensitizer zinc phthalocyanine and the immune checkpoint inhibitor anti-CTLA-4 antibody (Chen et al. 2020b). After the MN insertion, the co-loaded nanoparticles were delivered to the targeted tumor site and aggregated effectively around the tumor cells. Under the irradiation treatment, PDT first killed tumor cells and triggered the immune response. Then, anti-CTLA-4 antibodies played a role in antitumor immunotherapy.

Due to the simple fabrication technology, MNs can integrate the advantages of various therapeutic strategies in cancer treatment, such as chemo-photothermal therapy and photothermal-immunotherapy. These MN-based combined therapeutic strategies can not only increase the antitumor efficacy but also alleviate the side effect on the healthy tissues, thereby arousing widespread attention among researchers. Additionally, this integrated MN transdermal delivery platform can be a promising tool for other disease treatments.

References

Abdelghany S, Tekko IA, Vora L et al (2019) Nanosuspension-based dissolving microneedle arrays for intradermal delivery of curcumin. Pharmaceutics 11(7):308. https://doi.org/10.3390/pharmaceutics11070308

Alimardani V, Abolmaali SS, Tamaddon AM et al (2020) Recent advances on microneedle arrays-mediated technology in cancer diagnosis and therapy. Drug Deliv Transl Res 11:788. https://doi.org/10.1007/s13346-020-00819-z

Andersen TE, Andersen AJ, Petersen RS et al (2018) Drug loaded biodegradable polymer microneedles fabricated by hot embossing. Mater Sci Eng C Mater Biol Appl 195:57–61. https://doi.org/10.1016/j.mee.2018.03.024

Arya J, Henry S, Kalluri H et al (2017) Tolerability, usability and acceptability of dissolving microneedle patch administration in human subjects. Biomaterials 128:1–7. https://doi.org/10.1016/j.biomaterials.2017.02.040

Banchereau J, Palucka K (2018) Immunotherapy: cancer vaccines on the move. Nat Rev Clin Oncol 15(1):9–10. https://doi.org/10.1038/nrclinonc.2017.149

Bernelin-Cottet C, Urien C, McCaffrey J et al (2019) Electroporation of a nanoparticle-associated DNA vaccine induces higher inflammation and immunity compared to its delivery with microneedle patches in pigs. J Control Release 308:14–28. https://doi.org/10.1016/j.jconrel.2019.06.041

Bhatnagar S, Bankar NG, Kulkarni MV et al (2019) Dissolvable microneedle patch containing doxorubicin and docetaxel is effective in 4T1 xenografted breast cancer mouse model. Int J Pharm 556:263–275. https://doi.org/10.1016/j.ijpharm.2018.12.022

Boehm RD, Miller PR, Schell WA et al (2013) Inkjet printing of amphotericin b onto biodegradable microneedles using piezoelectric inkjet printing. JOM 65(4):525–533. https://doi.org/10.1007/s11837-013-0574-7

Cahill EM, Keaveney S, Stuettgen V et al (2019) Metallic microneedles with interconnected porosity: a scalable platform for biosensing and drug delivery. Acta Biomater 86:493–495. https://doi.org/10.1016/j.actbio.2018.09.007

Cai B, Xia W, Bredenberg S et al (2015) Bioceramic microneedles with flexible and self-swelling substrate. Eur J Pharm Biopharm 94:404–410. https://doi.org/10.1016/j.ejpb.2015.06.016

Carcamo-Martinez A, Mallon B, Dominguez-Robles J et al (2020) Plasmonic photothermal microneedle arrays and single needles for minimally-invasive deep in-skin hyperthermia. J Mater Chem B 8(25):5425–5433. https://doi.org/10.1039/d0tb00962h

Caudill CL, Perry JL, Tian S et al (2018) Spatially controlled coating of continuous liquid interface production microneedles for transdermal protein delivery. J Control Release 284:122–132. https://doi.org/10.1016/j.jconrel.2018.05.042

Chen DS, Mellman I (2013) Oncology meets immunology: the cancer-immunity cycle. Immunity 39(1):1–10. https://doi.org/10.1016/j.immuni.2013.07.012

Chen X, Prow TW, Crichton ML et al (2009) Dry-coated microprojection array patches for targeted delivery of immunotherapeutics to the skin. J Control Release 139:212–220. https://doi.org/10.1016/j.jconrel.2009.06.029
Chen J, Qiu Y, Zhang S et al (2016a) Dissolving microneedle-based intradermal delivery of interferon-α-2b. Drug Dev Ind Pharm 42(6):890–896. https://doi.org/10.3109/03639045.2015.1096282
Chen MC, Lin ZW, Ling MH (2016b) Near-infrared light-activatable microneedle system for treating superficial tumors by combination of chemotherapy and photothermal therapy. ACS Nano 10(1):93–101. https://doi.org/10.1021/acsnano.5b05043
Chen M, Quan G, Wen T et al (2020a) Cold to hot: binary cooperative microneedle array amplified photo-immunotherapy for eliciting antitumor immunity and abscopal effect. ACS Appl Mater Interfaces 12(29):32259–32269. https://doi.org/10.1021/acsami.0c05090
Chen SX, Ma M, Xue F et al (2020b) Construction of microneedle-assisted co-delivery platform and its combining photodynamic/immunotherapy. J Control Release 324:218–227. https://doi.org/10.1016/j.jconrel.2020.05.006
Chen Y, Yang Y, Xian Y et al (2020c) Multifunctional graphene-oxide-reinforced dissolvable polymeric microneedles for transdermal drug delivery. ACS Appl Mater Interfaces 12(1):352–360. https://doi.org/10.1021/acsami.9b19518
Chu KF, Dupuy DE (2014) Thermal ablation of tumours: biological mechanisms and advances in therapy. Nat Rev Cancer 14(3):199–208. https://doi.org/10.1038/nrc3672
Cole G, Ali AA, McCrudden CM et al (2018) DNA vaccination for cervical cancer: strategic optimisation of RALA mediated gene delivery from a biodegradable microneedle system. Eur J Pharm Biopharm 127:288. https://doi.org/10.1016/j.ejpb.2018.02.029
Courtenay AJ, Rodgers AM, McCrudden MTC et al (2019) Novel hydrogel-forming microneedle array for intradermal vaccination in mice using ovalbumin as a model protein antigen. Mol Pharm 16(1):118–127. https://doi.org/10.1021/acs.molpharmaceut.8b00895
Courtenay AJ, McAlister E, McCrudden MTC et al (2020) Hydrogel-forming microneedle arrays as a therapeutic option for transdermal esketamine delivery. J Control Release 322:177–186. https://doi.org/10.1016/j.jconrel.2020.03.026
Dangol M, Yang H, Li CG et al (2016) Innovative polymeric system (IPS) for solvent-free lipophilic drug transdermal delivery via dissolving microneedles. J Control Release 223:118–125. https://doi.org/10.1016/j.jconrel.2015.12.038
DeMuth PC, Garcia-Beltran WF, Ai-Ling ML et al (2013) Composite dissolving microneedles for coordinated control of antigen and adjuvant delivery kinetics in transcutaneous vaccination. Adv Funct Mater 23(2):161–172. https://doi.org/10.1002/adfm.201201512
Deng YL, Juang YJ (2014) Polydimethyl siloxane wet etching for three dimensional fabrication of microneedle array and high-aspect-ratio micropillars. Biomicrofluidics 8(2):026502. https://doi.org/10.1063/1.4871038
Dong L, Li Y, Li Z et al (2018) Au nanocage-strengthened dissolving microneedles for chemo-photothermal combined therapy of superficial skin tumors. ACS Appl Mater Interfaces 10(11):9247–9256. https://doi.org/10.1021/acsami.7b18293
Dragicevic N, Maibach H (2018) Combined use of nanocarriers and physical methods for percutaneous penetration enhancement. Adv Drug Deliv Rev 127:58–84. https://doi.org/10.1016/j.addr.2018.02.003
Du GS, Hathout RM, Nasr M et al (2017) Intradermal vaccination with hollow microneedles: a comparative study of various protein antigen and adjuvant encapsulated nanoparticles. J Control Release 266:109–118. https://doi.org/10.1016/j.jconrel.2017.09.021
Du H, Liu P, Zhu J et al (2019) Hyaluronic acid-based dissolving microneedle patch loaded with methotrexate for improved treatment of psoriasis. ACS Appl Mater Interfaces 11(46):43588–43598. https://doi.org/10.1021/acsami.9b15668
Duong HTT, Yin Y, Thambi T et al (2018) Smart vaccine delivery based on microneedle arrays decorated with ultra-pH-responsive copolymers for cancer immunotherapy. Biomaterials 185:13–24. https://doi.org/10.1016/j.biomaterials.2018.09.008

Gerstel MS (1976) Drug delivery device. US Patent 3,964,482
Gittard SD, Ovsianikov A, Chichkov BN et al (2010) Two-photon polymerization of microneedles for transdermal drug delivery. Expert Opin Drug Deliv 7(4):513–533. https://doi.org/10.1517/17425241003628171
González-Vázquez P, Larraeta E, Mccrudden M et al (2017) Transdermal delivery of gentamicin using dissolving microneedle arrays for potential treatment of neonatal sepsis. J Control Release 265:30–40. https://doi.org/10.1016/j.jconrel.2017.07.032
Haj-Ahmad R, Khan H, Arshad MS et al (2015) Microneedle coating techniques for transdermal drug delivery. Pharmaceutics 7(4):486–502. https://doi.org/10.3390/pharmaceutics7040486
Hao Y, Dong ML, Zhang TY et al (2017) Novel approach of using near-infrared responsive PEGylated gold nanorod coated poly(l-lactide) microneedles to enhance the antitumor efficiency of docetaxel-loaded MPEG-PDLLA micelles for treating an A431 tumor. ACS Appl Mater Interfaces 9(18):15317–15327. https://doi.org/10.1021/acsami.7b03604
Hao Y, Chen Y, He X et al (2020) Near-infrared responsive 5-fluorouracil and indocyanine green loaded MPEG-PCL nanoparticle integrated with dissolvable microneedle for skin cancer therapy. Bioact Mater 5(3):542–552. https://doi.org/10.1016/j.bioactmat.2020.04.002
Hardy JG, Larraneta E, Donnelly RF et al (2016) Hydrogel-forming microneedle arrays made from light-responsive materials for on-demand transdermal drug delivery. Mol Pharm 13(3):907–914. https://doi.org/10.1021/acs.molpharmaceut.5b00807
Hare JI, Lammers T, Ashford MB et al (2017) Challenges and strategies in anti-cancer nanomedicine development: an industry perspective. Adv Drug Deliv Rev 108:25–38. https://doi.org/10.1016/j.addr.2016.04.025
He X, Sun J, Zhuang J et al (2019) Microneedle system for transdermal drug and vaccine delivery: devices, safety, and prospects. Dose Response 17(4):1559325819878585. https://doi.org/10.1177/1559325819878585
Held J, Gaspar J, Ruther P et al (2007) Systematic characterization of DRIE-based fabrication process of silicon microneedles. MRS Proc 1052:707. https://doi.org/10.1557/PROC-1052-DD07-07
Henry S, McAllister DV, Allen MG et al (1998) Microfabricated microneedles: a novel approach to transdermal drug delivery. J Pharm Sci 87:922–925. https://doi.org/10.1021/js990783q
Hou A, Quan G, Yang B et al (2019) Rational design of rapidly separating dissolving microneedles for precise drug delivery by balancing the mechanical performance and disintegration rate. Adv Healthc Mater 8(21):e1900898. https://doi.org/10.1002/adhm.201900898
Huang Z, Wei G, Zeng Z et al (2019) Enhanced cancer therapy through synergetic photodynamic/immune checkpoint blockade mediated by a liposomal conjugate comprised of porphyrin and IDO inhibitor. Theranostics 9(19):5542–5557. https://doi.org/10.7150/thno.35343
Ingrole RSJ, Gill HS (2019) Microneedle coating methods: a review with a perspective. J Pharmacol Exp Ther 370(3):555–569. https://doi.org/10.1124/jpet.119.258707
Ita K (2015) Transdermal delivery of drugs with microneedles-potential and challenges. Pharmaceutics 7(3):90–105. https://doi.org/10.3390/pharmaceutics7030090
Ito Y, Kashiwara S, Fukushima K et al (2011a) Two-layered dissolving microneedles for percutaneous delivery of sumatriptan in rats. Drug Dev Ind Pharm 37(12):1387–1393. https://doi.org/10.3109/03639045.2011.576426
Ito Y, Murano H, Hamasaki N et al (2011b) Incidence of low bioavailability of leuprolide acetate after percutaneous administration to rats by dissolving microneedles. Int J Pharm 407(1–2):126–131. https://doi.org/10.1016/j.ijpharm.2011.01.039
Izumi H, Aoyagi S (2007) Novel fabrication method for long silicon microneedles with three-dimensional sharp tips and complicated shank shapes by isotropic dry etching. IEEJ Trans Electr Electron Eng 2(3):328–334. https://doi.org/10.1002/tee.20147
Jin CY, Han MH, Lee SS et al (2009) Mass producible and biocompatible microneedle patch and functional verification of its usefulness for transdermal drug delivery. Biomed Microdevices 11:1195–1203. https://doi.org/10.1007/s10544-009-9337-1
Juárez MCP, Escobar-Chávez JJ, Rodrigo-Guadarrama-Escobar O (2021) Transdermal delivery systems for biomolecules. J Pharm Innov 17:319. https://doi.org/10.1007/s12247-020-09525-2

Juríek P, Zou H, Zhang S et al (2013) Design and fabrication of hollow out-of-plane silicon microneedles. Micro Nano Lett 8:78–81. https://doi.org/10.1049/mnl.2012.0903

Kapoor Y, Milewski M, Dick L et al (2019) Coated microneedles for transdermal delivery of a potent pharmaceutical peptide. Biomed Microdevices 22(1):7. https://doi.org/10.1007/s10544-019-0462-1

Ke CJ, Lin YJ, Hu YC et al (2012) Multidrug release based on microneedle arrays filled with pH-responsive PLGA hollow microspheres. Biomaterials 33(20):5156–5165. https://doi.org/10.1016/j.biomaterials.2012.03.056

Kearney MC, Caffarel-Salvador E, Fallows SJ et al (2016) Microneedle-mediated delivery of donepezil: potential for improved treatment options in Alzheimer's disease. Eur J Pharm Biopharm 103:43–50. https://doi.org/10.1016/j.ejpb.2016.03.026

Kennedy J, Larrañeta E, McCrudden MTC et al (2017) In vivo studies investigating biodistribution of nanoparticle-encapsulated rhodamine B delivered via dissolving microneedles. J Control Release 265:57–65. https://doi.org/10.1016/j.jconrel.2017.04.022

Kim M, Jung B, Park JH (2012) Hydrogel swelling as a trigger to release biodegradable polymer microneedles in skin. Biomaterials 33(2):668–678. https://doi.org/10.1016/j.biomaterials.2011.09.074

Lan X, She J, Lin DA et al (2018) Microneedle-mediated delivery of lipid-coated cisplatin nanoparticles for efficient and safe cancer therapy. ACS Appl Mater Interfaces 10(39):33060–33069. https://doi.org/10.1021/acsami.8b12926

Lan X, Zhu W, Huang X et al (2020) Microneedles loaded with anti-PD-1-cisplatin nanoparticles for synergistic cancer immuno-chemotherapy. Nanoscale 12(36):18885–18898. https://doi.org/10.1039/d0nr04213g

Lane ME (2013) Skin penetration enhancers. Int J Pharm 447(1–2):12–21. https://doi.org/10.1016/j.ijpharm.2013.02.040

Lee JW, Park JH, Prausnitz MR (2008) Dissolving microneedles for transdermal drug delivery. Biomaterials 29(13):2113–2124. https://doi.org/10.1016/j.biomaterials.2007.12.048

Lee SJ, Lee HS, Hwang YH et al (2019) Enhanced anti-tumor immunotherapy by dissolving microneedle patch loaded ovalbumin. PLoS One 14(8):e0220382. https://doi.org/10.1371/journal.pone.0220382

Leone M, Monkare J, Bouwstra JA et al (2017) Dissolving microneedle patches for dermal vaccination. Pharm Res 34(11):2223–2240. https://doi.org/10.1007/s11095-017-2223-2

Li J, Zeng M, Shan H et al (2017a) Microneedle patches as drug and vaccine delivery platform. Curr Med Chem 24(22):2413–2422. https://doi.org/10.2174/0929867324666170526124053

Li QY, Zhang JN, Chen BZ et al (2017b) A solid polymer microneedle patch pretreatment enhances the permeation of drug molecules into the skin. RSC Adv 7(25):15408–15415. https://doi.org/10.1039/c6ra26759a

Li W, Yang J, Luo L et al (2019a) Targeting photodynamic and photothermal therapy to the endoplasmic reticulum enhances immunogenic cancer cell death. Nat Commun 10(1):3349. https://doi.org/10.1038/s41467-019-11269-8

Li Y, Zhang H, Yang R et al (2019b) Fabrication of sharp silicon hollow microneedles by deep-reactive ion etching towards minimally invasive diagnostics. Microsyst Nanoeng 5:41. https://doi.org/10.1038/s41378-019-0077-y

Li D, Hu D, Xu H et al (2021) Progress and perspective of microneedle system for anti-cancer drug delivery. Biomaterials 264:120410. https://doi.org/10.1016/j.biomaterials.2020.120410

Lin S, Cai B, Quan G et al (2017) Novel strategy for immunomodulation: dissolving microneedle array encapsulating thymopentin fabricated by modified two-step molding technology. Eur J Pharm Biopharm 122:104. https://doi.org/10.1016/j.ejpb.2017.10.011

Lin S, Quan G, Hou A et al (2019) Strategy for hypertrophic scar therapy: improved delivery of triamcinolone acetonide using mechanically robust tip-concentrated dissolving microneedle array. J Control Release 306:69–82. https://doi.org/10.1016/j.jconrel.2019.05.038

Ling MH, Chen MC (2013) Dissolving polymer microneedle patches for rapid and efficient transdermal delivery of insulin to diabetic rats. Acta Biomater 9(11):8952–8961. https://doi.org/10.1016/j.actbio.2013.06.029

Liu S, Zhang S, Duan Y et al (2019) Transcutaneous immunization of recombinant Staphylococcal enterotoxin B protein using a dissolving microneedle provides potent protection against lethal enterotoxin challenge. Vaccine 37(29):3810–3819. https://doi.org/10.1016/j.vaccine.2019.05.055

Loizidou EZ, Williams NA, Barrow DA et al (2015) Structural characterisation and transdermal delivery studies on sugar microneedles: experimental and finite element modelling analyses. Eur J Pharm Biopharm 89:224–231. https://doi.org/10.1016/j.ejpb.2014.11.023

Lopez-Ramirez MA, Soto F, Wang C et al (2020) Built-in active microneedle patch with enhanced autonomous drug delivery. Adv Mater 32(1):e1905740. https://doi.org/10.1002/adma.201905740

Lutton RE, Larraneta E, Kearney MC et al (2015) A novel scalable manufacturing process for the production of hydrogel-forming microneedle arrays. Int J Pharm 494(1):417–429. https://doi.org/10.1016/j.ijpharm.2015.08.049

Luzuriaga MA, Berry DR, Reagan JC et al (2018) Biodegradable 3D printed polymer microneedles for transdermal drug delivery. Lab Chip 18(8):1223–1230. https://doi.org/10.1039/c8lc00098k

Ma G, Wu C (2017) Microneedle, bio-microneedle and bio-inspired microneedle: a review. J Control Release 251:11–23. https://doi.org/10.1016/j.jconrel.2017.02.011

Maaden K, Heuts J, Camps M et al (2018) Hollow microneedle-mediated micro-injections of a liposomal HPV E743-63 synthetic long peptide vaccine for efficient induction of cytotoxic and T-helper responses. J Control Release 269:347–354. https://doi.org/10.1016/j.jconrel.2017.11.035

Maitra J, Shukla VK (2014) Cross-linking in hydrogels – a review. Am J Polit Sci 4(2):25–31. https://doi.org/10.5923/j.ajps.20140402.01

Marwah H, Garg T, Goyal AK et al (2016) Permeation enhancer strategies in transdermal drug delivery. Drug Deliv 23(2):564–578. https://doi.org/10.3109/10717544.2014.935532

McAlister E, Dutton B, Vora LK et al (2020) Directly compressed tablets: a novel drug-containing reservoir combined with hydrogel-forming microneedle arrays for transdermal drug delivery. Adv Healthc Mater 10:e2001256. https://doi.org/10.1002/adhm.202001256

McAllister DV, Wang PM, Davis SP et al (2003) Microfabricated needles for transdermal delivery of macromolecules and nanoparticles: fabrication methods and transport studies. Proc Natl Acad Sci U S A 100:13755–13760. https://doi.org/10.1073/pnas.2331316100

Mcgrath MG, Vrdoljak A, O'Mahony C et al (2011) Determination of parameters for successful spray coating of silicon microneedle arrays. Int J Pharm 415:140–149. https://doi.org/10.1016/j.ijpharm.2011.05.064

Mehta P, Haj-Ahmad R, Rasekh M et al (2017) Pharmaceutical and biomaterial engineering via electrohydrodynamic atomization technologies. Drug Discov Today 22:157–165. https://doi.org/10.1016/j.drudis.2016.09.021

Migdadi EM, Courtenay AJ, Tekko IA et al (2018) Hydrogel-forming microneedles enhance transdermal delivery of metformin hydrochloride. J Control Release 285:142–151. https://doi.org/10.1016/j.jconrel.2018.07.009

Miller PR, Narayan RJ, Polsky R (2016) Microneedle-based sensors for medical diagnosis. J Mater Chem B 4(8):1379–1383. https://doi.org/10.1039/c5tb02421h

Mina R, Niclas R, Zandi SR et al (2016) Flexible and stretchable microneedle patches with integrated rigid stainless steel microneedles for transdermal biointerfacing. PLoS One 11:e0166330. https://doi.org/10.1371/journal.pone.0166330

Naguib YW, Kumar A, Cui Z (2014) The effect of microneedles on the skin permeability and antitumor activity of topical 5-fluorouracil. Acta Pharm Sin B 4(1):94–99. https://doi.org/10.1016/j.apsb.2013.12.013

Nair VS, Elkord E (2018) Immune checkpoint inhibitors in cancer therapy: a focus on T-regulatory cells. Immunol Cell Biol 96:21–33. https://doi.org/10.1111/imcb.1003

Naito C, Katsumi H, Suzuki T et al (2018) Self-dissolving microneedle arrays for transdermal absorption enhancement of human parathyroid hormone (1-34). Pharmaceutics 10(4):215. https://doi.org/10.3390/pharmaceutics10040215

Norman JJ, Choi SO, Tong NT et al (2013) Hollow microneedles for intradermal injection fabricated by sacrificial micromolding and selective electrodeposition. Biomed Microdevices 15:203–210. https://doi.org/10.1007/s10544-012-9717-9

Ono A, Azukizawa H, Ito S et al (2017) Development of novel doubledecker microneedle patches for transcutaneous vaccine delivery. Int J Pharm 532:374–383. https://doi.org/10.1016/j.ijpharm.2017.08.110

Park JH, Allen MG, Prausnitz MR (2006) Polymer microneedles for controlled-release drug delivery. Pharm Res 23(5):1008–1019. https://doi.org/10.1007/s11095-006-0028-9

Pei P, Yang F, Liu J et al (2018) Composite-dissolving microneedle patches for chemotherapy and photothermal therapy in superficial tumor treatment. Biomater Sci 6(6):1414–1423. https://doi.org/10.1039/c8bm00005k

Peng T, Huang Y, Feng X et al (2020) Dissolving microneedles loading TPGS biphasic functionalized PLGA nanoparticles for efficient chemo-photothermal combined therapy of melanoma. Adv Ther 3(6):1900190. https://doi.org/10.1002/adtp.201900190

Prausnitz MR (2017) Engineering microneedle patches for vaccination and drug delivery to skin. Annu Rev Chem Biomol Eng 8:177–200. https://doi.org/10.1146/annurev-chembioeng-060816-101514

Prausnitz MR, Langer R (2008) Transdermal drug delivery. Nat Biotechnol 26(11):1261–1268. https://doi.org/10.1038/nbt.1504

Qin W, Quan G, Sun Y et al (2020) Dissolving microneedles with spatiotemporally controlled pulsatile release nanosystem for synergistic chemo-photothermal therapy of melanoma. Theranostics 10(18):8179–8196. https://doi.org/10.7150/thno.44194

Raphael AP, Crichton ML, Falconer RJ et al (2016) Formulations for microprojection/microneedle vaccine delivery: structure, strength and release profiles. J Control Release 10(225):40–52. https://doi.org/10.1016/j.jconrel.2016.01.027

Rodriguez A, Molinero D, Valera E et al (2005) Fabrication of silicon oxide microneedles from macroporous silicon. Sensors Actuators B Chem 109(1):135–140. https://doi.org/10.1016/j.snb.2005.03.015

Rui YE, Yang J, Li Y et al (2020) Fabrication of tip-hollow and tip-dissolvable microneedle arrays for transdermal drug delivery. ACS Biomater Sci Eng 13:2487–2494. https://doi.org/10.1021/acsbiomaterials.0c00120

Sala M, Diab R, Elaissari A et al (2018) Lipid nanocarriers as skin drug delivery systems: properties, mechanisms of skin interactions and medical applications. Int J Pharm 535(1–2):1–17. https://doi.org/10.1016/j.ijpharm.2017.10.046

Seah BC, Teo BM (2018) Recent advances in ultrasound-based transdermal drug delivery. Int J Nanomedicine 13:7749–7763. https://doi.org/10.2147/IJN.S174759

Seok HY, Suh H, Baek S et al (2014) Microneedle applications for DNA vaccine delivery to the skin. Methods Mol Biol 1143:141–158. https://doi.org/10.1007/978-1-4939-0410-5_10

Seong KY, Seo MS, Hwang DY et al (2017) A self-adherent, bullet-shaped microneedle patch for controlled transdermal delivery of insulin. J Control Release 265:48. https://doi.org/10.1016/j.jconrel.2017.03.041

Shakya AK, Lee CH, Gill HS (2018) Coated microneedle-based cutaneous immunotherapy prevents Der p 1-induced airway allergy in mice. J Allergy Clin Immunol 142(6):2007–2011.e2003. https://doi.org/10.1016/j.jaci.2018.08.017

Sharma S, Hatware K, Bhadane P et al (2019) Recent advances in microneedle composites for biomedical applications: advanced drug delivery technologies. Mater Sci Eng C 103:109717. https://doi.org/10.1016/j.msec.2019.05.002

Song G, Jiang G, Liu T et al (2020) Separable microneedles for synergistic chemo-photothermal therapy against superficial skin tumors. ACS Biomater Sci Eng 6(7):4116–4125. https://doi.org/10.1021/acsbiomaterials.0c00793

Sung H, Ferlay J, Siegel RL et al (2021) Global Cancer Statistics 2020: GLOBOCAN estimates of incidence and mortality worldwide for 36 cancers in 185 countries. CA Cancer J Clin 71(3):209–249. https://doi.org/10.3322/caac.21660

Tekko IA, Chen GY, Dominguez-Robles J et al (2020a) Development and characterisation of novel poly (vinyl alcohol)/poly (vinyl pyrrolidone)-based hydrogel-forming microneedle arrays for enhanced and sustained transdermal delivery of methotrexate. Int J Pharm 586:119580. https://doi.org/10.1016/j.ijpharm.2020.119580

Tekko IA, Permana AD, Vora L et al (2020b) Localised and sustained intradermal delivery of methotrexate using nanocrystal-loaded microneedle arrays: potential for enhanced treatment of psoriasis. Eur J Pharm Sci 152:105469. https://doi.org/10.1016/j.ejps.2020.105469

Tian Z, Cheng J, Liu J et al (2019) Dissolving graphene/poly(acrylic acid) microneedles for potential transdermal drug delivery and photothermal therapy. J Nanosci Nanotechnol 19(5):2453–2459. https://doi.org/10.1166/jnn.2019.15884

Tomomo T (2018) Puncture performance of sharpen microneedles by using inclined contact UV lithography. Microsyst Technol 24:3589–3599. https://doi.org/10.1007/s00542-018-3780-y

Tseng PL, Shih TY, Tien CC et al (2009) Photo-thermal-based nonablative dermal remodeling device. Thirteenth International Conference on Miniaturized Systems for Chemistry and Life Sciences, Jeju, Korea, 1–5 Nov 2009

Turner JG, White LR, Estrela P et al (2020) Hydrogel-forming microneedles: current advancements and future trends. Macromol Biosci 21:e2000307. https://doi.org/10.1002/mabi.202000307

Uddin MJ, Scoutaris N, Klepetsanis P et al (2015) Inkjet printing of transdermal microneedles for the delivery of anticancer agents. Int J Pharm 494:593–602. https://doi.org/10.1016/j.ijpharm.2015.01.038

Vinayakumar KB, Hegde GM, Nayak MM et al (2014) Fabrication and characterization of gold coated hollow silicon microneedle array for drug delivery. Microelectron Eng 128:12–18. https://doi.org/10.1016/j.mee.2014.05.039

Vora LK, Donnelly RF, Larraneta E et al (2017) Novel bilayer dissolving microneedle arrays with concentrated PLGA nanomicroparticles for targeted intradermal delivery: proof of concept. J Control Release 265:93–101. https://doi.org/10.1016/j.jconrel.2017.10.005

Waghule T, Singhvi G, Dubey SK et al (2019) Microneedles: a smart approach and increasing potential for transdermal drug delivery system. Biomed Pharmacother 109:1249–1258. https://doi.org/10.1016/j.biopha.2018.10.078

Wang C, Ye Y, Hochu GM et al (2016) Enhanced cancer immunotherapy by microneedle patch-assisted delivery of anti-PD1 antibody. Nano Lett 16(4):2334–2340. https://doi.org/10.1021/acs.nanolett.5b05030

Wang M, Han Y, Yu X et al (2020) Upconversion nanoparticle powered microneedle patches for transdermal delivery of siRNA. Adv Healthc Mater 9(2):e1900635. https://doi.org/10.1002/adhm.201900635

Wilke N, Hibert C, O'Brien J et al (2005a) Silicon microneedle electrode array with temperature monitoring for electroporation. Sensors Actuators A Phys 123-124:319–325. https://doi.org/10.1016/j.sna.2005.05.017

Wilke N, Mulcahy A, Ye SR et al (2005b) Process optimization and characterization of silicon microneedles fabricated by wet etch technology. Microelectron J 36:650–656. https://doi.org/10.1016/j.ejpb.2021.06.005

Wu B, Fu J, Zhou Y et al (2020) Tailored core–shell dual metal-organic frameworks as a versatile nanomotor for effective synergistic antitumor therapy. Acta Pharm Sin B 10(11):2198–2211. https://doi.org/10.1016/j.apsb.2020.07.025

Xu B, Jiang G, Yu W et al (2017) H2O2-Responsive mesoporous silica nanoparticles integrated with microneedle patches for the glucose-monitored transdermal delivery of insulin. J Mater Chem B 5:8200–8208. https://doi.org/10.1039/C7TB02082A

Xu Q, Li X, Zhang P et al (2020) Rapid dissolving microneedle patch for synergistic gene and photothermal therapy of subcutaneous tumor. J Mater Chem B 8(19):4331–4339. https://doi.org/10.1039/D0TB00105H

Yan L, Raphael AP, Zhu X et al (2014) Nanocomposite-strengthened dissolving microneedles for improved transdermal delivery to human Skin. Adv Healthc Mater 3(4):555–564. https://doi.org/10.1002/adhm.201300312

Yang P, Lu C, Qin W et al (2020) Construction of a core-shell microneedle system to achieve targeted co-delivery of checkpoint inhibitors for melanoma immunotherapy. Acta Biomater 104:147–157. https://doi.org/10.1016/j.actbio.2019.12.037

Ye Y, Wang J, Hu Q et al (2016) Synergistic transcutaneous immunotherapy enhances antitumor immune responses through delivery of checkpoint inhibitors. ACS Nano 10(9):8956–8963. https://doi.org/10.1021/acsnano.6b04989

Ye Y, Yu J, Wen D et al (2018) Polymeric microneedles for transdermal protein delivery. Adv Drug Deliv Rev 127:106–118. https://doi.org/10.1016/j.addr.2018.01.015

Zaric M, Lyubomska O, Touzelet O et al (2013) Skin dendritic cell targeting via microneedle arrays laden with antigen-encapsulated poly-D,L-lactide-co-glycolide nanoparticles induces efficient antitumor and antiviral immune responses. ACS Nano 7(3):2042–2055. https://doi.org/10.1021/nn304235j

Zhang W, Gao J, Zhu Q et al (2010) Penetration and distribution of PLGA nanoparticles in the human skin treated with microneedles. Int J Pharm 402(1–2):205–212. https://doi.org/10.1016/j.ijpharm.2010.09.037

Zhang Y, Yu J, Kahkoska AR et al (2019) Advances in transdermal insulin delivery. Adv Drug Deliv Rev 139:51–70. https://doi.org/10.1016/j.addr.2018.12.006

Zhao X, Li X, Zhang P et al (2018) Tip-loaded fast-dissolving microneedle patches for photodynamic therapy of subcutaneous tumor. J Control Release 286:201. https://doi.org/10.1016/j.jconrel.2018.07.038